METRIC AND U. S. UNITS OF MEASUREMENT

Units

1 kilometer (km) = 1000 meters (m)

1 meter (m) = 100 centimeters (cm)

1 centimeter (cm) = 0.39 inch (in)

1 mile (mi) = 5280 feet (ft)

1 foot (ft) = 12 inches (in)

1 inch (in) = 2.54 centimeters (cm)

1 square mile (mi^2) = 640 acres (a)

1 kilogram (kg) = 1000 grams (g)

1 pound (lb) = 16 ounces (oz)

1 fathom = 6 feet (ft)

Conversions

When you want to convert:	multiply by:	to find:
Length		
inches	2.54	centimeters
centimeters	0.39	inches
feet	0.30	meters
meters	3.28	feet
yards	0.91	meters
meters	1.09	yards
miles	1.61	kilometers
kilometers	0.62	miles
Area		
square inches	6.45	square centimeters
square centimeters	0.15	square inches
square feet	0.09	square meters
square meters	10.76	square feet
square miles	2.59	square kilometers
square kilometers	0.39	square miles
Volume		
cubic inches	16.38	cubic centimeters
cubic centimeters	0.06	cubic inches
cubic feet	0.028	cubic meters
cubic meters	35.3	cubic feet
cubic miles	4.17	cubic kilometers
cubic kilometers	0.24	cubic miles
liters	1.06	quarts
liters	0.26	gallons
gallons	3.78	liters
Mass and Weight		
ounces	28.35	grams
grams	0.035	ounces
pounds	0.45	kilograms
kilograms	2.205	pounds

Temperature

To convert from Centigrade (°C) to Fahrenheit (°F), multiply by 1.8 and add 32.

To convert from Fahrenheit (°F) to Centigrade (°C), subtract 32 and divide by 1.8.

Principles of
Glacial Geomorphology
and Geology

Principles of Glacial Geomorphology and Geology

I. Peter Martini
Michael E. Brookfield
Steven Sadura

Prentice Hall
Upper Saddle River, NJ 07458

Library of Congress Cataloging-in-Publication Data

Martini, I. P. (Ireneo Peter)
  Principles of glacial geomorphology and geology / I. Peter Martini, Michael E. Brookfield, and Steven Sadura.
      p.   cm.
  Includes bibliographical references (p.).
  ISBN 0-13-526518-5
  1. Glaciers.   I. Brookfield, M. E. (Michael E.)   II. Sadura, Steven.   III. Title.

GB2403.2 .M37 2001
551.31 — dc21   00-052465

Senior Editor: *Patrick Lynch*
Editorial Assistant: *Sean Hale*
Assistant Managing Editor: *Beth Sturla*
Production Editor/Composition: *G & S Typesetters, Inc.*
Marketing Manager: *Christine Henry*
Marketing Assistant: *Erica Clifford*
Art Director: *Jayne Conte*
Cover Design: *Bruce Kenselaar*
Manufacturing Manager: *Trudy Pisciotti*
Manufacturing Buyer: *Michael Bell*

© 2001 by Prentice-Hall, Inc.
Upper Saddle River, New Jersey 07458

Printed in the United States of America
10 9 8 7 6 5 4 3 2 1

ISBN 0-13-526518-5

Prentice-Hall International (UK) Limited, *London*
Prentice-Hall of Australia Pty. Limited, *Sydney*
Prentice-Hall Canada Inc., *Toronto*
Prentice-Hall Hispanoamericana, S.A., *Mexico*
Prentice-Hall of India Private Limited, *New Delhi*
Prentice-Hall of Japan, Inc., *Tokyo*
Pearson Education Asia Pte. Ltd.
Editora Prentice-Hall do Brasil, Ltda., *Rio de Janeiro*

Contents

Preface

We live in an interglacial period during which glacier ice covers 11% of the continents. A further 11% of the ground is permanently frozen, 12% of the surface water is frozen, and ice surrounds us in the atmosphere. This sphere of ice (cryosphere) influences all our activities. Changes in average global temperatures of just a few degrees Celsius or changes in insolation at mid-latitudes can either move us into a warmer interglacial time or, conversely, plunge us back into a full glacial period. In either case humankind will have to make tremendous efforts to adjust to changing climatic conditions: including dryer conditions in some places, wetter conditions in others, and changes in sea level. All this is nothing new; it has gone on for the last 2.5 million years of recent geological time, and has occurred several times before during the last 3 billion years of Earth's history. The so-called ice ages occur when the average temperature on Earth is so low (approximately what we have today) that small temperature changes (a few degrees Celsius) may force alternating periods of glacial advance (glacial periods) and retreat (interglacial periods). We are now living in the most recent interglacial period, called the Holocene: a time of rapid change in the last 10,000 years.

Changes bring difficulties, but also opportunities if we are prepared for them. This book aims to help the reader understand the processes and history of glaciation: how glaciers form and move, what effect they have, when and where they have affected the Earth, and the consequences of ice ages. The approach is to analyze the workings of present glaciers and learn how to "read" the sediments and landforms left by previous glaciers. To do this we first need to understand how glaciers form and act: this is the field of glaciology. Then we need to analyze how glaciers and the meltwater derived from them interact with the substrate: how they erode it, how and where they deposit sediments, and what landscape they develop; this is the field of glacial geomorphology. Finally, we need to establish what kind of sediment and rock past glaciers have left behind, and the history of glaciation they record; this is the field of geology. These three approaches constitute what we call here the "Principles of Glacial Geomorphology and Geology." We use the term "principles" because the fundamental processes and their effects are the major focus of the book, rather than detailed analysis of any particular region or environment.

This book is designed for science and nonscience students alike who have an interest in natural sciences and in understanding how nature and humanity have been tremendously impacted by glacial events. It is designed for anyone who has completed a first-year university geology or geomorphology course, and/or some high school science courses. For this reason, the subject matter is approached in a scientific manner, but using a minimum amount of mathematics. For those wanting to pursue a topic further, numerous up-to-date references are reported in the two volumes by Menzies (1995, 1996), the book by Benn and Evans (1998), and for periglacial settings, that of French (1996).

Geology and geomorphology are visual subjects. So we have included numerous photographs and diagrams, many of which are "classics" in this field. To keep the cost of the book down, black-and-white photographs have been used. However, stunning color pictures are available in books like that by Andersen and Borns (1994) and on numerous World Wide Web (WWW) sites on the Internet. The site *http://instaar.colorado.edu/* provides good information and gateways to many other sites dealing with glacial geomorphology and geology. A WWW site companion to this textbook will update information with the help of the readers. Any feedback, constructive criticism, suggestions, information, diagrams, or stunning photographs you have to offer will be welcome and duly acknowledged. Please contact us via our website at www.prenhall.com/glacial.

I. P. Martini
M. E. Brookfield
S. Sadura
University of Guelph, Ontario, Canada

To our students who enjoyed and suffered through it all, and helped in focusing this text.

Introduction

CHAPTER 1

Introduction

GLACIOLOGY, GEOMORPHOLOGY, AND GEOLOGY

Nowadays everyone is, or should be, concerned about the environment, and the effect humanity has on it. Expanding populations create ever-increasing needs for resources such as sand and gravel, minerals, and energy, and they increasingly contaminate soil, water, and air. Conversely, natural phenomena such as flooding and rise in sea level increasingly constitute hazards. People at high latitudes and altitudes need to better understand the processes that generated those landscapes and materials. People at lower latitudes and altitudes may be affected by local climate changes, floods, and global sea level changes brought about by glaciation. This means understanding cold-climate processes, principally those related to glacial environments and frozen ground. To achieve this, the subject matter of this book is subdivided into three major parts: glaciology, glacial geomorphology and geocryology, and glacial geology.

Glaciology is the study of ice in all of its forms. Here the study of glaciers is emphasized because they brought about tremendous changes on Earth's surface both during the Quaternary and in earlier times (Fig. 1.1). To understand a glacier we need to understand its component parts and its behavior. Therefore, we need to analyze the snow and ice that compose it. Ice is a mineral with a simple composition (H_2O), which forms on water, underground, in the atmosphere (primarily as snow and hail), and as the transformation of snow in glaciers. Snow can be considered a wind-blown sediment — for example, deposited in snowdrifts — which through changes (metamorphism) may form ice masses (metamorphic ice rock) in high altitude and latitude areas (above the snowline). An accumulation of snow and metamorphic ice is called a **glacier** if it is thick enough to be able to move under its own weight.

Glacial geomorphology is the study of interactions between the glacier and the land surface, and of the landforms and sediments of areas presently or formerly occupied by a glacier, as well as of adjacent areas indirectly affected by the glacier. Geocryology is the study of frozen ground.

Glacial geology is the study of sediments formed by glaciers, and of the history of areas molded directly or indirectly by glaciers. This includes the study of not only the lithology, texture, shape, sedimentary structure, and architecture of such deposits, but also their physical and temporal distribution, that is, their stratigraphy. Here emphasis is placed on the study of Pleistocene deposits, but some reference is also made to glacigenic deposits formed in ancient times, such as during the Permian-Carboniferous, Devonian, Ordovician, and Precambrian (Fig. 1.1).

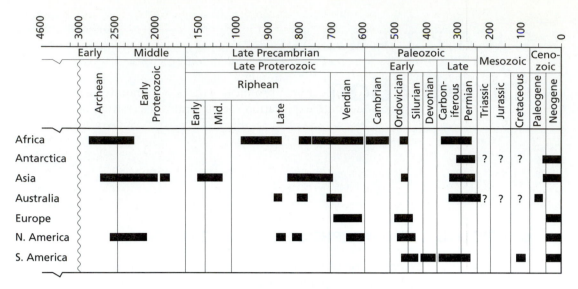

FIGURE 1.1   Diagram showing glacial record on Earth. (From Eyles and Eyles 1992.)

GUIDING CONCEPTS: GRAVITY AND HEAT

This book is not encyclopedic; rather, it tries to aid the reader in achieving a basic understanding of a few fundamental concepts to explain what has happened on Earth during cold periods (ice ages).

The basic concept followed is that whatever is happening, or has happened, at Earth's surface is driven by **gravity** and **heat** (forms of energy).

1. Gravity does not change much, but its components change according to topography, and it determines mass redistributions on Earth, including the flow of glacial ice.

2. The heat of Earth's surface changes through time and space, depending on the conditions that modulate either the total amount of energy input or its distribution on Earth's surface. Heat has a profound effect on ice because H_2O is close to its crystallization (and melting) point at Earth's surface temperatures. A slight drop in average global temperature may significantly contribute to the formation of large new glaciers, and thus to a new full-blown ice age.

3. Heat and gravity combined affect ice a great deal, whether it is found in the atmosphere, inside a glacier, or in the ground. For example, the melting and crystallization point of ice decreases as the pressure increases. This concept is fundamental in understanding the formation and behavior of glaciers, the formation of glacial sediments and landforms, the development and persistence of ground ice, and for planning human activities in cold regions without major damage and costly environmental modifications.

4. Changes in temperature distribution (climate) in the last 2.6 Ma had a profound effect on the distribution of glaciers, and on the development of certain species, like humankind, and the disappearance of others, like mastodons. Similar events have occurred throughout Earth's history. Cold "ice ages" have alternated in the past with warm "greenhouse ages," due to terrestrial and extraterrestrial factors that changed the amount and distribution of heat at Earth's surface. The study of what is happening today, and what has happened in the last 2.6 Ma, and in more ancient times, has not only academic but practical implications. The distribution of modern peat and ancient coal deposits, for example, is related to specific climatic conditions, with the peat and coal of cold periods having markedly different energy properties from those formed under warmer conditions. Glaciofluvial sands and gravels are excellent construction materials, and glaciolacustrine (lake) clay is used for making bricks and other ceramic products. Traces of minerals dispersed over continents by glaciers facilitate exploration for mineralized substrate in areas now covered by forests and wetlands, as in the discovery of diamonds in northeastern Yukon Territory (Canada) in 1993. Finally, the study of the past may help us understand, predict, and prepare for future changes, which are likely to include a warming trend in the short term (on the order of a few tens of thousands of years) and a return to a fully developed cold glacial time in the long run (on the order of hundreds of thousands of years). Indeed, we are living in an interglacial period.

Glaciology

Glaciology "is the science of natural ice in all its diversity" (SHUMSKII 1964).

The basic properties of ice have to be considered first in order to better understand the behavior of glaciers and their ability to mold the landscape and generate sediments, as well as to understand the effect of ice that forms underground and over rivers, lakes, and seas.

Ice Properties

WATER

Water (H_2O) makes Earth a unique solar planet. It can be found in nature in various forms (phases): vapor, liquid, and solid. Ice is the solid phase of H_2O.

The water molecule (H_2O) is composed of two positive univalent hydrogen ions (H^+) asymmetrically bonded (**covalent bond**) to one bivalent negative oxygen ion (O^{-2}) (Fig. 2.1A). This asymmetry leads to small residual negative charges at the oxygen end of the molecule and slight positive charges at the hydrogen end. The water molecule is thus slightly charged (**polar**) (Fig. 2.1B). In a mixture of water molecules, the hydrogen end of one molecule attracts the oxygen end of another molecule, forming a weak electrostatic bond called a **hydrogen bond.** The hydrogen bond is about 6% as strong as the covalent bond between oxygen and hydrogen within each molecule. Hydrogen bonds are easily formed and broken by reducing or adding energy, such as heat, to the system. In the vapor phase (water vapor) the H_2O molecules move independently, temporarily forming hydrogen bonds with other molecules and breaking them. The energy of the system is too high for the molecules to stick to one another in any sort of permanent position. In the liquid phase (water), the energy of the system is lower, and the H_2O molecules can form long chains (Fig. 2.1C) packed tightly against one another. Nonpermanent bonds may form between chains. In the solid phase (ice), the energy of the system is lower still and the H_2O molecules can get closer and form semipermanent bonds with the surrounding molecules, usually generating a crystal structure.

ICE MINERALS

Ice differs from water in structure, physical properties, and isotopic composition.

Ice Varieties

There are **nine** known varieties of ice: one amorphous (glass) and the others crystalline or behaving as such (Fig. 2.2).

1. Amorphous ice does not occur naturally, but can be produced experimentally by condensing water vapor supercooled to temperatures below –120°C. It has the properties of a glass, but devitrifies when heated to –70°C.
2. Crystalline ice (ice I) has two polymorphs at the surface of the earth, that is, two solid phases with the same chemical composition but different structures.

9

a. A *high-temperature polymorph* of crystalline ice is hexagonal with a ditri-gonal pyramidal habit. It exists at temperatures between 0°C and about −70°C and at pressures as high as approximately 2 kilobars (kb), or 2000 atmospheres (atm) (2047 atm to be exact) (Fig. 2.2). This is the type of ice that we are all familiar with and that forms glaciers, where the pressure may reach up to about 250 atm.

b. A *low-temperature polymorph* is cubic. It exists at temperatures below about −70°C, under very high vacuum conditions. On Earth, it is found only high in the troposphere.

Other ice varieties (Fig. 2.2) exist at pressures greater than about 2 kb, and at various temperatures, some as high as +200°C. Those that exist at temperatures higher than 0°C are called **hot ices.** They exist only at enormous pressures, well be-yond any that can be physically exerted at Earth's surface. Very high pressure ices behave as solids, as they are composed of water molecules fixed in space. What keeps water molecules immobile at such high pressures are chemical attraction

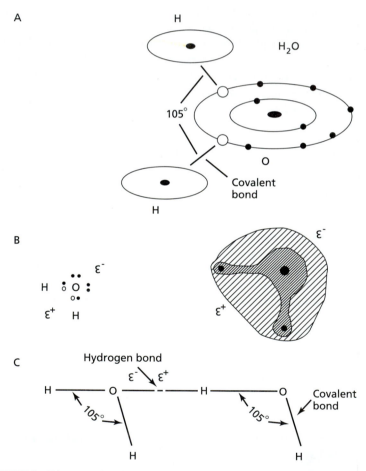

FIGURE 2.1   Diagrams showing structure of H_2O. *(A)* Molecule of water where hydrogen (H^+) shares electrons with oxygen (O^{-2}) (strong covalent bond) (black ellipse = nucleus; black dots = electrons; small open circles = shared electrons between oxygen and hydrogen). *(B)* Isolated molecules of water in space showing residual positive (+) and negative (−) charges. *(C)* Long chains of H_2O molecules joined by hydrogen bonds due to residual charges.

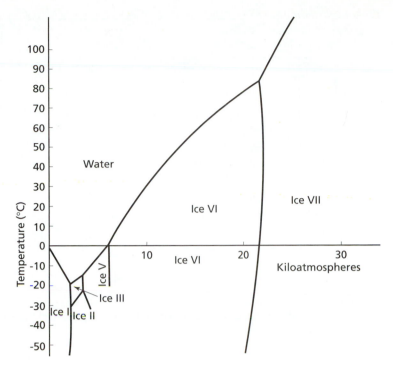

FIGURE 2.2   Phase diagram of water showing the ranges of pressure and temperature under which ice types and fluid phase are stable (1 kiloatmosphere = 1 kilobar = 1000 atm). (After Shumskii 1964.)

forces at the surfaces of other minerals. Essentially these ices are films a few molecules thick, chemically bonded to other minerals. Although these ice polymorphs are not important for the formation and behavior of glaciers, they may affect properties of rocks, such as susceptibility to weathering and permeability to water.

Crystallography

Common ice (the type dealt with in this book) is a mineral, and therefore has a relatively stable crystalline structure within a certain range of pressure and temperature. Ice is made up of water molecules held together by relatively weak hydrogen (electrostatic) bonds in a tetrahedral arrangement. Each oxygen ion is surrounded at equal distances (0.275 nm) by four other oxygens situated at the corners of a regular tetrahedron (four-sided three-dimensional shape) (Fig 2.3). These tetrahedral building blocks can be arranged in space to form crystals with either hexagonal or cubic symmetry. Glacier ice and other types of natural ice show hexagonal (trigonal) symmetry. This structure has been determined by X-ray diffraction analysis of the position of the oxygen atoms in space. The hydrogen atoms are too small, and not static enough above −70°C, to be recorded by X-ray diffraction techniques.

The dimensions of a unit cell of ice at 0°C are 0.448 nanometer (nm = 10^{-9} m) along the a-axis, and 0.731 nm along the c-axis (Fig. 2.3), from which the crystallographic constant of ice at 0°C = c/a = 1.632 can be calculated.[1]

Physical Properties

The properties of a pure ice crystal are usually defined at the equilibrium conditions of the pressure melting point. This is 0°C at 1 atm, but it can be lower at higher

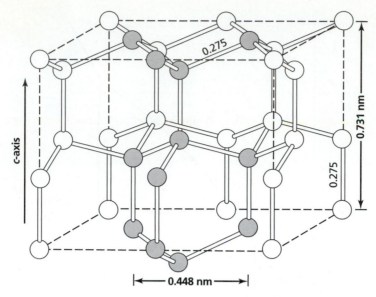

FIGURE 2.3 Diagram of structure of ordinary ice (hexagonal). Oxygens are indicated by circles, and hydrogen bonds are indicated by rods. The unit cell is highlighted by shading of the oxygen atoms.

pressures; nevertheless, it is standard practice to use the symbol 0°C to indicate that the system is at the pressure melting point, even when the actual temperature is less than 0°C.

The **melting point** decreases with increasing pressure in the following ways.

For hydrostatic pressure:

$$dT/dp = -0.00752°C/atm \text{ (at the pressure melting point)}$$

where

dT = small change in absolute temperature
dp = small change in pressure

For oriented pressure:

$\Delta T = -0.0044°C/atm$ (for a pressure up to 30 kg/cm², which is the force equaling the breaking point of ice in compression at the pressure melting point)

where

ΔT = change in absolute temperature

By increasing the total (confining and oriented) pressure, the melting point may be reduced to an approximate minimum of −22°C (at about 2047 atm). Further increase in pressure leads to the transformation of ice I into other types of ice with decreasing volume (compression), and not to melting (Fig. 2.2).

Specific Gravity and Volume. The *specific gravity* of ice is lower than that of water, whereas its *volume* (for a given mass) is higher. This is because ice has a three-dimensional structure, which is more open than that of water, that is instead composed of well-packed chains of H_2O.

Phase	Specific Gravity	Volume
Ice (at 0°C, 1 atm)	0.91689 g/cm^3	1.0906 cm^3/g
Water (at 0°C, 1 atm)	1 g/cm^3	1 cm^3/g

The **specific heat** is the amount of heat necessary to raise the temperature of 1 g of ice by 1°C.

$$C_p = 0.5057 + 0.001863T \text{ cal/g °C}$$

where

C_p = specific heat of ice at constant pressure
T = temperature

Note that one kilocalorie (kcal = 1000 cal) is the amount of heat necessary to raise the temperature of 1 kg of water from 14°C to 15°C, at 1 atm. Note also that one calorie is equal to 1/1000 kcal, and that 1 BTU (British thermal unit) is equal to 0.252 kcal.

The **thermal conductivity (K)** is a measure of how readily heat is transmitted through a substance. The thermal conductivity for ice is:

$$K = 0.0051(1 + 0.0015T) \text{ cal/cm}^2 \text{ sec °C}$$

The thermal conductivity measured parallel ($K\!/\!/$) to the principal crystallographic axis (c-axis) of ice is greater than that measured perpendicular ($K\perp$) to it:

$$K\!/\!/ \text{ varies between 0.0050 and 0.0052 cal/cm}^2 \text{ sec °C}$$

$$K\perp = 0.0051 \text{ cal/cm}^2 \text{ sec °C}$$

The **heat content (enthalpy)** is the amount of heat needed to bring a body to a given state by means of the irreversible isobaric (unchanging pressure) process. Ice (at 1 atm and 0°C) has a heat content (H_i) of 72 cal/g, and water (at 1 atm and 0°C) has a heat content (H_w) of 152 cal/g.

The **latent heat** is the heat absorbed or freed by a unit weight (such as 1 g) of water or ice during phase transition at the pressure melting point (Fig. 2.4).

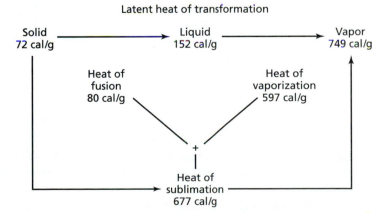

FIGURE 2.4   Model of heat exchanges during H$_2$O phase changes.

Melting (ice → water) requires a latent heat of fusion (L) at 0°C of:

$$L = H_w - H_i = 80 \text{ cal/g}$$

Evaporation (water → vapor) needs the latent heat of vaporization (V) at 0°C of:

$$V = 597 \text{ cal/g}$$

Below 0°C, the latent heat of vaporization increases by 0.00537 cal/g for each degree drop in temperature.

Sublimation (ice → vapor) requires the latent heat of sublimation (S)

$$S = L - V = 677 \text{ cal/g}$$

Note that theoretically the crystallization of 1 g of water vapor at 0°C produces 677 calories of heat, which is sufficient to melt 8.5 g of ice.

During the transformation of ice into water, ice into vapor, and water into vapor, energy is required from the surrounding environment, and therefore, the environment cools. This process is used to reduce body temperature by sweating, or to cool off water inside a bottle by wrapping the bottle in spongy material, wetting it, and letting it evaporate. Conversely, the transformation of water into ice releases energy to the environment.[2]

The **mechanical properties** of an ice crystal depend on the weakness of the electrostatic hydrogen bonds and the geometry of the space lattice. Like any other crystalline substance, ice may behave in an elastic, plastic, or brittle fashion. Factors that determine this behavior are temperature and intensity and rate of force application.

1. When ice is near the melting point, it shows plastic behavior. Lowering the temperature makes it difficult for atoms in the crystal structure to move around; the structure becomes more stable and progressively more elastic and brittle. As a by-product, the **hardness** of ice crystals increases with decreasing temperature. The brittle hardness (measured with Mohs' scale for mineral hardness) of ice at various temperatures is shown below.

Phase	Specific Gravity	Volume
Ice (at 0°C, 1 atm)	0.91689 g/cm³	1.0906 cm³/g
Water (at 0°C, 1 atm)	1 g/cm³	1 cm³/g

Temperature (0°C)	Ice Hardness (Mohs' Scale)	Reference Minerals
0	1.5	Talc/gypsum
−15	2–3	Gypsum-calcite
−30	3–4	Calcite-fluorite
−40	4	Fluorite
−78.5	6	Orthoclase

So, glacial ice alone at or near the pressure melting point would not be capable of abrading any bedrock material except talc and gypsum, whose Mohs' hardness is 1 and 2, respectively.[3]

2. The nature of ice deformation depends on the amount and rate of applied force and associated stress.[4]

a. When an increasing force is rapidly applied to a body, elastic deformation occurs, followed eventually by breakage.

b. If a given force is applied to a body slowly over a long time, the elastic limit decreases and the body deforms permanently in a plastic way. As the body deforms plastically, a constantly decreasing force is required over time to keep the body in a given deformed state. The time (**relaxation time**) required to reduce the stress needed to maintain a given deformation to the eth power ($e = 2.718$) is a measure of the relative importance of elastic and plastic properties of the substance. Consequently, a long relaxation time is related primarily to elastic deformation, whereas a short relaxation time is related primarily to plastic deformation.

The crystallographic structure of ice is elongated in the direction of the c-axis because of the morphology and makeup of the unit cell (Fig. 2.3). The number of bonds parallel to the c-axis differs from those perpendicular to this axis. This leads ice minerals to show different properties parallel and perpendicular to it. Under a given condition, the relaxation time of ice crystals can vary greatly (from about 8 to 90 minutes) depending on the direction of force application relative to the crystallographic axes. So a given force applied parallel to the c-axis elastically bends the crystallographic basal planes until the breaking point is reached (Fig. 2.5A). The same force applied perpendicular to the c-axis makes the basal planes slip like a deck of cards (Fig. 2.5B), and produces plastic deformation by elongating the crystal in that direction (Fig. 2.5C). This slippage occurs along **gliding planes** perpendicular to the c-axis. The crystals are weaker in this direction because fewer bonds need to be broken in the unit cell (Fig. 2.3). These bonds break and are reformed at the submicroscopic scale, generating what macroscopically appears as plastic behavior.

Viscosity is the internal friction due to molecular cohesion, that is, a property that enables a substance to withstand a shearing stress. For water, the viscosity is 0.0179 poise at 0°C, and this decreases with increasing temperature to a minimum of 0.0028 poise at 100°C.[5] Ice crystals are anisotropic for viscosity; that is, they have variable viscosity in different crystallographic directions. Viscosity is about 1000 times greater parallel to the c-axis than perpendicular to it. The viscosity perpendicular to the c-axis is 10^{10}–10^{11} poise, and parallel to the c-axis it is 10^{14}–10^{15} poise. Ice viscosity decreases with increasing temperature near the pressure melting point, due to partial melting.

Strength is resistance of a substance to deformation under specific environmental conditions, that is, the ability to withstand differential stress. It can also be considered the amount of unit volume change when pressure (stress) increases by one unit. The crushing strength of ice is slightly lower perpendicular to the c-axis than parallel to it. The strength of ice parallel to the c-axis varies from about 31 to 33 kg/cm², whereas the strength perpendicular to the c-axis varies from about 20 to 25 kg/cm².

Radiant energy, such as solar energy, can penetrate and break bonds inside transparent ice crystals. Because ice crystals are hexagonal, a six-sided bubble of water can form inside the mineral, along planes perpendicular to the c-axis. These are called **ice flowers.** The importance of this is that water impurities are generated inside the crystal, weakening its structure.

Isotopic composition of water varies in terms of hydrogen isotopes (the ratio between the heavy (^2H) and light (^1H) isotopes on average is about 1:6500) and in terms of the oxygen isotopes ^{18}O, ^{17}O, and ^{16}O. Ice crystals that form from water of different isotopic composition do not differ significantly in physico-chemical prop-

FIGURE 2.5   Diagrams showing deformation of an ice crystal under applied forces (black arrows). *(A)* When the applied force is parallel to the *c*-axis, the basal planes of the ice crystal bend. *(B)* When the applied force is perpendicular to the *c*-axis, the basal planes of the ice crystal slide like a deck of cards (black arrows indicate applied stress). *(C)* Force applied perpendicular to the *c*-axis causes slippage along glide planes.

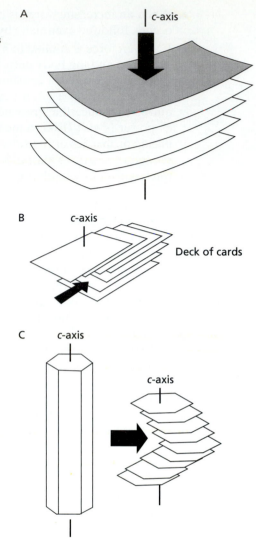

erties, although the concentration of certain isotopes in ice accumulations can provide proxy data for past climatic conditions.

ICE AS A SEDIMENT AND ROCK

Loose snow is an accumulation of ice crystals, which when solidified can be considered a rock. Ice rock can be found:

1. above other rocks: for example, in glaciers;
2. within other rocks: for example, as cement or lenses in soil; and
3. on water: for example, an ice shelf.

Ice is a peculiar rock because it is directly related to climate, and on Earth it is mostly restricted to the Quaternary, except in Greenland and Antarctica, where some older ice may exist. Apart from these two characteristics, ice rock behaves like any other rock and is an important part of the earth's crust. It forms facies that have constant characteristics and composition and are bound to particular geological processes.

The two major forms of ice rock are:

1. congelation ice formed from freezing of water, and
2. metamorphic glacier ice formed through changes of snow.

As with any other rock, ice rock contains impurities that modify the properties of single ice crystals. The principal types of impurities are gas, liquid, salt, and insoluble solid inclusions (other rocks). Some of these impurities are incorporated as the ice forms, whereas others are added later. Impurities do not affect the ice mineralogy, but they do affect the properties of ice accumulations.

Gas Impurities

Gas impurities are mostly air, but CH_4, H_2S, CO_2, and other gases also are incorporated (Fig. 2.6). In snow, as much as 99% of the volume can be air, in firn (last-year snow) up to 56%; but in ice most of the air has been eliminated. The transformation of the sediment snow into metamorphic glacier ice mostly involves the progressive elimination of air impurities.

Gases, mostly air, affect ice in various ways. Because of their lower thermal conductivity, air bubbles (Fig. 2.6A) reduce the thermal conductivity of ice rock. Because they contain a lot of air, snow packs, for example, are very good insulators.

Air bubbles reduce the transparency of ice and its permeability to radiant energy (which becomes scattered at the interface between ice and air). So, trapped air bubbles reduce the melting of ice at depth, although concen-

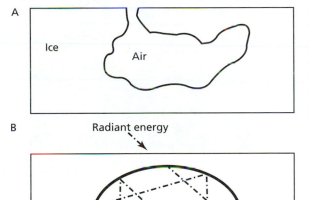

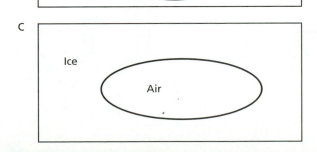

FIGURE 2.6   Diagrams showing position of gas impurities in ice. *(A)* Air bubbles in connection with the atmosphere decrease the thermal conductivity of the system. *(B)* Air bubbles in ice scatter incoming radiation, and thus concentrate energy near the surface (dashed arrow indicates the path of scattered energy). *(C)* Air bubbles isolated from the atmosphere may become overpressured.

tration of energy, and the resultant melting, may be enhanced near the surface (Fig. 2.6B).

If air within the ice is not in communication with the outside atmosphere, it participates in the transmission of stresses (Fig. 2.6C). In this way, ice masses can become suspended on overpressured[6] air bubbles, and can move readily and at relatively high speed.

Liquid Impurities

The most important liquid impurity in ice is water. Water forms in ice because of heat input, increased pressure that reduces the ice melting point, and salt impurities that also lower the ice melting point.

Input of heat into ice that is much colder than the pressure melting point simply heats it. Near the melting point, water appears at the crystal boundaries (Fig. 2.7). At the melting point, a dynamic equilibrium is established between the solid and liquid phases (Fig. 2.8). As some of the ice melts, it cools the surrounding environment to below the melting point so that more energy is necessary to start melting again, and so on. Any added energy (heat) is used to melt ice, and the ice-

FIGURE 2.7   At the melting point, water develops first at the ice crystal boundaries. *(A)* Photomicrograph of a thin section of ice with impurities and water (arrows) between the crystals. Bubbles are artifacts of the thin-sectioning process. *(B)* Diagrammatic representation of water developing between ice crystals.

A
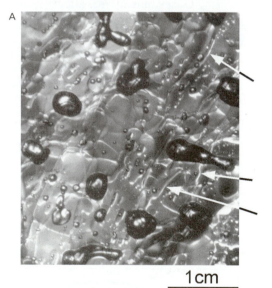
1cm

B
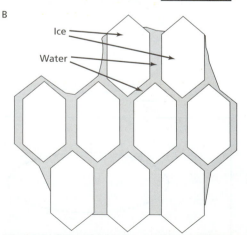
Ice
Water

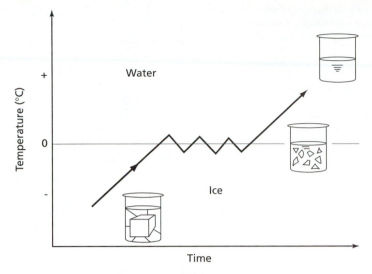

FIGURE 2.8 Diagram showing that addition of heat through time causes ice to warm up and melt. As long as a mixture of water and ice exists, the temperature of the system is kept at the pressure melting point (0°C at 1 atm). Only after all the ice has been melted will the water warm up.

water mixture remains at the melting point (0°C at 1 atm) until all the ice has been melted; only then does the water heat up.

During this process, melting begins at the crystal boundaries; so the interstitial water is concentrated there. At the surface of snow or ice accumulations, penetrating radiant energy may melt ice crystals along basal planes, and within ice crystals some water may be found as ice flowers. Meltwater may fill the pores of snow and firn and can constitute up to 60% of the entire volume. If the water content increases above this level, ice crystals lose contact with one another, and slush (ice crystals floating in water) is formed.

The very top (about 10 m) of a glacier can experience large changes in water content due to diurnal or seasonal melting and freezing cycles. This top zone is called the **active layer.** Most of the atmospheric energy added to glaciers melts ice at the surface. Only a very small amount is transmitted into deeper parts of the glacier by **conduction** along a very weak temperature gradient (approximately 1°C per 1400 m). Heat is mainly transferred into a glacier by downward percolation and re-freezing of water, which liberates the latent heat of fusion (melting) at depth.

Pressure changes occur within a glacier as its thickness or flow rate changes. When there is very little heat exchange, the processes are near **adiabatic** (changes in pressure or volume without changes in heat). This means that *the amount of water in a given volume of ice at the pressure melting point is proportional to the pressure at that point.* For example, under the weight of an ice layer 750 m thick, the ice at the pressure melting point contains 3.6 g of liquid water per cubic decimeter, while at a depth of 1500 m it contains 7.2 g/dm^3. The slightest pressure change leads to re-freezing of water or further melting of ice.

Salt impurities

Salt impurities occur in all natural ice accumulations because all natural waters contain salts. Salt can also be blown in or dissolved from solid rock inclusions within the ice. The main effect of salt impurities is to lower the freezing point of

FIGURE 2.9   Phase diagram of salt (NaCl) + water solution. T1 is the temperature at which the first ice crystals (in black) form; T2 is a lower crystallization temperature corresponding to higher salt content; and E is the eutectic point. (Note that the volume of the black crystals in the beakers is shown to increase because, contrary to other minerals, the volume of ice is greater than the volume of equivalent water.)

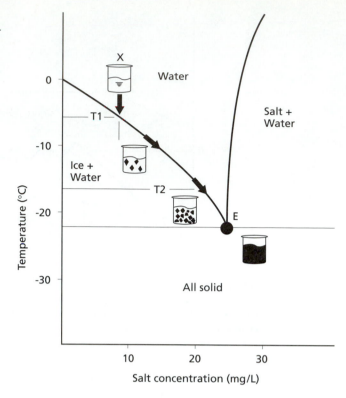

water in proportion to the concentration of salt dissolved in it. Figure 2.9 shows which phase is stable in solutions with different concentrations of salt at different temperatures. If a given solution (X, Fig. 2.9) begins to cool, pure ice crystals start forming when the pressure melting point is reached (T1, Fig. 2.9). There is no space in the crystal structure to incorporate extraneous ions; so the salt concentration in the residual fluid progressively increases, and its freezing point is lowered along the boundary between the water and the water + solid phase. As the temperature is lowered to a new freezing point of the solution (T2, Fig. 2.9), more ice crystals form and the residual fluid becomes more concentrated. A slush of crystals and brine is formed, until at the eutectic point (E, Fig. 2.9) everything freezes. So the presence of salt may locally change the water content of a glacier and therefore its behavior.

Mineral impurities have several effects on ice-rock properties depending on the types of minerals present and their concentrations (Fig. 2.10). (1) Minerals are sources of salts; so their presence may lead to the generation of water. (2) Minerals have a larger coefficient of absorption of radiant energy than ice; so at the surface of the glacier, small isolated rock particles promote local melting and increased plasticity of ice deposits (Fig. 2.10A). (3) Large or thick particle accumulations on the surface prevent melting by shielding the ice from radiant energy (Fig. 2.10B). (4) Isolated rock particles at the base of the glacier can induce high local stresses,[7] leading to pressure melting and decreased ice strength (Fig. 2.10C). Conversely, numerous rock impurities within the glacier increase the strength of the ice-impurity mixture because rocks are not as plastic as ice (Fig. 2.10D).

METAMORPHISM OF SNOW

Metamorphism means change. Any material undergoing change leads to a minimum free energy state and an increase in entropy. That is, the tendency is to oblit-

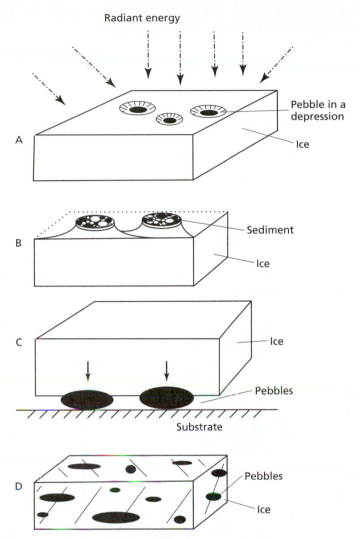

FIGURE 2.10   Diagrams showing effects rock impurities have on ice. *(A)* Melting occurs around small rock clasts at the surface as they absorb radiant energy. *(B)* Thick deposits of sediment or large rock fragments locally protect the ice from melting. *(C)* Melting occurs at depth due to increased stress over isolated clasts where the pressure melting point is reached. *(D)* Increased number of strong rock fragments increases the rigidity of the whole deposit.

erate distinct entities and increase the random order of matter in the universe. All minerals and rocks, including snow and ice, undergo metamorphism in a primarily solid phase, under high (for the particular system) pressure and temperature. The change is analyzed here first as it affects a single entity such as a snowflake (an ice crystal), and then an assemblage of entities. Rates of change and metamorphic processes active under various ambient conditions are also discussed.

Concept of Free Energy

Free energy is that portion of energy from any natural process or body that can be converted to useful work. Absolute free energy cannot be measured for any chemical element or compound; only differences in free energies can be measured.

The most stable natural chemical configuration is one that leads to the lowest free energy for the system; thus *spontaneous chemical reactions* occur only when the free energy changes are negative.

$$\Delta G = H - T\Delta S$$

where

    ΔG = change in free energy
      H = enthalpy (heat content)
      T = temperature
      S = entropy (a measure of disorder)

Total free energy can be subdivided into bulk energy (the energy of all molecules in a body) and surface free energy (the energy of molecules at the surface of a body) (Fig. 2.11). Surface ions, atoms, or molecules of a solid are not completely surrounded by other ions, atoms, and molecules of the body. They are not as tightly bound to the solid as are the inner ions, atoms, and molecules; thus they have an *excess* surface free energy. This means that they can react with the surrounding environment and readily undergo chemical or structural change, thereby releasing en-

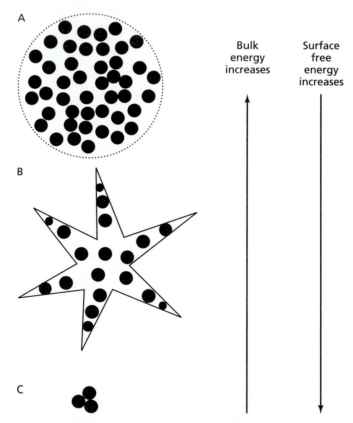

FIGURE 2.11   Diagrams indicating changes in free energy related to the number of molecules in the crystal and the crystal shape. *(A)* Aggregation of many molecules (black) within spherical crystal; this disposition shows the lowest number of molecules at the surface in relation to those in the main body, hence lowest surface free energy. *(B)* Snowflake-like body with many molecules at the surface. *(C)* Incipient, unstable aggregation of molecules; all molecules are at the surface, so all react with the surroundings.

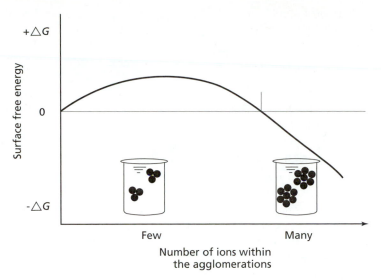

FIGURE 2.12 Diagram of changes in surface free energy (ΔG) during homogeneous nucleation. During early stages of nucleation when few ions aggregate, surface free energy increases. This goes against the natural reaction trend, rendering the process difficult.

ergy in other forms. The magnitude of surface energy is directly proportional to the surface area of the solid. For large crystals, bulk energy is much greater than surface free energy (Fig. 2.11A), and surface free energy can be neglected in thermodynamic calculations. For small or sharp-cornered crystals, though, and particularly during initial stages of crystallization, surface free energy is high enough that it cannot be neglected (Fig. 2.11C).

Surface free energy controls nucleation and crystal growth. There are two types of nucleation: homogeneous and heterogeneous.

Homogeneous nucleation may occur when particles of ice or other substance are not present in the solution. It develops when chance collision of ions forms a crystal embryo large enough (with sufficient bulk energy) to continue to grow, and thus decrease its surface free energy. Homogeneous nucleation rarely occurs in natural solutions at their standard freezing point, because of the high reactivity of the early agglomeration of a few ions. During the early stages of agglomeration, due to random impacts of ions, the system undergoes an unsustainable increase in surface free energy (Fig. 2.12). The temperature of the system must be lowered well below the freezing point (supercooled) for homogeneous nucleation to work.

Heterogeneous nucleation occurs around a pre-existing seed of ice or other matter. The seed provides sufficient initial bulk energy for an ice crystal to start growing on it. As the crystal grows, there is a corresponding decrease in surface free energy.

From this it follows that small crystals or the points of a skeletal crystal have excess surface free energy and are more soluble than larger, rounder crystals. Thus, natural crystal accumulations modify (metamorphose) following two major pathways (Fig. 2.13).

1. Single crystals tend to reduce the surface-area-to-volume ratio to a minimum and tend to become more spherical (a sphere has the smallest specific surface of all geometric figures) (Fig. 2.13A).

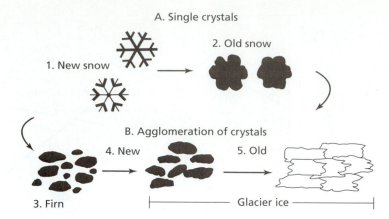

FIGURE 2.13 Schematic representation of metamorphism of ice. Transformation of single crystals *(A)* and agglomerations of crystals *(B)* during transformation of snow into glacier ice.

2. A polycrystalline agglomeration has different sizes of crystals. Larger crystals have smaller surface free energy, are more stable, and grow at the expense of smaller crystals, which dissolve (Fig. 2.13B).

Transformation of Snow into Glacier Ice

Snow is transformed into glacier ice in two environments:

1. In a cold environment where the ice minerals are consistently below the pressure melting point and where sublimation and solid-state transformations prevail

2. In a warm environment where the ice minerals are at the pressure melting point and some liquid water is present throughout

The rate of transformation is lower in (1) than in (2). This can be shown by following the transformation of snow with depth within a glacier (Fig. 2.14). Note that this transformation leads to a progressive decrease in porosity of the ice, hence in the elimination of interstitial fluids (generally air and water).

Stage 1. New, fresh snow in a cold environment consists of skeletal, hexagonal crystals (Fig. 2.13A). The accumulated snow is fluffy and light, with a density of about 0.05 to 0.1 g/cm^3 and a porosity of about 95%. The pores are filled with air, making snow a good insulator, a property used by animals, such as polar bears for their dens, and by Inuit to build ice huts (igloos).

Freshly deposited snow contains an enormous internal surface area and it can exchange moisture readily with interstitial fluids. Sublimation occurs in cold conditions, and local melting and infiltration in warm conditions. In both cases, the most reactive molecules of the skeletal crystals are located at the points, that is, those less well held by the crystal itself and having more exposure to the surrounding fluids, hence higher excess surface free energy. Through sublimation or melting (breaking of a number of crystalline bonds), molecules are removed from the points and tucked away into the saddles where they are surrounded by a larger number of crystal-bond molecules. In this position, crystals have lower surface free energy and therefore the molecules are held more tightly. This leads to a progressive rounding of the crystals, and because the long skeletal arms are eliminated, the particles become smaller, though denser. Furthermore, due to the weight of the snow deposit itself, compaction occurs and some crystals are crushed.

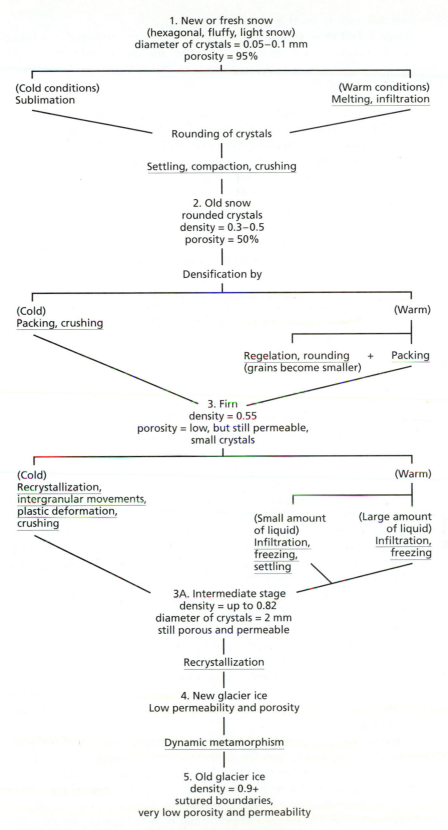

FIGURE 2.14   Summary of stages of snow-ice metamorphism.

25

All this leads to what is called **old snow.**

Stage 2. Old snow contains rounded crystals of ice arranged in relatively close packing. The density may reach 0.3–0.5 g/cm³ and the porosity about 50% of the original snow. Old snow is found in glaciers as well as in regions below the snowline with seasonal snowfall.

1. In cold conditions, old snow becomes denser as continued sublimation and crushing rearranges the crystals and packs them tighter. Densification is achieved slowly and at considerable depth. For example, in central Greenland, sublimation, crystal breakage, and packing increase snow-pack density to about 6 to 7 m depth. A further increase in density below this takes place through plastic deformation of the particles.

2. In warm conditions, old snow becomes denser mostly by pressure melting of crystals at their sharp edges, followed by infiltration of this water and regelation (refreezing) in more protected areas (saddles and lower-pressure areas around the minerals). Densification in warm conditions is achieved rapidly and at relatively shallow depth.

Under both warm and cold conditions these changes produce **firn.**

Stage 3. Firn is a German word meaning "last-year snow." Firn deposits contain granular, rounded crystals about 1 mm in size and with a density of about 0.55g/cm³ (Figs. 2.13B, 2.14). Firn deposits show a relatively low porosity compared to snow, but are still very permeable.

Up to the firn stage, modifications mostly occur by single crystals becoming rounder and smaller. Further transformation of the firn is achieved through interaction between crystals. The smaller crystals, having greater surface free energy and thus being less stable, are disrupted, and their molecules are used to expand larger crystals that have lower surface free energy and thus are more stable (Figs. 2.13B, 2.14). This reduces the intercrystalline space in the deposit still further, thus its porosity and permeability.

In cold conditions, further densification of the firn is achieved primarily through continued intergranular movements, plastic deformation of crystals into voids, crushing, and recrystallization. Recrystallization occurs primarily in the solid state. Molecules are transferred from one crystal to another a few at a time without breaking down (melting) of the crystals. In warm conditions, the firn stage is very short lived because infiltration and refreezing of meltwater produce rapid changes (Fig. 2.14).

Substage 3A. All of the above leads to an *intermediate firn-glacier ice stage,* where the average crystal size increases to about 2 mm. The deposit is still somewhat porous, but has low permeability, and some of the water or air trapped in the residual voids may become overpressured.

From this point on, the distinction between cold and warm conditions becomes less important. Although a certain amount of water is always present in a glacier with ice at the pressure melting point, recrystallization is the predominant process of transformation and densification of the ice. This leads to the formation of **new glacier ice** (Figs. 2.13B; 2.14).

Stage 4. New glacier ice contains larger crystals than firn and forms an almost impermeable deposit, although some isolated pores remain. In warm conditions, some water molecules can still migrate through water films around the crystals.

Further changes of new glacier ice occur through **dynamic metamorphism;** that is, physical deformation of the deposit is caused by its own weight, temperature fluctuations, and external forces. Strong directed forces induce stresses that cause melting at some crystal-to-crystal contacts, and water is forced to migrate into lower-pressure areas where it refreezes. Accordingly, ice crystals melt in some directions and grow in others, forming large elongated crystals.

The product of these processes is **old glacier ice.**

Stage 5. Old glacier ice is the final stage of metamorphism of snow and ice. The crystals are large, elongated, and sutured against one another (Figs. 2.13B, 2.14). The deposit reaches densities of about 0.9 g/cm^3 and is essentially impermeable, although some water molecules can still be transmitted along a continuous water phase (water-molecule film around crystals).

SNOW HAZARDS AND RESOURCES

Snow seasonally covers the northern two-thirds of North America and Eurasia and perennially covers most of Greenland, Antarctica, and many high mountains at lower latitudes. As such, snow represents a large resource to people in northern climates and in mountainous regions, but it also presents potential costs and hazards. As a resource, it provides water for direct use or for hydroelectricity, material for winter transportation (commercial and recreational snow routes), and thermal insulation for crops and wildlife. Major expenses relate to the prevention of snow deposition on buildings and roads and the removal of these deposits when they do form. Prevention of thick snow accumulation in unwanted places requires proper design of roads, buildings, and bridges, and, in many areas, construction of snow-protection structures such as snow fences (Fig. 2.15). Every year billions of dollars are spent by states and cities to physically and/or chemically remove snow from roads.

Properties of snowcover such as depth, density, temperature, and hardness are affected by topography and vegetation and other natural or artificially made ob-

FIGURE 2.15   Photograph showing snow fence used to capture drifting snow.

structions. The snowcover is further modified by snow metamorphism, with changes occurring, for example, in layering, strength, porosity, and permeability. These properties vary widely in the landscape, and their mapping is of safety, environmental, and economic importance.

Hazards

Hazards for Animals   Some animals are better adapted than others for life in snow-covered areas; for example, the snowshoe hare has large feet in relation to its body weight, and can readily move over most snowcovers. The adaptation of other animals is to switch food sources; for example, caribou switch from ground mosses and lichens to tree branches and lichens. Other animals migrate away from areas of thick snowfall. When migration is not possible, such as for domesticated animals on ranches such as those in the midwestern states of North America, snowfalls during calving time may result in high newborn mortality rates. In addition, thick snowfall may lead to increased costs of beef production because cattle lose weight unless extra feed on the range is provided.

Snowcover also affects animals indirectly; for instance, migrating great snow geese may not mate if snow lasts two or more weeks after they arrive at their arctic breeding grounds.

Hazards for Plants   When blown about by wind, fresh, cold snow can abrade materials. Trees can be damaged by snow blasting. Their branches can also be broken by the weight of snow after large snowfalls. Indeed the shape of trees is in part dictated by the amount of snow in the area. The bending of branches by heavy snow, however, has the indirect benefit of making branches available as food to animals.

Compacted snow, either by natural (wind, avalanches) or artificially induced (snowmobile trails) means, has an increased thermal conductivity (about 12 times that of undisturbed snow). This may allow the roots of perennial plants to freeze and be damaged by desiccation or frost heave. The resultant decrease in porosity, permeability, and the exchange of gases with the underlying soil also has a detrimental effect on bacterial activity (soil bacteria can be reduced 100-fold), reducing soil productivity.

Hazards for Human Constructions   Snowfall and snowcover can be dangerous for humans: for instance, by generating **whiteouts** (loss of visibility) along roads during storms (blizzards are common in nonforested areas), disastrous avalanches, and snow bridges dangerous for mountaineers. Thick, heavy snowfalls can also damage power lines as a result of loading, hence indirectly causing hazardous or costly conditions.

Avalanche   Abundant snowfall on steep mountainsides may lead to avalanches. An avalanche is a rapid downslope movement of a large mass of snow. It is a mass flow, carrying the denser fluid-sediment mixture through the less-dense fluid (air) body. In many respects it is similar to other density flows such as cold air that sinks under warm air, sediment-laden stream flows that dip under lake waters, or sediment masses that slump underwater and may transform into turbidity currents.

Avalanches are typically composed of fine snow dust (powder avalanche, with densities up to 15 kg/m^3) or dry snow (dry-flowing avalanche, with densities up to 150 kg/m^3) or wet snow (wet-flowing avalanche, with densities up to 400 kg/m^3). Usually these different types occur together and grade into one another. For instance the lower part of a powder avalanche is normally composed of a dry-flowing

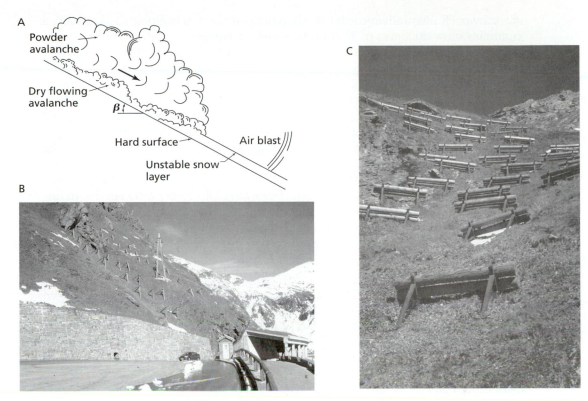

FIGURE 2.16 Representation of a type of avalanche and photographs of protective constructions. *(A)* Mixed snow avalanche. *(B)* Snowshed along an alpine road, Austria. *(C)* Slatted fencing used as an avalanche retaining structure, Austria. (Drawing after Gray and Male 1981.)

avalanche (Fig. 2.16A). Furthermore, as the avalanche gathers speed, friction and collisions among particles generate heat and stresses, which can transform the dry-flowing avalanche into a wet-flowing avalanche.

Avalanches develop when stress exceeds strength at some point in a thick snow pack. This generally occurs on slopes between 30° and 45°. Avalanches seldom occur on slopes less than 25°, or greater than 60°. In the latter case, frequent small snow slides are triggered, negating the possibility of thick snow accumulation.

In relation to the initiation of an avalanche, it is interesting to note that freshly fallen, cold snow behaves as a cohesive material because the irregular, freshly deposited crystals stick together. As soon as crystal rounding occurs through metamorphism, however, the snow behaves as a less cohesive, granular material and loose snow avalanches may occur. Initiation of movement is also enhanced by the formation of discontinuities (layering) within the snowpack, such as the formation of deep hoar through condensation of water vapor. In this case, and if the snow is compacted or otherwise metamorphosed, detachment can occur along these discontinuities and slabs may move and generate slab avalanches. Avalanches are more common on the leeward than on the windward sides of slopes, because more snow is deposited there. Furthermore, in the Northern Hemisphere, north-facing slopes are more hazardous in full winter, because they are colder and likely to develop surface- and deep-hoar layers along which slabs can slide. In the spring, however, south-facing slopes are more dangerous because they receive greater insolation, thus increased melting and weakening of the snowpack. Ground conditions under the snowpack also influence the development of snow movement. Essentially, a minimum snow thickness of 30 cm is required for any movement on a smooth substratum, 50 cm on average treeless ground, and 120 cm on treeless ground with some significant roughness.

Once initiated, the avalanche accelerates downslope, reaching speeds up to 70 m/s for powder ones, 60 m/s for dry-flowing avalanches, and 30 m/s for wet-flowing avalanches. Once they reach flat terrain, avalanches slow down and stop but not before traveling some distance, at times crossing valleys and climbing the opposite flanks. Damages to trees and buildings are caused by the direct impact of the moving mass, whose pressure can reach 30 kPa (1 atm = 101.3 kPa) for powder avalanches and 300 u for flow avalanches. Damage is also caused by the air blast that precedes the main body of the avalanche, by localized lift forces associated with turbulent motion, and finally by the weight and sintering of the deposited snow.

Avalanches cause property damage, communications interruptions, and are responsible for numerous deaths each year. People caught in avalanches usually die of suffocation and hypothermia. Generally a person who is uninjured during transportation can survive only for about 30 minutes once buried in the compacted snow of an avalanche. The best protection is to avoid areas where avalanches occur, but if this is not possible, protective constructions can be built along roads and railways or other remedial actions taken. Snow sheds (Fig. 2.16B) or deflection walls can be constructed, slopes can be kept forested or armored with permanent steel fences (Fig. 2.16C), or snow can be removed from dangerous areas before it becomes too thick. Several methods are used for snow removal, among them explosions and shell bombardment. However, there is no foolproof method, and bad weather may not permit remedial action. This, plus the fact that more skiers and mountaineers can reach increasingly more remote areas (for example, by helicopter), leads to numerous fatalities by avalanches each year.

Snowmelt and Flooding   Snowmelt can be of considerable benefit in some areas, as it provides irrigation waters. However, because it occurs rapidly during a short period, it can cause extremely damaging floods. Streams fed primarily by snowmelt display a characteristic nival hydrograph (Fig. 2.17). They are characterized by very

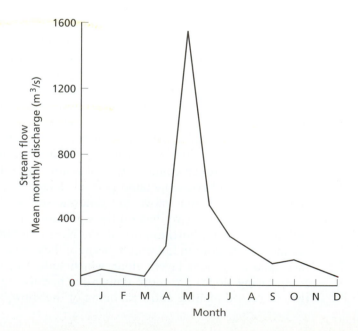

FIGURE 2.17   Annual hydrograph of a river with a typical nival regime, northern Ontario, Canada.

low to no flow during the winter; short-duration, very high flows during spring melt; and relatively low flows during the summer and fall.

Costs

For those who live in northern latitudes or mountainous areas, snow represents a major cost. Snow removal (by mechanical, chemical, or thermal means) from roads is a large municipal budget item, most notably in cities such as Chicago and Buffalo in the United States that are located at the end of one of the Great Lakes and receive high snowfall. The wind picks up moist air from the unfrozen lakes and dumps the moisture as snow as it cools upon reaching land (Fig. 2.18). In 1967, 58 cm of snow fell in Chicago in a two-day period; snowstorms dumping 120 to 180 cm at a time, capable of generating local snowdrifts 7 to 8 m deep are not uncommon in Buffalo. Furthermore, the costs are related not just to snow removal, but also to business shutdowns due to unexpected or large snowfalls and thus to loss of productivity. A large snowfall in 1978 paralyzed activity in the Boston area, and had an estimated cost of about 1 billion dollars to regional commerce.

At the end of 1996, around Christmas, unexpected large snowstorms created havoc in British Columbia and in the western United States (Washington and Oregon). In British Columbia, up to 35 cm of snow was dumped in a few days on Vancouver, a major city, and a similar amount on Vancouver Island, surpassing a record that had lasted for 60 years. This area near sea level usually receives high rainfall from the moist Pacific air, but in 1996 a cold front led to the unusual snowfall. Highways were closed, trapping people in their cars, and the army was called in to help with the rescue. Vancouver airport had to be closed and flights were canceled because de-icing systems for the planes were inadequate (by the time part of the plane was de-iced, another part had re-iced). Subsequently, in the last two days of the year, normal warmer conditions were re-established, with considerable rainfall and snowmelt. This led to dangerous, damaging floods, particularly in the northwestern United States where more than 50 counties, including some in the semiarid state of Nevada, were declared disaster areas.

The best way to protect structures and roadways is to plow away the snow as soon as it has fallen. This, however, is not always possible and is too costly; and thus

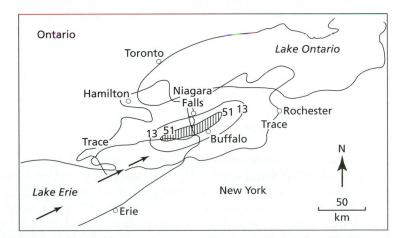

FIGURE 2.18   Map showing a typical snowfall deposition pattern produced by lake effect in the area of Buffalo, N.Y. (arrows indicate direction of storm track; contour lines = snow isopach in centimeters). (After Gray and Male 1981.)

several other methods (use of de-icing substances, heating of surfaces) are practiced to melt and remove snow and ice.

Salts like NaCl are efficient in reducing the melting point of the ice/salt mixture down to about $-21°C$ (Fig. 2.9). Below that point, the mixture remains solid. Normally in very cold climate conditions, snow and ice are not melted; instead, as much as possible they are removed and granular material (sand) is added to roughen the road surface. $CaCl_2$ depresses the melting point down to about $-51°C$, but it costs two or three times as much as NaCl and is seldom used.

The amount of de-icing material put on a roadway depends on the amount of traffic and the temperature of the road surface (which usually is somewhat different from the air temperature). These substances are required to bring the systems near to the melting point, then the pressure imparted by the vehicular traffic and the frictional heat generated by the tires triggers more melting. Furthermore, only about 30 to 50% of the ice needs to be melted, because the resultant slush is then physically removed from the roadway by the traffic itself. Thus, heavily traveled highways are sparingly salted, whereas secondary streets remain ice- and snow-bound for a long time unless considerable salting is done. The amount of melting is related to the amount of salt applied; so to maximize the effect, salt is usually spread along the center of the road. The ice melts rapidly there, exposing the pavement; thus more radiant energy is absorbed. This raises the temperature of the system and fosters further and more rapid melting of the entire road surface. The timing of salt application is also important. Usually it is done during or soon after a snowfall so that the snow cannot be compacted by traffic, but instead is partially melted by it, remaining granular and slushy, and easily removed by snowplows.

The use of de-icing substances improves road safety and in many places it is necessary. However, it has some costly and damaging consequences. The salts corrode cars and any other steel objects, and damage concrete structures as well. Use of corrosion inhibitors with the salt is inefficient unless used in relatively high concentrations, making their costs prohibitive. Furthermore, salts added to the roadways eventually find their way into the surrounding fields, damaging plants and reducing soil fertility. They can also contaminate shallow ponds, small streams, and groundwater. It has been demonstrated that dumping salt-laden snow collected by municipal plowing in local landfill sites can result in mobilization of heavy metals like mercury and lead via complexation with the Cl^- ions. As more chloride ions are added, there is the possibility of finding more highly mobile $PbCl^+$, for instance, in landfill leachate.

In areas that must be kept free of snow and ice and where de-icing substances cannot be used for safety reasons or because it is too cold, heating systems are installed. These may be embedded pipes, electrical cables, or overhead infrared lamps. The high cost of setting up and maintaining these permanent structures is justified only in special conditions such as building entrances, some sidewalks, some parking areas, toll plazas, ramps, roofs, some bridge approaches, and perhaps parts of some airport runways.

Rural areas incur major costs to remove snow from buildings, roadways, and ditches to allow for water discharge. Additional costs relate to substances that are spread on fields. For example, coal dust is sometimes spread on early snowfalls to accelerate melting and allow the sugar beet harvest to be completed. The dark dust particles absorb solar energy and accelerate melting. An alternative practice used to melt snow on turfgrass is to use dark-colored fertilizer.

In resort areas and in mountainous areas crossed by roads, considerable costs are incurred managing snow to prevent dangerous avalanches.

Benefits

Insulation Snow insulates the ground, inhibiting deep frost penetration, thus preventing damage to plants (crops in particular) or buried pipelines. Snow is a very efficient insulating material, particularly when fresh. Furthermore, snow-cover increases the albedo (radiation reflected away from the surface) and reduces the net heat exchange between the ground and atmosphere. The insulating properties of a thick blanket of snow moderate the local ambient temperature, allowing plants and small animals near the ground surface to survive. Indeed, plants can start growing under a relatively thin snowpack where some light can penetrate. The snowcover protects both wanted and unwanted (weeds, fungi, some insects) organisms. For instance, fungi can survive under the snowcover and damage plants. However, most winter plant-kill is due to freezing, desiccation, and frost heaving. Snow and ice cover on fields may not kill plants (unless large amounts of toxic gases accumulate) as much as the subsequent snowmelt and wet conditions that are generated.

It is a practice in some northern countries to accumulate snow around the base of fruit trees for two reasons: to protect the roots of the tree from frost, and to ensure sufficient moisture in the spring.

Transportation Travel on fresh snow is difficult for animals and people alike. Snowshoes, skis, and specialized vehicles such as snowmobiles are needed for cross-country travel. Most snow vehicles are built, however, to be used efficiently and safely on roads made by compacted snow. Such winter snow routes are an essential feature of northern countries for activities such as geophysical surveying, logging, building pipelines, and moving materials to and from mine sites as well as to construction sites for hydroelectric dams. Winter roads link many northern communities, and goods are transported overland and locally across the artificially thickened ice cover of rivers and lakes. The thickening of the ice cover (formation of ice bridges) is done either by removing the snow, allowing ice to thicken by congelation of the underlying water, or by pumping water onto the ice surface where it freezes. Winter roads of packed snow provide both relatively smooth surface to travel on and some, but not total, protection for the underlying vegetation, peat, and permafrost.

Resources

Recreation Snow is an excellent resource for recreational purposes. Various physical properties of snow make it suitable for activities such as skiing, snowshoeing, dogsled travel, snowmobiling, and camping. To ensure that snow is present in the right amount in certain areas, it is also generated artificially using increasingly more effective and reliable snowmaking equipment — another source of revenue related to snow.

There has been considerable research on the friction between snow and skis, and on the shape and composition of skis, to arrive at an optimum compromise between speed, maneuverability, and durability. The introduction of high-speed skiing has led to detailed analysis of the problem of friction. This problem is more complex for skiing than for ice-skating, where high stresses are generated by the skater's weight over the narrow, sharp blades; hence the skater floats over a thin layer of meltwater. In addition, ice in competitive skating rinks is made to a specified temperature, that is, hardness. For skiing, the conditions are less controlled because snow, unlike artificial ice, ranges from very cold, freshly deposited snow, to hard, cold snow, to various degrees of soft, partially melted snow. Assuming no external factors such as sinking or plowing of the skis into the snow, the principal factors that affect friction are brittle deformation of snow under cold conditions, plastic deformation of warmer substratum, and thickness of the water film under warmer conditions still. Essentially, minimum friction is achieved at intermediate conditions where a small amount of water is generated and the skier floats on it; under colder conditions, the hard crystals gouge into the skis and may be hard to break; under warmer conditions, too much water is present and the adhesion forces between the ski surface and the watery substratum may increase friction. Ski manufacturers try to minimize the friction between the ski and the snow, producing a stable (non-vibrating) frictionless instrument. In addition, waxes are used to reduce friction and match the ski base to the snow conditions. Hard waxes are used for cold conditions so that the ice crystals do not gouge too deeply into the ski surface. Softer, water-resistant waxes are used for intermediate conditions so that both the snow and the ski base can deform plastically, and soft aquifuge waxes are used under warmer conditions to reduce the adhesion forces between water and skis. It is also important to realize that the local conditions at the bottom of the skis are very different under static or slow-moving conditions than under fast-moving conditions. In the latter case, heat is generated by impact and friction. The frontal part of the ski runs on dry substratum where brittle deformation occurs and some heat is generated. As the ski passes over that part of the substratum, some of the crystals melt, leading to a smoothing of their surfaces and, at the same time, generating some water molecules that allow the central and tail part of the ski to glide (Fig. 2.19). In this case, it is possible that different waxes may need to be used on different parts of the ski. Note that at high speeds the skier's weight is an important variable, but not a determinant one

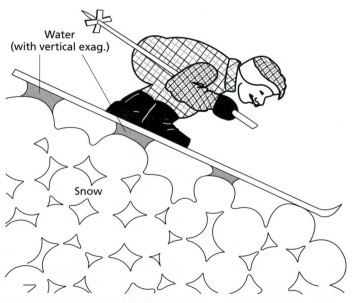

Water
(with vertical exag.)

Snow

FIGURE 2.19   Diagrammatic representation of friction between ski and snow and formation of a thin water film. (After Colbeck 1992.)

in achieving maximum speed. Rather, friction is most important, and, everything else being equal, ski size and waxing are critical. Accordingly, downhill skis are relatively wide (for stability), long (generally 2 m), and made of materials that dampen their vibration (for stability and continuous contact with the snow). Skis used for high-speed trials are solid, wide, and long to achieve maximum stability, low friction, and thus high speed.

Source for Water   Snow provides large amounts of water during spring melt and is a vital resource for irrigation that needs to be managed. For example, in parts of the great plains of southwestern Canada and adjacent states of the United States, the absence of a sizable snowfall during the winter may mean drought during the subsequent summer. For this reason, snow is actually harvested in a variety of ways. Snow-management practices consist of erecting snow fences, planting vegetation barriers (strategically located rows of trees), snow plowing (snow is piled up into parallel ridges, generating space for new snow to accumulate between them), stubble management (leaving tall crops [sunflowers, corn] or their stubble in the field to intercept and accumulate snow), and surface modification (such as digging benches where snow can be trapped and snowmelt can be retained).

Snow supplies an abundant, renewable source of clean drinkable water for homes, and for use in agriculture. Water from snowmelt also serves to recharge reservoirs that store water for generation of hydroelectricity.

Snow supplies one-third of the world's irrigation water, irrigating approximately 185 million hectares. This is very important for the arid and semiarid lee slopes of major mountain chains throughout the world. The snow collects in upland watersheds and supplies snow-fed mountain streams during the early parts of the growing season between mid-April and mid-June. Water generated from snow usually contains low levels of dissolved solids (especially salts), making it highly suitable for agricultural purposes. Five large countries (China, India, the United States, Pakistan, and parts of the former USSR) also cultivate vast semiarid areas, irrigating them from rivers fed from distant mountain snow. Even in monsoon areas, such as in parts of India, crop production during the dry season is possible only because of water derived from the mountains.

Snowmelt contributes more water to streams than comparable rainfall events because snow accumulates during periods of lower evapotranspiration, and snowmelt occurs when the soil may be still frozen and little water infiltrates the ground. However, the production of snowmelt occurs rapidly and, in some arid countries, it must be artificially retained for use during the growing season.

ENDNOTES

1. Older books may use Ångstroms (Å = 10^{-8} cm) instead of nanometers; so it is useful to know that 1 nm is equal to 10 Å.

2. Citrus fruit growers make use of this phenomenon to protect fruit from freezing when air temperature falls below 0°C. By continually spraying their trees with water during the cold spell, the latent heat released during the transformation of water into ice (80 cal/g) may be enough to prevent the temperature of the fruit itself from falling below 0°C. In this way, slight freezing of the skin can actually save the crop.

3. The Mohs' scale indicates that a mineral with a given hardness value, for example, 4, can scratch a mineral with lower hardness, 3, and can be scratched by a mineral with a larger hardness number, 5.

4. Stress (τ) is a force (F) applied over an area (a), so $\tau = F/a$.

5. Poise is a unit of absolute viscosity, equivalent to 1 dyne per s/cm^2; a dyne is the force required to accelerate 1 g at a rate of 1 cm/s^2.

6. Overpressure in a fluid is the pressure in excess of that of a column of such a fluid open to the atmosphere. It is achieved by the fluid being trapped within the rock and in part sustaining the weight of the rock itself.

7. Small particles have a small surface area; so if a force (such as the weight of a glacier) is exerted on them, the stress (force/area) increases to a level where locally the pressure melting point is reached. The same force distributed over a larger surface could be insufficient to cause pressure melting to occur.

CHAPTER 3

Glaciers

LOCALIZATION OF GLACIERS

Glaciers are bodies of ice thick enough to move under their own weight. They develop unique internal characteristics (**facies**) that reflect the local environment. These facies occur in different parts of a glacier, from fresh snow at the top to old glacier ice at the bottom (Fig. 3.1A). It is possible to apply the stratigraphic principle of **Walther's law,** which states that the lateral order of facies seen on the surface of a sedimentary environment can also be seen vertically in sections cut through the deposit, provided there are no major unconformities (Fig. 3.1B). The converse is also true. So, expanding the internal vertical succession of a glacier laterally, the various facies can be seen on the glacier surface (Fig. 3.1B). New and old snow exist upslope from the annual snowline, firn upslope from the firn line, and glacier ice downslope from the firn line.

The **snowline** is the altitude above which there is a net accumulation of snow; that is, some of the snow precipitated during the winter survives during the subsequent summer. Below the snowline, regardless of the amount of winter snowfall, no snow is preserved during the subsequent summer. On a glacier, the snowline generally coincides with the firn line, firn being last year's snow. The **firn line,** also called the **equilibrium line,** separates the **accumulation zone** from the **ablation zone** (Fig. 3.1B). The **accumulation zone** lies above the firn line and has a yearly net addition of ice. The **ablation zone** (wastage area) lies below the firn line and undergoes a yearly net loss of ice. In some cases, meltwater generated at the surface in the accumulation zone flows onto the ablation zone and refreezes, forming a surface layer of congelation ice or **superimposed ice** (Fig. 3.1B).

Glaciers form above the snowline, but can expand below it as well; so the formation and distribution of glaciers depend both on the elevation of the snowline and the temperature of the region. In turn, snowline elevation is controlled by summer temperature and the amount of winter snowfall. The colder the summer temperature, the lower the snowline will be; furthermore, the snowline is lower in regions with moist maritime climates (high snowfall) and is higher in regions with dry continental climates (low snowfall).

Because of temperature changes, the average elevation of the snowline drops from the equator to the poles. But this drop is not uniform because moisture conditions, hence snowfall, change from south to north (Fig. 3.2). Atmospheric moisture conditions are, in part, related to temperature, because colder air can contain less water vapor, but also to other factors like availability of moisture sources and trade-wind patterns. Snowfall decreases in the dry conditions of the so-called **horse latitudes** (approximately latitude 25° N to latitude 30° N), and in the cold deserts of the

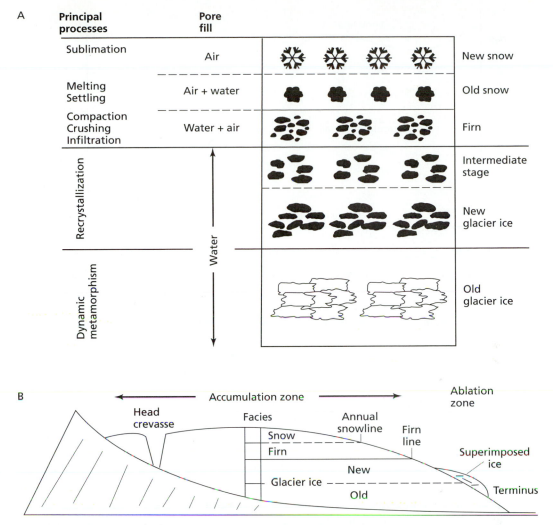

FIGURE 3.1 Diagrammatic representation of a valley glacier. *(A)* Idealized vertical facies sequences in a glacier. *(B)* Internal and surficial ice facies distribution in an ideal valley glacier.

high-polar regions. The amount of fallen snow influences snowline altitude because, even above the snowline, some snow precipitated during the winter melts during the summer. If very little snow falls during the winter, as in the horse latitudes, then permanent snowfields occur at higher altitudes than expected. Conversely, in areas of high snowfall, the altitude of the snowline may be lower than expected.

The regional (average) snowline thus drops regularly from the equator to about 25° N, then rises to about 30° N because of lower-than-average snowfall, and then drops regularly to about 45° N (Fig. 3.2). North of 45° N, the drop in snowline altitude is steeper, because the line of maximum snowfall intersects and rises above the regional snowline (Fig. 3.2). The altitude of the snowline depends on the balance between winter accumulation and summer melting. In the north the preserved snow is the unmelted portion of the maximum snowfall of the area. In the south the preserved snow is only a portion of smaller snowfalls for the area: the maximum amount of snow falls below the snowline, but all that snow melts during the subsequent summer. This process can be seen on many famous ski slopes that develop

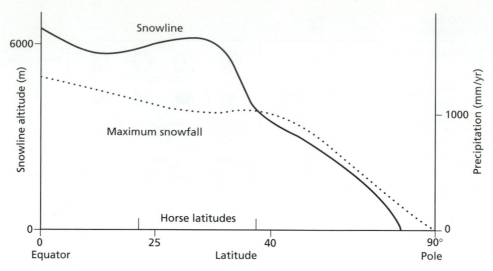

FIGURE 3.2 Diagram showing changes in elevation of the snowline and of the line of maximum snowfall along an idealized transect from the equator to the poles. (After Sudgen and John 1976.)

thick snow in winter but are grass covered in summer. As a glacial age develops, large glaciers could first form around 45–50° N latitude because that is where the snowline and the maximum snowfall line intersect. Farther north, because of lower moisture and despite lower temperatures, the amount of snowfall decreases. The large Pleistocene ice sheets of North America (Fig. 3.3A) did indeed start spreading from mid-latitudes where the snowline intersected the topographic elevation (Fig. 3.3B). Along the same transect, present glaciers are located north of 65° because only there does the snowline intersect the ground surface. In Europe, the Pleistocene glaciers nucleated at higher latitudes than in North America because the warm currents of the **Gulf Stream** displaced climate belts northward. In conclusion:

1. At the equator, snow falls only at high altitudes. The snowline is very high, much of the snow melts, and only small glaciers can form at very high altitudes.

2. At middle latitudes, around 45°–50° N, much snow remains throughout the summer because the maximum snowfall occurs above the snowline. Large glaciers can develop and flow long distances below the snowline.

3. In polar regions, the snowline is at sea level. But the low humidity and low snowfall allow only small net accumulation. Hence glaciers acquire and lose little matter throughout the year, and are not very active.

CLASSIFICATION OF GLACIERS

Glaciers can be classified according to their dimensions and relation to the underlying bedrock (morphological classification), their internal temperature regime (thermal classification), and the type of snow and ice (facies) exposed at their surface during the summer (facies classification).

Morphological Classification

There are several morphological classifications. The one used here assigns glaciers to the somewhat arbitrary classes of **ice sheets, valley glaciers,** and **lowland glaciers.** Each of these classes can be subdivided further, and for completeness we include

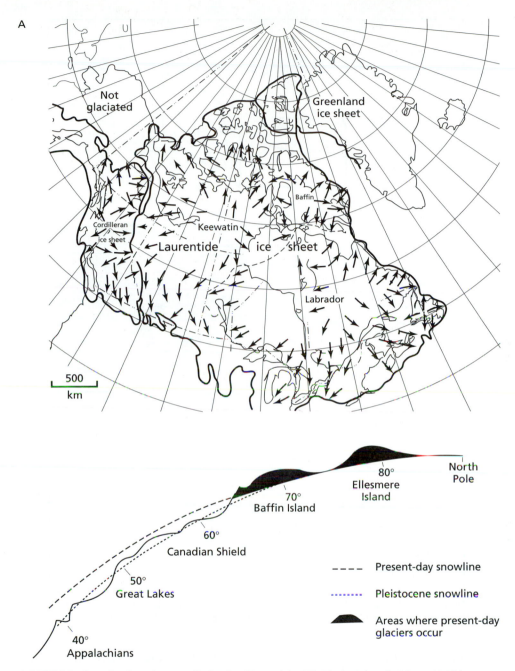

FIGURE 3.3 Localization of centers (Labrador, Keewatin) of North American ice sheets at mid-latitudes, where the Pleistocene snowline intersected Earth's surface. *(A)* Map showing Keewatin and Labrador centers of Pleistocene glaciation and direction of ice movement (arrows). *(B)* Schematic profile along a north-south line from the North Pole to latitude 40° N, showing altitudes of present and Pleistocene snowlines in North America. (After Stearn, Caroll, and Clark 1979.)

accumulations of ice and snow that persist over several summers, although they may not experience any detectable movement.

Ice sheets (Fig. 3.4) are broad, unconfined masses, which generally flow in an irregular radial fashion from one or more central ice domes. They are only partially affected by the morphology of the underlying bedrock surface. The following divi-

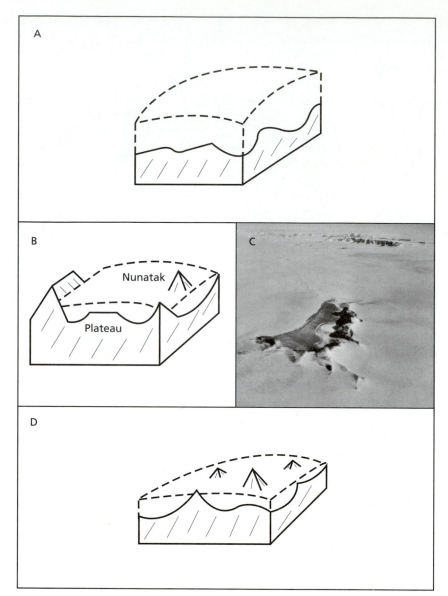

FIGURE 3.4   Schematic diagrams and photograph of ice sheets. *(A)* Continental; *(B)* and *(C)* plateau; *(D)* highland. (Photograph from Post and LaChapelle 1971.)

sions show a progressive decrease in size and thickness, and an increasing control of land morphology on their shapes and behaviors.

Continental ice sheets (Fig. 3.4A) are shieldlike domes of continental dimensions, larger than 25,000 km^2 and thick enough to cover most irregularities of the land surface, although a few mountaintops **(nunataks)** may emerge locally. Modern-day examples can be found in Greenland and Antarctica.

Ice caps are smaller than 25,000 km^2, dome-shaped or flat, and with common nunataks. An example is Barnes Ice Cap in Baffin Island, Canada.

Plateau glaciers (Fig. 3.4B,C) contain smaller, central nourishment areas. They are generally flat, occur on highland plateaus, and may show tonguelike ice cascades around their rims. Nunataks occur around the margins. Examples can be found in Iceland and the Norwegian Mountains.

Highland ice sheets (Fig. 3.4D) are broad, with undulating surfaces somewhat controlled by the underlying bedrock surface morphology. They show numerous isolated, randomly distributed nunataks. The Canadian Rocky Mountains contain good examples.

Valley glaciers develop in well-defined valleys, where ice flow is confined, and acquire elongated, streamlike forms (Fig. 3.5).

Ice streams (Fig. 3.5A) are elongated, fast-moving ice masses within a continental ice sheet.

Reticular glaciers (Fig. 3.5C,D) are a transitional form between ice sheets and fully developed valley glaciers. Their flow is channeled by underlying bedrock valleys whose form is outlined by numerous aligned nunataks.

Outlet glaciers (Fig. 3.5E,F) descend from ice sheets through confining bedrock valleys.

Alpine glaciers are confined to bedrock valleys, and have several forms depending on the number and size of the tributaries. Simple valley glaciers have one main valley; dendritic glaciers have a main glacier joined by various smaller ones; hanging glaciers develop when tributaries join discordantly, with smaller ones cascading down onto major ones; composite valley glaciers are formed by various glaciers that join and maintain their identities.

Cliff and reconstituted glaciers (Fig. 3.5G) form where steep slopes separate thin glaciers into two parts. The cliff portion forms at the head of a slope too steep to hold ice. The ice cascades down to the reconstituted (reformed) portion at the base of the cliff.

Wall-side glaciers (Fig. 3.5H) are ice tongues that extend downvalley from the accumulation zone, but are not confined by valley walls. They are either tongues marginal to ice sheets or remnants of retreating alpine glaciers no longer large enough to fill the whole valley (misfit glacier), as is the case in many glaciers in the European Alps today.

Cirque glaciers (Fig. 3.5B) are oval to circular masses of ice confined to a relatively small bedrock alcove. These glaciers develop near the snowline and experience significant accumulation and ablation.

Apron glaciers are thin masses of ice and snow on mountainsides.

Lowland glaciers represent a mixed group of glaciers whose common characteristic is their location on flat areas (Fig. 3.6).

Piedmont glaciers (Fig. 3.6A,B) form where confluent valley glaciers spread out onto flatter areas at any altitude. They have low surface slopes, great breadth, and distinctive multilobate margins. Intermontane glaciers spread out within enclosed mountain valleys.

Expanded foot glaciers (Fig. 3.6C,D) form where a valley glacier fans out over a wider plain or trunk valley. Typical examples occur in Alaska.

Fringing glaciers (Fig. 3.6E,F) are low-lying ice belts around seacoasts. They are truncated by ice cliffs of constant height and are probably relics of shelf ice that once filled marine bays.

Stagnant parts of glaciers (Fig. 3.6G,H) are debris-laden terminal zones that do not move unless pushed by the main glacier. They develop hummocky, pitted surfaces. A typical example is the terminal part of the Malaspina Glacier in Alaska where a forest grows on the thick debris covering the ice.

Thermal (Geophysical) Classification[1]

Glaciers contain a surficial **active layer** that is directly affected by seasonally changing atmospheric conditions, and a lower body that is only indirectly affected by surface conditions. The internal temperature distribution within glaciers is used to

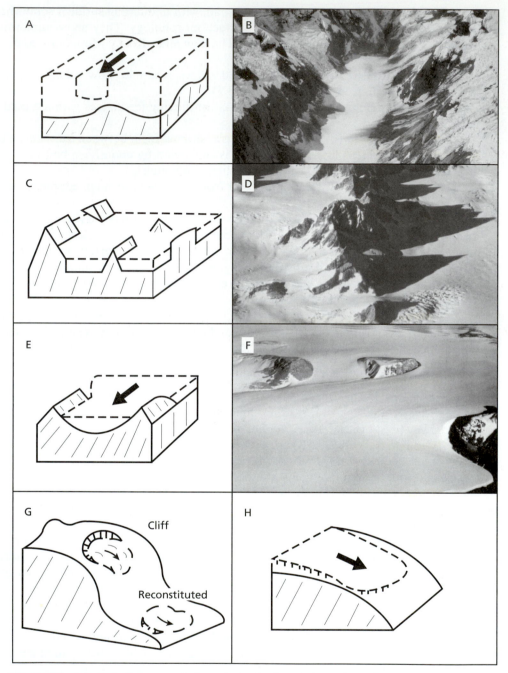

FIGURE 3.5   Schematic diagrams and photographs of valley glaciers. *(A)* Ice stream; *(B)* cirque glacier; *(C)* and *(D)* reticular glacier; *(E)* and *(F)* outlet glacier; *(G)* cliff and reconstituted glacier; *(H)* wall-side glacier. (Arrows indicate glacial flow.) (Photographs from Post and LaChapelle 1971.)

divide them into temperate and polar types, with the latter further subdivided into high-polar and subpolar varieties.

Temperate glaciers have warm ice throughout; that is, the temperature inside the glacier is at or near the pressure melting point. The upper active layer may be

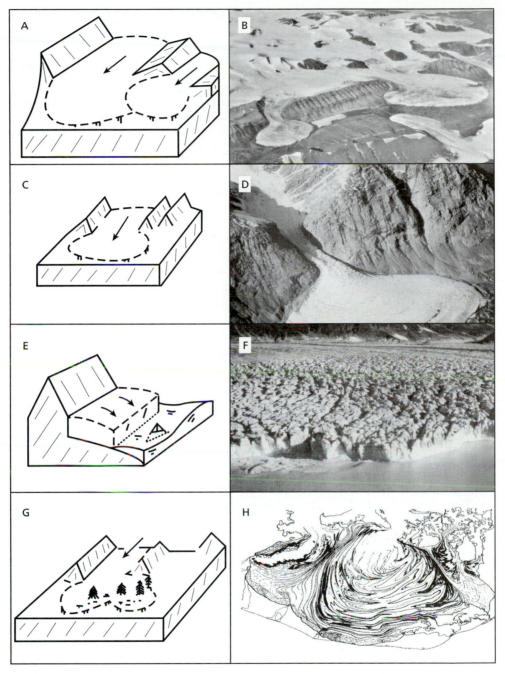

FIGURE 3.6   Schematic diagrams and photographs of lowland glaciers. *(A)* and *(B)* Piedmont glacier; *(C)* and *(D)* expanded foot glacier; *(E)* and *(F)* fringing glacier; *(G)* and *(H)* stagnant glacier. (Arrows indicate direction of glacial flow.) (Photographs from Post and LaChapelle 1971.)

relatively thick, and its upper few meters may melt a lot during the summer. Large amounts of meltwater percolate downward to refreeze at depth, releasing enough latent heat of transformation (80 cal/g) to keep the glacier at the pressure melting point throughout. Because of this, the deeper ice, subjected to higher pressure, is colder than the shallower ice (Fig. 3.7A). Therefore, any geothermal heat that

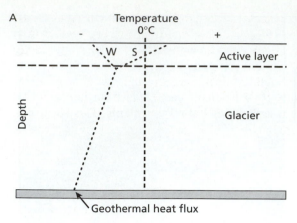

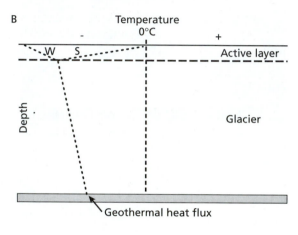

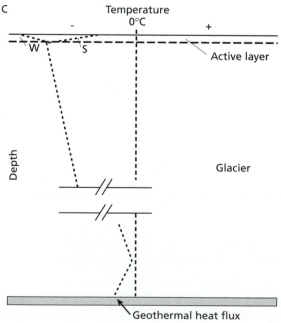

FIGURE 3.7   Temperature profiles within glaciers. Dotted lines indicate the internal temperature profile; S = summer profile and W = winter profile; arrows indicate input from the geothermal gradient. *(A)* Temperate glacier with the main body of ice at the pressure melting point. *(B)* Thin polar glacier with the main body of ice below the pressure melting point. *(C)* Thick polar glacier with the main glacier body below the pressure melting point and a lower layer of ice at the pressure melting point.

reaches the earth's surface below the glacier (approximately 38 cal/cm^2/year on average; sufficient to melt 0.6 cm of ice at the pressure melting point per year) cannot be conducted upward and melts ice at the ground/glacier interface. This may lead to overpressured fluids, which reduce friction at the base of the glacier and promote movement and basal slippage. Slippage, in turn, can, through friction, generate more heat (on average approximately 10–20 cal/cm^2/year), which further increases the basal melting, thus water content, enhancing basal movements. This basal slippage contributes to basal glacial erosion and formation of several geomorphic features such as striae.

Polar glaciers contain cold ice in significant parts of their inner layer, and their active layer may be thin (Fig. 3.7B). They can be subdivided into two groups.

High-polar glaciers have thick firn in their accumulation zone. Their active layer is very thin; hence there is little water generated and, therefore, no possibility of warming up the interior ice.

Subpolar glaciers contain firn down to a depth of 10–20 m in the accumulation zone. Their active layer is thin, but sufficient meltwater can be generated during the summer to percolate into and slightly warm the upper part of the glacier body, but not to raise it to the pressure melting point.

Polar glaciers contain thick firn layers in the accumulation zone, because the metamorphism of snow occurs slowly under cold conditions, and the transformation into glacier ice can occur only at high pressures. Furthermore, these glaciers are generally frozen to the bedrock and cannot slide. Their temperature profile, with colder upper layers and warmer lower ones (polythermal glaciers), allows geothermal heat to be dispersed into the glacier by conduction. In some cases, polar glaciers become thick enough for basal layers to reach the pressure melting point; that is, they become warm-based (or wet-based) (Fig. 3.7C). In these cases, the lower layers behave as temperate glaciers and basal sliding is possible but difficult because the thick areas may be surrounded by wide areas of thinner ice still frozen to the ground.

Facies Classification[2]

This classification is based on the distribution of snow and ice types (facies) observable at the surface of a glacier during the summer (Figs. 3.8, 3.9). Three main facies occur in the accumulation zone. The *dry-snow facies* has cold snow and old snow and little or no melting. The *percolation facies* experiences some melting, but not enough meltwater is generated to percolate the whole snowpack and raise it to the pressure melting point. The *soaked facies* has a lot of meltwater capable of percolating through the snowpack and raising it to the pressure melting point. Two main facies occur in the ablation zone, *glacier ice* and *superimposed ice*. Glaciers can be classified during the summer by estimating the proportion of the surface area covered by each facies. No specific glacier categories have yet been established based on facies.

Comparison Between Classifications

There are equivalent features in the various classifications (Table 3.1).

Each classification presents advantages and disadvantages. The morphological classification is descriptive and nongenetic, and it can be used with remote sensing. The disadvantage is that it has arbitrary class limits and gives only general information on the internal makeup and behavior of the glaciers.

The advantage of the thermal classification is that once the temperature profile of a glacier is known, its behavior and erosive power can be predicted. The

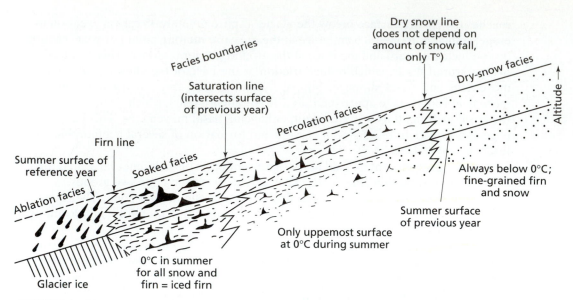

FIGURE 3.8 Diagram showing distribution of snow and ice facies at the surface of glaciers. (0°C indicates pressure melting point.) (After Benson and Thompson 1987.)

FIGURE 3.9 Map of distribution of snow and ice facies on the Greenland Ice Sheet.

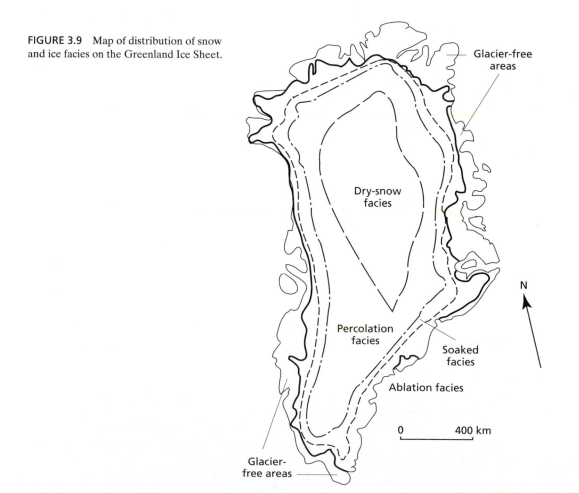

TABLE 3.1   Equivalent Glacier Classes in the Thermal and Facies Classifications

Thermal Classification	Facies Classification	Geographic Distribution
Temperate glacier	Glacier-ice and soaked zones	Low altitudes and middle latitudes
Subpolar	Percolation zone	Mid-high altitudes and high latitudes
High-polar	Dry-snow zone	High altitudes and latitudes

disadvantage is that it is difficult, time consuming, expensive, and often impractical to obtain temperature profiles of glaciers.

The facies classification provides some insight into the internal characteristics of the glaciers from observable surface features. These features can be observed directly or through remote sensing. This classification has no defined classes for glaciers; rather the glaciers can be subdivided according to a not-yet-formalized percentage of coverage of the various facies.

MASS BUDGET OF GLACIERS

Climatic conditions determine whether a glacier can form or not. Relative rates of accumulation and ablation of snow and ice determine how fast the ice mass shrinks or grows, and this, in turn, determines the dynamics of the glacier.

Accumulation relates to addition of ice to the glacier. The material comes primarily from snowfall and secondarily from hail, freezing rain, and, locally, snow avalanches. These contributions vary regionally according to cyclone paths, altitude, prevailing winds, regional climate (continental or marine), and topography. Feedback between glaciers and their surroundings also affects accumulation of the glacier itself.

Ablation relates to removal of ice from the glacier. Ice can be removed in a liquid phase (by melting, primarily at the surface and at the bottom of glaciers), vapor phase (evaporation-sublimation), and solid phase (calving, wind-blown snow) (Fig. 3.10). The relative importance of these processes varies with the type of glacier.

Melting is important in temperate glaciers that end on land. It is caused by direct short-wave solar radiation, indirect long-wave radiation reflected by the atmosphere, conduction of heat from the atmosphere (enhanced by air convection and turbulence), and latent heat of transformation released during condensation.

Radiation input into the ice mass is affected by season, time of day, latitude, altitude, cloud cover, exposure (attitude), humidity, and albedo (reflectivity) of the materials. Because of the high albedo of snow (from 0.8 for freshly deposited snow to 0.3 for dust-covered snow), up to four-fifths of the direct radiation on a glacier is reflected away.

Conduction of heat from the air to the ice is controlled by temperature differences and the temperature gradient near the interface. Under calm conditions, the still air and underlying ice quickly reach equilibrium. During windy conditions, new warm air may be continually brought into contact with the ice. The resulting steep temperature gradient increases the rate of heat conduction and melting of ice.

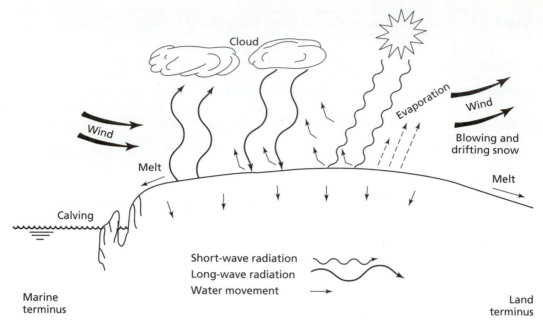

FIGURE 3.10 Diagrammatic representation of conditions and energy fluxes at the surface of a glacier that affect the melting of ice.

Release of the latent heat of transformation occurs when water vapor in contact with ice or snow condenses as it cools to the dew point. Condensation at the pressure melting point (0°C at 1 atm) releases a large amount of latent heat of transformation. Each gram of water vapor upon freezing releases about 680 calories, sufficient to melt about 7.5 grams of ice or snow if it is already close to the melting point (note that the specific heat of moist air is greater than that of dry air at the same temperature). *Rain* does not melt ice efficiently because it is often close to freezing. In addition, almost as much ice is formed as melted because the latent heat of transformation between ice and water is only about 80 cal/g. Rain appears more effective in melting snow than it actually is because of the compaction and settling it causes in the snow pack.

Sublimation needs a lot of heat, about 608 cal/g for ice at its melting point. Because the amount of water vapor that can be acquired by cold air is small, the amount of ice removed by sublimation from glaciers is small. Nevertheless, this is the main way that ice is removed from polar glaciers.

Calving is the physical removal of blocks of ice from a glacier. It occurs when the glacier terminates into a deep body of water (sea or lake) and the ice becomes buoyant. This occurs when about 75–90% of the ice is submerged and starts to float. Marine and lacustrine currents carry away the blocks of ice that break from the terminus of the glacier as icebergs.

The **mass budget** of a glacier is the ratio between accumulation and ablation. It varies over time and from place to place and causes changes in glacier thickness, speed of flow, length, and dynamic response. Budget estimates are usually done annually (Fig. 3.11), though they can be averaged over a longer time. A glacier with a positive budget gains more material than it loses (Fig. 3.11A). It responds by advancing, increasing the surface area, and in this way increasing its melting potential. A glacier with a negative budget loses more than it gains (Fig. 3.11B). It responds by retreating, shrinking in surface area, and decreasing its melting potential. For any glacier, the net mass budget of the accumulation zone is always positive, whereas that of the ablation zone is always negative, and that at the firn line is at equilibrium

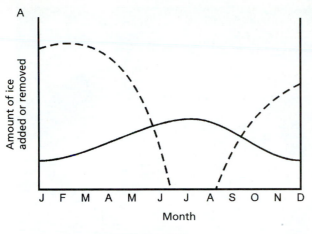

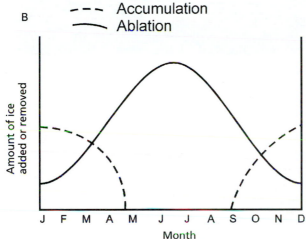

FIGURE 3.11   Diagrams showing annual budgets of a glacier. *(A)* Positive budget (more material is added than removed); *(B)* negative budget.

(because of this, the firn line is also called the equilibrium line). Depending on the overall ice budget, the area of the accumulation and ablation zones may shrink or expand, and the position of the firn line may change.

The mass budget is related to ambient temperature and snowfall. However, the dynamic response of glaciers to climatic changes is not instantaneous, and for large glaciers there may be a significant time delay of up to hundreds of years. That is, a period of higher snowfall in the accumulation zone may not trigger an expansion of the ablation zones until much later. Furthermore, the rate of response may differ in adjacent glaciers or zones within an ice sheet: at any given time some may surge, others retreat, and still others may not change at all (Fig. 3.12). So, not all surges and sudden retreats recorded in the glacial landscape can be directly related to discrete changes in climatic conditions. They may instead be due to different dynamic time-responses of various glaciers or parts of ice sheets to one climatic event.

MOVEMENT OF GLACIERS

The fundamental characteristic of a glacier is that it is capable of moving on its own. A glacier flows to reach and maintain a gravitational equilibrium form, as material is moved from the accumulation zone to the ablation zone (Fig. 3.13).

FIGURE 3.12   Photograph showing different behavior of adjacent glaciers subjected to the same climatic conditions. Surging glacier with a steep terminus at right and a retreating glacier with flat terminus at left. (From Post and LaChapelle 1971.)

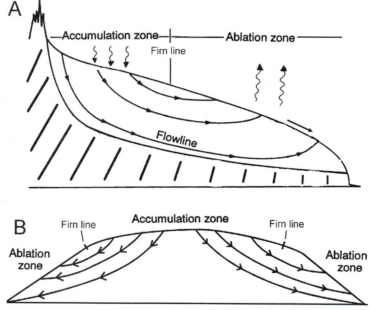

FIGURE 3.13   Diagrams of idealized glaciers showing zones of accumulation, ablation, and the firn line. *(A)* Valley glacier; *(B)* ice sheet. (After Sudgen and John 1976.)

Rates of Movement

Glaciers move at different speeds depending on local environmental conditions. Polar glaciers move very slowly, as do the debris-laden margins of temperate glaciers. Temperate glaciers usually move at a few centimeters per day; however, a few steep glaciers may reach velocities ranging from 0.3–0.5 cm/day to 3.0–6.0 m/day. Sporadically fast moving (surging) glaciers may move very fast: outlet glaciers in

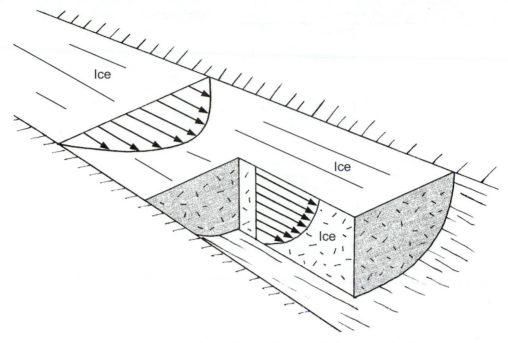

FIGURE 3.14 Idealized section of a valley glacier showing horizontal and vertical velocity profiles (arrows indicate velocity).

Greenland may move up to 40 m/day, and some Alaskan glaciers can reach up to 80 m/day. Furthermore, glaciers move at variable speeds in their different parts and during different seasons. They move faster in winter in the accumulation zone because of increased weight of the snow; they move faster during the summer in the ablation zone because of increased melting.

Increased precipitation in the accumulation zone does not cause a glacier to speed up immediately everywhere. Changes in the rate of movement happen when the pressure of the ice in the upper reaches becomes large enough to overcome the resistance of slower-moving parts of the glacier lower down. So glaciers move neither uniformly nor steadily.[3] Glaciers, in fact, often move in a pulsating and spasmodic fashion; long periods of slow movement alternate with rapid advances.

Measurements of glacier surface movement can be made by placing targets on the surface and periodically noting their position relative to a fixed point away from the glacier. Drilling vertical holes, placing flexible rods in them, and noting their deformation and displacement over time is a way to measure internal movement of the glacier.

The measured speed profiles taken in valley glaciers show how the underlying topography affects glacier movement, and how the movement is partitioned in different parts of the glacier itself. Because of friction with the walls and floor of the valley, a glacier moves slower at its flanks and base and faster at its center and surface (Fig. 3.14). A glacier flowing in a regular trough-shaped (U-shaped) valley shows a regular U-shaped horizontal velocity distribution across its width (Fig. 3.15A). In rapidly moving glaciers, and during surges, the velocity profile at the surface of the glacier shows a box shape with shear near the valley walls (Fig. 3.15B). In most cases, this shear does not occur at the transition between the ice and the bedrock at the walls, but within the ice near the walls where it is weaker. Irregularly shaped valleys modify the surficial velocity profiles (Fig. 3.15C).

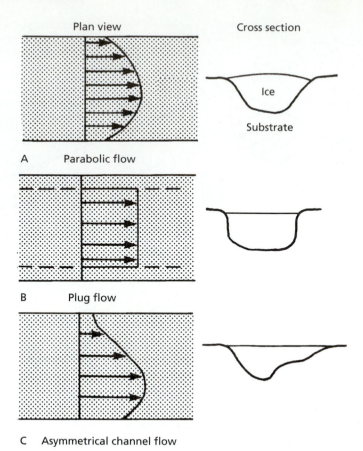

C   Asymmetrical channel flow

FIGURE 3.15   Diagrams showing surface movement of valley glaciers. *(A)* Flow in a regular U-shaped valley shows a parabolic velocity profile. *(B)* Block flow of a surging glacier. *(C)* Flow in an irregularly shaped valley. (After Sharp 1960.)

The vertical velocity profile shows that the maximum speed of a glacier is near the top (Figs. 3.14, 3.16). This displacement is the sum of movements within the ice mass and slippage of the glacier at its base. The basal slippage is important because it is a significant factor in glacial erosion.

Forces Involved in Glacier Movement

Though a glacier moves primarily due to the internal stresses associated with the weight of its material, external forces and temperature variations also affect it.

External forces mostly affect ice cover on bodies of water and glaciers with sea terminations. They are associated with wind, waves, tides, and channelized water flows. These forces compress, stretch, and tear the ice, causing it to crush, arch, crack, and fracture. Large pieces of ice can break from floating glaciers to form **icebergs.** Land glaciers are affected locally by wind, earthquakes, volcanoes, and other geotectonic phenomena.

Internal temperature fluctuations cause glacier ice to expand and contract, leading to intense tensile stresses that cause slight flexures and cleavage. In winter, cold penetrates downward into the glacier, causing contractions that may crack the

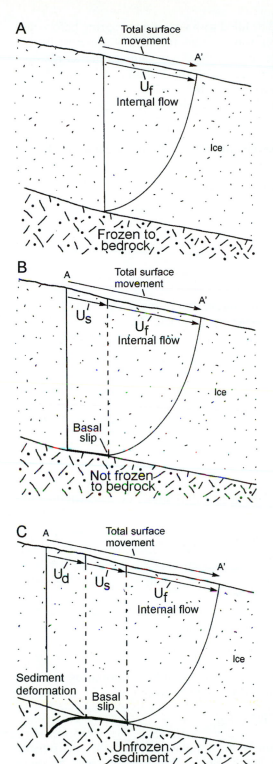

FIGURE 3.16   Vertical velocity profiles of a glacier. *(A)* Frozen to the bed. *(B)* With warm ice at the base and sliding on a rigid substratum. *(C)* With warm ice sliding on deformable sediment bed. (U_f = deformation component related to internal flow; U_s = deformation component related to basal slip of glacier; U_d = deformation component related to moving basal sediment layer.) (After Boulton 1996.)

topmost part of the ice. The cracks may be filled with water in the spring before they can close, and congelation ice (freezing water) veins form. Subsequently, further warming and expansion in the summer causes these ice veins to act as resistant wedges, and strong tangential stresses are created. If the system is unbalanced, such as on a slope, this leads to increased movement of the glacier.

The **natural weight of snow and ice** leads to development of internal stresses in the glacier as follows.

1. When lateral expansion is not possible, the weight of the ice generates hydrostatic pressure that leads to uniform compression.

$$\Delta V = 34.10^{-6} pV$$

where

ΔV = change in volume
p = pressure in kg/cm^2
V = initial volume in cm^3

Because of this, air is forced out of the interstitial pores of snow and firn, or it can become overpressured in isolated bubbles within glacier ice.

2. If the glacier is an unbalanced system, the stresses associated with the weight lead to lateral deformation. The vertical pressure (p) created by the weight (ρg) at depth (z) is

$$p = \rho g z$$

where

ρ = density of ice
g = acceleration due to gravity

If the relationship between pressure and deformation is linear, the coefficient of lateral pressure (E) is

$$E = \xi/1 \times \xi$$

where

ξ is the Poisson coefficient for ice = 0.361

thus:

$$E = 0.565$$

This means that 56.5% of the vertical pressure in ice would be transmitted horizontally, and the lateral pressure component ($p1$) at depth z would be

$$p1 = 0.565 \, \xi g z$$

This relationship is valid only in the top 35 m of a glacier where the relationship between pressure and deformation is linear. Below that, the pressure becomes hydrostatic.

Different types of deformation occur in different parts of the glacier due to temperature and pressure of the ice. Brittle deformation (fracture) dominates in cold surface layers, and plastic deformation is dominant at depth.

Mechanism of Glacier Movement

The movement of a glacier is a complex phenomenon and occurs through several observable mechanisms, some more important than others, in different parts of the glacier.

1. Rotation of individual ice crystals past one another (Fig. 3.17A) is possible in snow and firn zones.

2. Movement of water downslope and inside the glacier (Fig. 3.17B) transfers material downglacier, and ice crystals are elongated downslope as the water refreezes on surfaces least affected by pressure.

3. Slippage of the glacier over bedrock (Fig. 3.17C) occurs only in glaciers with ice at the pressure melting point and with enough overpressured water at the base to reduce friction between the ice and substrate. Soft, deformable, water-saturated sediments at the bases of glaciers enhance basal slippage.

4. Internal slippage along fractures can occur in any part of the glacier, but is most effective toward the terminus, where thrust faults develop (Fig. 3.17D).

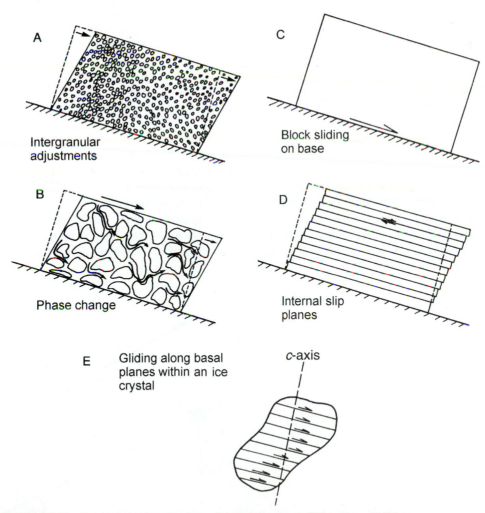

FIGURE 3.17 Illustrations of mechanisms of glacial movement. (After Sharp 1960.)

5. Slippage along basal planes of ice crystals can occur everywhere (Fig. 3.17E). In old glacier ice, large crystals are formed strongly elongated downslope. This is a form of movement and hence a contribution to flow.

Hypotheses of Movement

A slab of ice placed on an inclined surface constitutes an unbalanced system when subjected primarily to the force of gravity. In a two-dimensional system, its weight (weight = ρgz) can be resolved into two component vectors of force: one perpendicular to the bed ($\sigma = \rho gz \cos\alpha$) and one parallel to the bed ($\tau = \rho gz \sin\alpha$) (Fig. 3.18):

where

> ρ = density of ice
> g = force of gravity
> τ = force component tangential to bed
> σ = force perpendicular to the bed
> z = thickness of ice
> α = slope

When $\tau < \sigma$, the ice block is stable. If α or z increases, τ increases. For $\tau = \sigma$, the elastic limit near the bed is reached and downhill creep (very slow deformation) begins. If $\tau > \sigma$, the glacier deforms faster and moves appreciably downhill.

Several hypotheses try to explain how the glacier moves. One, the kinematic hypothesis, describes the movement without regard to the forces involved. The others, hydrodynamic and plastic hypotheses, make basic assumptions about the glacier ice and pressure stress.

The **kinematic hypothesis** assumes first that glacier movement is geometrically similar to the laminar flow of a viscous fluid. A second assumption is that for every point of the glacier surface in the accumulation zone there is a corresponding point in the ablation zone. These points are connected by flow lines, which are the paths of particles moving through the glacier. Following the concept of laminar flow, the flow lines never intersect, and the entire glacier may be subdivided into elementary flow tubes along these lines. A **steady state** is achieved if an identical amount of ice passes through every cross section of a given flow tube in the same amount of time.

FIGURE 3.18   Diagrammatic representation of idealized setting of a slab of ice on an inclined surface (symbols explained in text).

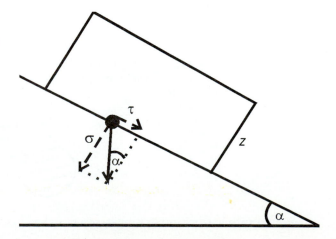

The flow lines descend into the glacier in the accumulation zone and emerge in the ablation zone; thus material flows downward with respect to the surface of the glacier in the accumulation zone, parallel to the surface at the firn zone, and upward in the ablation zone (Fig. 3.13A). Furthermore, in the accumulation zone of valley glaciers, the movement also shows a component directed away from the sidewalls; at the firn line it is straight downvalley; and in the ablation zone, it shows a component of movement from the center toward the sides. This is caused by differences in mass balance between the sides and center of the glacier. In the accumulation zone, more material is added to the sides than to the center due to snowdrifts, avalanches, and rock debris. In the ablation zone, more ice is removed from the sides because of the large absorption of energy closer to the valley walls, which increases melting and runoff. All things considered, under normal circumstances the glacier should be moving directly downhill fastest at the firn line.

In ice sheets, the flow lines are inferred from the equilibrium profile in the peripheral area and extend outward from the area of maximum snowfall to the margin of the glacier (Fig. 3.13B). Within this zone, the same considerations made for valley glaciers stand, except for the effect of valley walls, which do not exist in ice sheets. The kinematic hypothesis is useful for describing and understanding

1. changes in the surface form of the glacier as a function of movement and melting rates;
2. the distribution of material transported on and in a glacier;
3. the effects of climate changes on a glacier; and
4. the possible shape of the glacier bed, inferred by observing the glacier surface and speed.

The **hydrodynamic hypothesis** and the **plastic flow hypothesis** make the assumption that glacier ice behaves, respectively, as a Newtonian viscous fluid or a plastic fluid. In fact, glacier ice can be considered a pseudoplastic material; that is, under certain circumstances, it behaves almost like a viscous fluid, under others, like a plastic material. A Newtonian viscous fluid starts deforming immediately upon application of a force, and continues to do so as long as the same stress is applied (Fig. 3.19). If the stress is increased, the rate of strain (deformation) is proportionally increased so there is a linear relationship between stress and strain in a fluid, with the slope of the curve reflecting the viscosity of the fluid. A plastic material does not deform as soon as a force is applied. Instead, no visible deformation occurs until the stress is greater than a certain value (plastic limit), at which point the material deforms macroscopically. At this critical point, for an infinitesimal (infinitely small) increase in stress there will be a finite strain; thus the rate of deformation is infinite (Fig. 3.19).

Glacier ice at the pressure melting point (0°C at 1 atm) does not deform until the stress reaches a value of about 0.1 kg/cm^2, when the elastic limit is reached. This initial deformation is very slow, and the relationship between stress and strain is quasi-linear, similar to that of a Newtonian fluid with a viscosity higher than that of water (Fig. 3.19). The plastic limit of ice is reached at a stress of between 1 and 1.5 kg/cm^2, at which point the glacier undergoes (macroscopic) deformation. The relationship between stress and strain is no longer linear, and eventually the rate of strain (deformation) becomes infinite, similar to a plastic material (Fig. 3.19).

The **hydrodynamic hypothesis** assumes that ice behaves as a Newtonian viscous fluid. This assumption is in part reflected in the statement, "a glacier is like a river of ice." The morphology of some alpine glaciers does resemble that of a frozen

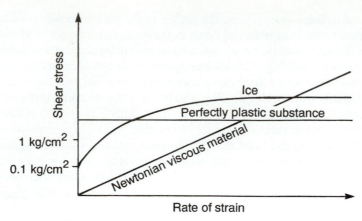

FIGURE 3.19 Graph showing behavior of different kinds of substances subjected to shear stress. Ice is a pseudoplastic material. (After Sharp 1960.)

river, with tributaries and all. However, very slow deformation of ice under a stress of less than 1 kg/cm^2 so as not to exceed the plastic limit is observed only in snowpacks and during the transfer of meltwater. Any other shearing mechanism breaks the tenets of a viscous fluid material, because for an infinitesimal increase in height there will be a finite deformation; thus the rate of deformation will be infinite. The hydrodynamic hypothesis, therefore, does not satisfactorily explain the movement of glaciers.

Plastic flow hypothesis assumes that glacier ice behaves as a plastic or pseudoplastic body with a critical yield stress (τ_{crit}) of 1 bar. The interplay of various stresses in a valley glacier creates three types of movement, which, combined, explain the total movement of the glacier. These are

1. plastic flow,
2. block slipping along the bed, and
3. internal slipping along faults.

Plastic Flow   The glacier must have a certain thickness (z) before it can reach the plastic limit of 1 to 1.5 kg/cm^2 (at the pressure melting point) and flow down a slope (α). For movement to occur, the shear stress parallel to the bed (τ) must exceed the pressure perpendicular to it (σ). Conversely, if perfect plasticity is assumed, the thickness of a glacier at equilibrium ($\tau_{crit} = 1$ bar) can be calculated from $z = \tau_{crit}/\rho g \sin\alpha$ by just measuring the surface slope of the glacier.

The plastic movement of a glacier can be analyzed for uniform slope or changing slope. If the slope (α) is uniform, the glacier moves when it becomes thick (z) enough to exceed plastic equilibrium ($\tau_{crit} = 1$ bar $= 1$ kg/cm^2) for any given slope angle α (Table 3.2). A very thick glacier can move on a flat surface; but in this case, the movement is driven not by the slope of the substrate but by the slope of the glacier surface. Glaciers tend toward an equilibrium form under given ambient conditions. If the slope is constant, to maintain equilibrium conditions the glacier must change thickness (z), and thus must advance for an excess of precipitation or retreat for an excess of ablation.

Natural slopes are rarely uniform, and change along the glacier path. Ice thickness cannot change instantaneously to satisfy local conditions for plastic equilibrium; so the glacier moves faster on steeper slopes, creating both tension in adjacent flatter, slower-moving, upper reaches and compression in flatter, slower-

TABLE 3.2 Ice Thickness Necessary for Detectable Movement

Slope	Slight Movement ($\tau \sim 1 \text{ kg/cm}^2$)	Noticeable Movement ($\tau > 1.5 \text{ kg/cm}^2$)
45°	1.54 m	15.4 m
10°	6.28 m	62.4 m
1°	62.5 m	625.1 m

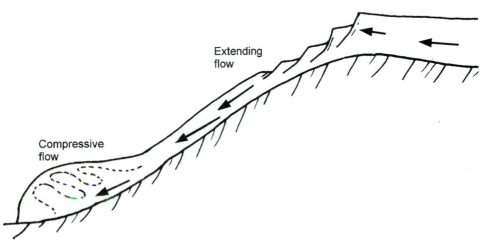

FIGURE 3.20 Schematic longitudinal section of a glacier with indications of areas of different flow velocities (represented by length of arrows) in a valley with variable slopes.

moving, lower reaches (Fig. 3.20). This results in the generation of **extending flow** in the steeper reaches and **compressive flow** in the flatter ones. The sum of these flow types constitutes the **plastic flow** of glaciers.

Extending flow helps in moving slower upstream parts of the glacier by pulling them. It also tends to thin the glacier out to reach plastic equilibrium. The fastest way to do this is to form **crevasses** (Box 3.1).

Compressive flow helps in moving the glacier by pushing it on flatter reaches. A by-product of this is folding and thickening of the ice until that portion of the glacier can flow under its own weight (Fig. 3.20). Thus, in this case, the glacier tends to reach plastic equilibrium by thickening, and the fastest way to do so is by folding. Typical features associated with compressive flow are compressive waves (**ogives**).

Block Slipping Along the Bed Whole reaches of a glacier may slide (block slippage) along its bed when pushed in flatter areas or if it is too thick in steep areas. Block slipping requires decoupling of the ice from its substratum. One way to get this decoupling is by **regelation slip,** whereby excess water is generated at the base of the glacier. This water reduces basal friction, facilitating block slipping.

The regelation-slip process works as follows. Most glaciers have irregular beds with resistant protuberances (Fig. 3.21). Glaciers can exert pressure on the upstream ends of the protuberances, causing the ice to reach and exceed the pressure melting point. If so, the produced meltwater flows along the pressure gradient around the obstructions until it refreezes at the low-pressure downstream ends.

BOX 3.1   Crevasses

Crevasses develop at the surface of glaciers, where large tensional stresses generated during extending flow exceed the brittle strength of ice. The depth of crevasses is generally less than 30–50 m. Various types of crevasses form, influenced by the distribution of stresses, which, in turn, are affected by the morphology of the substratum. In a valley glacier the main types of crevasses are bergschrund, transverse, marginal, and splaying (Fig. Box 3.1.1).

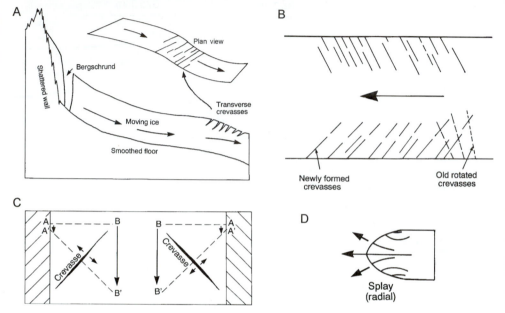

FIGURE BOX 3.1.1   Schematic representation of various crevasse types. *(A)* Plan view and cross section showing transverse crevasses. *(B)* Plan view of marginal crevasses (long dashes = newly formed, short dashes = rotated crevasses). *(C)* Plan view of marginal crevasses showing diagonal tensional stresses generated by differential ice movement at the valley sides (A–A′) and center (B–B′). *(D)* Plan view of splay crevasses associated with diverging ice flows (arrows). (After Sharp 1960.)

The *bergschrund* is the crevasse that forms at the head of a valley glacier where it pulls away from the headwall. The crevasse does not actually form at the ice/substratum contact, but within the ice itself (Fig. Box 3.1.1A). Sediment removed from the headwall by freeze-thaw processes may fall into the bergschrund, to be incorporated in the glacier.

Transverse crevasses form where a faster part of the glacier pulls away from a slower part, as in the case at a break in valley slope (Fig. Box 3.1.1A).

Marginal crevasses form in response to tensional stresses created by the speed difference between the middle of the glacier and its sides. Under these conditions, fractures form, rotate, and new fractures are formed (Fig. Box 3.1.1B,C). This creates a jumbled-up morphology at the surface of the glacier, which is also called serac (ice falls). One of these serac zones is very famous because most Mount Everest climbers must cross it on their ascent.

Splaying crevasses form primarily where a glacier expands in wider parts of the valley or at the end of the valley. The tensional stresses associated with radial flow create the fractures (Fig. Box 3.1.1D).

(continues)

BOX 3.1   (*continued*)

Crevasses constitute a hazard to climbers, particularly where they are hidden by weak snow bridges. From a geomorphological and geological point of view, they are important because they are indicators of the stress patterns within the glacier. This pattern can also be reconstructed in deglaciated landscapes by mapping the distribution of the poorly sorted sediments that accumulated in such fractured zones (**crevasse fillings**) and, upon melting of the ice, formed elongated isolated hills.

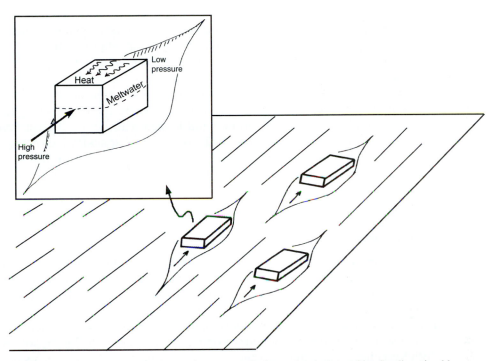

FIGURE 3.21   Diagrams of small protuberances at the base of a glacier and ice flowlines (molding) around them. High stress on the upflow side of the obstruction results in melting as the pressure melting point is reached. A low-pressure area is generated on the downflow side of the obstruction, resulting in refreezing. Heat derived from refreezing is conducted upflow through the rock.

Upon refreezing, the released latent heat flows preferentially through the obstructions (which are invariably characterized by higher thermal conductivity than the ice around them). If the obstructions are small enough, some of this heat reaches the upstream ends, heating and melting more ice. The plasticity of the icy mass increases, and the released water can locally become overpressured. These conditions reduce friction, allowing the glacier to slide on its base. This regelation-slip process occurs only if the glacier base is near the pressure melting point. It cannot occur in cold glaciers frozen to the ground. So block sliding occurs only in temperate or very thick, wet-based, polar glaciers. Note that during plastic flow, forces are expended in overcoming friction; during block slipping, forces are expended in accelerating the mass. Therefore, very rapid glacier movement is indicative of block slipping. Block slipping may foster basal glacial erosion.

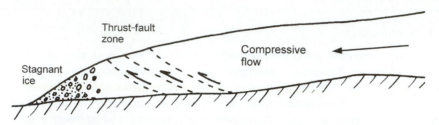

FIGURE 3.22   Schematic longitudinal cross section showing thrust-fault (shear) zone at a glacier terminus.

Internal Slipping Along Faults   Compressive flow against stagnant or sluggish parts of the glacier and sudden acceleration due to block slipping may increase stresses enough to exceed the brittle strength of ice. Thrust faults (low-angle reverse faults or shear planes) can form and release excess pressure in the direction of least resistance, upward to the surface (Fig. 3.22). Thrust faults can occur in any part of the glacier but are most common toward the terminus, where the glaciers may stagnate because they are frozen to the ground and loaded with debris. Thrust faults rarely occur singly, but rather form thrust-fault zones (shear zones), which carry ice and debris from the bottom of the glacier to its surface, mixing various types of sediments transported by the glacier and contributing to the formation of debris deposits (moraines) at the terminus.

Movement of Ice Sheets

In the previous analysis, a valley glacier was used as a model. The same mechanisms and general principles apply to ice sheets, although the rate of flow is generally controlled by slope of the ice surface, not the slope of the bed. Where ice sheets thin out at their margins, they are affected by bed morphology as well.

There are several hypotheses on movement of ice sheets. Two are mentioned here: the extrusion-flow hypothesis and the omnidirectional gravity-flow hypothesis.

Both hypotheses state that ice sheets can be considered as large bubbles of ice that try to reach gravitational equilibrium. An ice sheet therefore expands under its own weight when net accumulation exceeds net ablation, and retreats when ablation exceeds accumulation.

The *extrusion-flow hypothesis* focuses on the fact that thick glaciers can have warm basal ice that deforms plastically more readily than the cold ice above. The basal part of the glacier is subjected to higher pressure under the thicker parts of the glacier and lower pressure toward the margins (Fig. 3.23A). The more deformable, warm ice in the lower part of the glacier is thus extruded along a pressure gradient from the center to the margins beneath the colder upper glacier ice (Fig. 3.23B). The main problems with this hypothesis are internal friction and the shape of the ice sheet. If the true shape of the ice sheet is considered, rather than the usual vertically exaggerated conceptual diagrams, then it is obvious that pressure differences due to changes in glacier thickness are very small for ice sheets that extend over tens or even hundreds of kilometers (Fig. 3.23C). Internal friction is more than enough to resist stresses caused by these pressure differences. Extrusion flow would occur only if a glacier were higher than wide, like a large bar of ice placed vertically (Fig. 3.23D), but this is impossible for an ice body whose lateral extent is hundreds of times greater than its height.

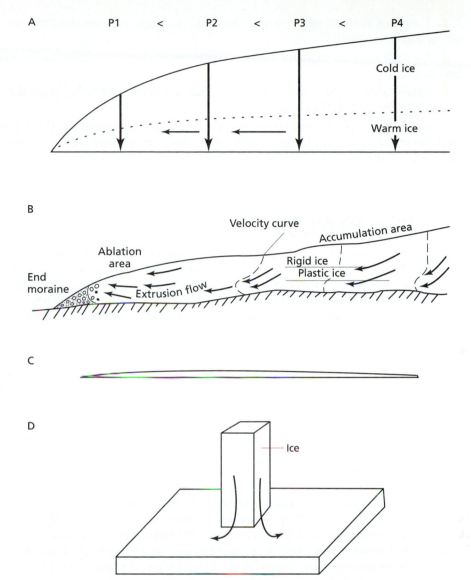

FIGURE 3.23 Diagrams illustrating extrusion-flow hypothesis. Theoretical distribution of pressures at the base of a thick polar glacier and hypothesized lateral flow of warm ice along the pressure gradient. *(A)* Extreme vertical-exaggeration diagram of a glacier showing pressure at the base and stratification of warm and cold ice (thick arrows = pressure, thin arrows = direction of flow, < = less than). *(B)* Theoretical movement and velocity profiles (dotted lines). *(C)* Realistic representation of an ice sheet with relatively small vertical exaggeration. *(D)* Vertical bar of ice tending to expand under its own weight (arrows indicate potential directions of expansion at the base of the block).

The *omnidirectional gravity-flow hypothesis* better explains the movement of ice sheets. It assumes that basal melting occurs at the base of thick glaciers, and thus basal friction is reduced and the ice can move. The upper, cold part of the glacier would ride on the lower, warmer part, and thus the ice sheet would move as an integrated whole.

Considering the geometry, mass budget, and type of ice in an ice sheet, the following can be concluded:

1. Large ice sheets move slowly at their centers where they receive little precipitation and do not melt.

2. Ice sheets are more active at the margins where they experience precipitation and melting.

3. Many polar glaciers are frozen to the bed and do not move by block slipping but only by internal deformation.

4. Parts of ice sheets move faster than others and form ice streams. Recent compilations of satellite radar images have revealed large ice streams that drain the central parts of the Antarctic Ice Sheet toward the sea. This indicates considerable movement and potential erosion even in the central portions of high-polar glaciers.

GLACIER HAZARDS AND RESOURCES

Most glaciers move slowly enough to pose little direct danger to properties and individuals. Indirect dangers are ice avalanches and floods associated with collapse of ice dams of meltwater lakes. On a global scale, glacier hazards are related to long-term increased melting of glaciers. This may lead to sea level rise, changes in ocean temperature, and ultimately to changes in atmospheric circulation and climate.

Hazards

The climate of the last 2500 years has been generally cool and moist. However, small (order of 1°C) variations in average global temperature have occurred, causing climatic warming during the eleventh and twelfth centuries, climatic deterioration from AD 1200 to 1500, and a colder spell (the Little Ice Age) from 1550 to 1840 during which glaciers readvanced. During the last century there has been a warming trend and overall retreat of glaciers; for example, calving glaciers retreated 14 km in Glacier Bay, Alaska, from 1907 to 1940, and mountain glaciers retreated 3 km in the Caucasus Mountains over the last 120 years. Thus, whereas medieval buildings and farm fields and roads were locally overridden by advancing ice in areas such as the European Alps, Caucasus, Karakorum, and Norway, the damage directly associated with glacier movement is minimal today. Nevertheless, past experience stands as a warning that similar events may occur in the future, and villages and resort facilities constructed in areas previously occupied by glaciers may be destroyed. The decision to develop an area, however, is based on a profit-risk assessment, and in most cases, the odds for the foreseeable future are judged to be favorable.

Some danger exists from the sudden, fast movement of surging glaciers or from ice detaching from hanging glaciers and moving rapidly downvalley as **ice avalanches.** When the snout of a glacier located on a steep slope becomes unstable because of crevasses and melting, large amounts of ice can detach and move rapidly downhill in a sort of avalanche (Fig. 3.24A). This ice avalanche accelerates, pulverizes the ice, and may mix with water and mud to generate very dangerous mudflows. In some valleys, such as the Chamonix Valley in France, such events are relatively frequent but small. This is in contrast to other places such as the Cordillera Blanca of Peru, where glacier avalanches involving more than 3 million m^3 of ice have been reported. These ice avalanches may initiate mudflows comprising about 13 million m^3 of ice, water, and mud moving at speeds greater than 100 km/hr. Although unstable glaciers are monitored, at times the events are unexpected or larger than estimated. In 1970, for example, 20,000 people were reported killed by a single glacier avalanche in Peru. These avalanches are extremely damaging because of their direct

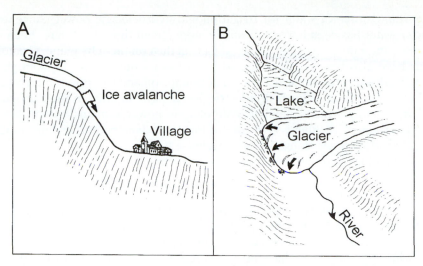

FIGURE 3.24   Diagrams of ice avalanches. *(A)* Start of an ice avalanche; *(B)* glacier damming a valley to create a lake (arrows indicate direction of movement of ice). (After Tufnell 1984.)

impact and, beyond the area they cover, because of the windblast they produce. Ice avalanches may locally form a reconstructed glacier at the bottom of the valley, possibly damming the drainage and leading to glacier floods.

Glacier-related floods generally develop when ice dams break and the impounded water moves downvalley, at times catastrophically. These dammed glacial lakes form in various ways, but most commonly they develop when an advancing glacier cuts off and dams a side valley (Fig. 3.24B), or when, like in Iceland, subglacial volcanoes induce large, subglacial melting. Temporary lakes may develop behind moraines, or where the valley is closed by ice- or rock-avalanches. Some historical floods reached a discharge of about 3000–6000 m³/s, to about 200,000 m³/s, the largest recorded being in Iceland. Glacier-related floods may be extremely damaging to agriculture, property, and people, as is commonly occurring in the Cordillera Blanca of Peru, where, for instance, one-third of the city of Huaraz was destroyed and about 6000 people killed in 1941. These floods can also cause damage far from their point of origin. For example, in 1929 an ice flood that began 400 km away destroyed the city of Abadan in southern Asia and adjacent agricultural fields.

To avoid these disasters, people have tried, with mixed success, to lower the rising lake level once the valleys have been dammed by a glacier. In Switzerland a first trial was made by cutting a tunnel through the ice to stop the impounded lake from exceeding a certain elevation. Water flowing though the tunnel, though, melted the ice, undermining the glacier dam so that it eventually gave away, and a flood was generated anyway, albeit smaller than it would have been if the lake had been allowed to deepen further. For slowly advancing glaciers, water is pumped onto the ice itself, melting the glacier into blocks that can be transported away by the rivers before the dam can form. The use of water is more efficient and less costly than other means (such as the use of explosives) of destroying the integrity of the glacial tongue.

Resources

Glaciers can be considered a resource as well. In the past, glaciers were used to keep material refrigerated, or as a source of ice for refrigeration. Mostly they are an important source of water for irrigation and hydroelectric power. Since antiquity, glacier water has been harvested, and in more recent times trials have been made to increase the rate of melting by spreading coal dust on them by plane. Grandiose designs have also been made to use icebergs from Antarctic or Alaskan glaciers to bring freshwater to semiarid areas for consumption and irrigation. Increasingly, glacial meltwater is used for hydroelectric power: for example, in the European Alps and in Norway. Much potential hydroelectric power remains untapped — for example, around Greenland — because of low local demand and technical difficulties in transporting power long distances across seas.

The other main use of a glacier as a resource is tourism: glaciers attract many summer visitors, as well as climbers and skiers. Some glaciers allow skiing to be a year-round activity.

Finally, it is interesting to note that although glaciers may create problems for transportation routes, the disappearance of glaciers may cause problems as well. In some Alpine valleys, routes used, for example, by animals to move between summer and winter grazing lands have been severed by glacial retreat. In the past, valleys were crossed on the ice; now, steep deglaciated slopes prevent animals from moving one flank to the other.

ENDNOTES

1. After Ahlmann 1948.
2. After Benson 1959.
3. Uniform flow occurs when, at a given time, the same amount of ice passes through cross sections of equal size in different parts of the glacier. Steady flow occurs when, at one locality, the same amount of ice passes through the same cross section at different times.

Glacial Geomorphology

The study of interactions between the glacier, other types of ice, and the substrate.

CHAPTER 4

Glacial Erosion

Glaciers are large and powerful agents of erosion and transportation that modify the morphology of entire regions.

POWER OF A GLACIER

Power[1] is the capacity of a system to perform work. The power of a glacier to move material is a function of its thickness and speed.

The **total power of a glacier (Wt)** is the shear stress (τ) the glacier can exert, multiplied by its average velocity[2] (v):

$$Wt = \tau v$$

where (τ) can be considered constant, with an average value of 1 bar at the base of a glacier at the pressure melting point. Therefore, on a shallow, uniform slope, where the thickness and weight of the glacier do not change significantly, the total power (Wt) of the glacier is directly proportional ($\propto$) to average velocity (v).

$$Wt \propto v$$

This means that the faster the glacier moves, the more total power it has.

Glacier velocity is normally measured at its surface, and consists of two major components. One component of movement is related to the internal deformation of the ice mass downvalley, the other to slippage of the glacier at its base (Fig. 3.16). This basal slippage is the main component of movement that effectively erodes the substrate. Thus, only a portion of the total power of the glacier is expended for erosion. This portion is called the **effective erosional power (We)** of the glacier:

$$We = \text{part of } Wt$$

For example, a glacier near the firn line with a 1 km^2 surface, a width of 430 m, a thickness of 50 m, and a velocity through its cross section varying between 2 m/yr and 40 m/yr has a total power varying between 0.25 and 1.5 watts (W), but with an estimated effective erosional power ranging between 0.03 and 1.0 W. The ratio We/Wt changes largely in relation to the proportion of basal slip to internal ice deformation. Where the amount of basal slippage is high, such as in a surging glacier, We/Wt tends toward 1; and where slippage does not occur, such as in a cold-based glacier, the ratio tends toward zero. As a first approximation, *We/Wt is about 0.5 in temperate glaciers and 0.1 in cold-based polar glaciers.* When compared with rivers, which experience velocities five to six orders of magnitude greater, the erosional power of glaciers is higher because of larger basal shear stress due to greater thickness.

Steep valley glaciers in the Himalayas move very fast (up to 100 m/yr) and have a greater erosive power than glaciers of gently sloping valleys (as in the European Alps), which move only a few centimeters per year. Continental ice sheets show maximum effective erosive power toward their margins, where a more dynamic regime is established because of high snowfalls and glacier ice removal (melting or calving at the continental margin). Indeed, the *rate of erosion* is greater toward the margins of ice sheets than toward their centers where cold ice persists and little basal sliding occurs.

Temperate glaciers thus erode more effectively than cold glaciers. However, in terms of *total* erosion, the lower effective erosive power of cold glaciers may be partly compensated for by their longer life. For example, an average erosion rate of 0.1 mm/yr over a period of 3 million years would be capable of eroding features up to 300 m deep. If the rate of erosion is increased to 1 mm/yr over 6 million years (the time some large glaciers, such as the Antarctic Ice Sheet, occupy an area), valleys as deep as any developed under temperate glaciers can form. Furthermore, glaciers start, grow, and wane. Thus, areas near centers of glaciations experience erosion *under temperate conditions* during the early and later stages of glaciations, and continue to be eroded, albeit at a slower rate, under polar glacier conditions during maximum glaciation.

EROSION PROCESSES

Erosion occurs under a glacier through several processes:

1. Abrasion: scratching action
2. Plucking or quarrying: removal of large fragments from the substrate
3. Moving meltwater: abrasion, corrasion, and dissolution

Abrasion

Glacier ice cannot directly abrade other rocks, unless they are softer, like talc or gypsum, because its hardness is low (2 to 3 at −15°C, on Mohs' scale), even in cold glaciers. It needs rock fragments embedded in its base to abrade (Fig. 4.1). With

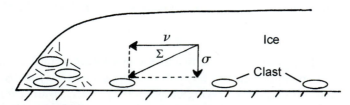

FIGURE 4.1   Schematic cross section of a glacier showing conditions necessary for abrasion. Abrasion occurs when the combined forces of weight (σ) and movement (v) of a glacier push a clast against the bedrock or against another clast, as indicated by the vector sum $\Sigma = v + \sigma$.

these, the glacier behaves as an enormous rasp, with ice as the matrix (glue) and hard rock fragments as the teeth. Thick continental glaciers acquire rock fragments through the process of subglacial freezing-and-thawing and quarrying. Valley glaciers also acquire rock fragments from the substrate and from valley walls.

Effective abrasion requires the rock fragments to be harder than the substrate, and they must stay or return to the base of the glacier through basal melting or glacial flow. Furthermore, significant abrasion occurs only when large clasts and/or a large number of particles occur at the base of a temperate, sliding glacier. Abrasion of both the bedrock and the moving, interacting particles may occur, and produces vast quantities of silt. If, instead, isolated clasts occur at the base, the weight of the glacier applied over their small cross-sectional area exerts local high stress (stress = force/area) such that local pressure melting occurs. This causes the clasts to retract inside the ice and lose contact with the substrate, and abrasion ceases.

Bulldozing and Quarrying (Plucking)

A glacier can exert large compressive forces on obstructions, and can also generate tensile stresses on the substrate when parts of the glacier freeze to the bottom and pull it away. When the substrate is loose or fractured, the glacier can **bulldoze** it. Furthermore, when the terminal part of the glacier is stagnant, the pressure exerted by the still-moving main body of the glacier may thrust basal material upward to the surface.

A glacier cannot quarry (pluck) solid rock, but it can remove fractured fragments. Repeated advances and retreats of the glacier load and unload the substrate, causing bending and fracturing of the rocks. Freeze-thaw processes enlarge these fractures and loosen the fragments. If this occurred only in permafrost areas in front of the glacier, the advancing glacier would simply erode the thin, uppermost, fractured part of the substrate. The glacier can, in fact, erode much deeper because freezing-and-thawing also occurs under the glacier itself; so a continuous breakdown and removal of newly exposed rocks takes place. Such subglacial freezing-and-thawing of the substrate depends on the thickness of the overlying glacier and the nature of the basal contact (Fig. 4.2). In some areas the basal ice is at the pressure melting point, and the glacier can slide at its base; but in other areas the ice is cold and frozen to the bottom. In this case the pressure melting point occurs below the ice/rock interface, and, if the rock is fractured, the upper frozen parts can be removed (pulled out, plucked) by the moving glacier and incorporated into the ice itself. Changes in ice thickness by snow accumulation, surface melting, crevassing, and other processes may displace the zones of bottom freezing-and-thawing back and forth; hence, considerable amounts of debris can be removed from the substrate.

Meltwater Erosion Under the Glacier

A large amount of water is generated at the base of temperate glaciers. This meltwater may flow through fractures, tunnels, and, at times, as thin sheets under the ice, and may also accumulate as large subglacial lakes like those discovered by geophysical means under thick polar glaciers in Antarctica. If the water is released suddenly from these lakes, powerful subglacial floods can be generated. Sediment-laden water flows can effectively abrade the substrate, and cold-water flows can in part

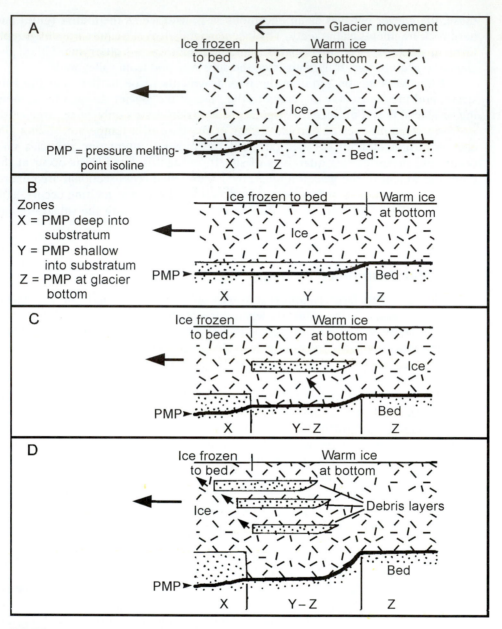

FIGURE 4.2 Schematic diagram showing glacial fluctuations leading to plucking (removal) of sediments or fractured bedrock at the bottom of a moving glacier. The point at which ice freezes to the bottom depends on the thickness of the glacier. *(A)* Glacier is frozen to the surface of the bottom sediments where it is thinner. Where ice is frozen to the bottom, the pressure melting-point (PMP) isoline, hence the potential detachment zone, is deeper into the sediments or rocks. *(B)* The glacier thins and the frozen bed expands upglacier. *(C)* The glacier thickens and advances, plucking out part of the substrate frozen to the ice. Such plucking occurs when glacier movement is faster than readjustment of temperature conditions at the base of the glacier. *(D)* After several cycles of thinning and thickening (shifting the position of the frozen bed), several fragments of the substrate have been plucked away. These fragments may be very thin or relatively thick depending on the depth of frozen ground and the rate of glacier movement. Once plucked, the material is moved inside the glacier along concave-upward shear zones (thrusts). (After Weertman 1961.)

dissolve rocks such as carbonates. During low winter discharge, in some meltwater streams, the solution load (order of 150 mg/L) far exceeds the suspended load (order of 50 mg/L). During summer, the suspended load prevails.

EROSIONAL FEATURES

Glacial erosional features are usually subdivided into small scale and large scale, the former due mostly to abrasion, and the latter due primarily to quarrying. Meltwater erosion may modify both. Furthermore, whole regions may develop glacially eroded landscapes that differ depending on whether they have been affected mostly by ice sheets or by valley glaciers.

Small-Scale Erosional Features

Striae are small-scale scratches produced by abrasion that vary in dimension and shape. They may be hairlike or more substantial. Striae develop and are preferentially preserved in fine-grained, brittle rocks such as quartzite and limestone, since friable rocks weather more rapidly and lose any striae they may have had.

Striae form parallel to the local flow direction of the glacier as rock fragments embedded in the basal ice are pressed against bedrock. Their continuity and variable depth of incision indicate the type of tools (size, hardness, concentration) carried at the base of the glacier and their behavior. Well-developed, continuous sets of striae are formed by a moving carpet of clasts. Discontinuous striae of variable depth and width are formed by isolated clasts that are readily retracted inside the ice by pressure melting. In these processes the clasts become striated as well.

There are many different types of striae.

Simple striae are sets of scratches of various lengths (Figs. 4.3A, 4.4A). Two or more sets of striae may intersect at various angles, with the different sets having different depths of incision (Fig. 4.3A). Intersecting sets of striae form when glaciers or parts of a glacier change their local flow directions.

Wedge-shaped striae are of various length and shape; some are triangular, others are ellipsoidal (Figs. 4.3B, 4.4B). They form due to differential abrasion as the abrading clasts score the bedrock progressively deeper until they are abruptly retracted inside the ice.

Nailhead striae have a pit generally at the downflow end (Fig. 4.3C). They form when a tool (clast) scratches progressively deeper into the bedrock until it is abruptly retracted inside the ice.

Rat-tail striae have minor longitudinal ridges downflow from an obstruction (Fig. 4.3D). They form by abrasion around an obstruction.

Fine scratches and *polished surfaces* (Fig. 4.3E) are generated by a moving mass of silt and sand.

Crescentic marks develop in brittle rocks, and are usually transverse to glacial flow.

Crescentic gouges are semilunate scours, concave upstream. They are formed as a fragment of rock is removed from between two fractures (Figs. 4.4C, 4.5A).

Crescentic fractures consist of single fractures without removal of any material. The fractures are concave and dip 60–90° in the glacier-downflow direction (Figs. 4.4D, 4.5B).

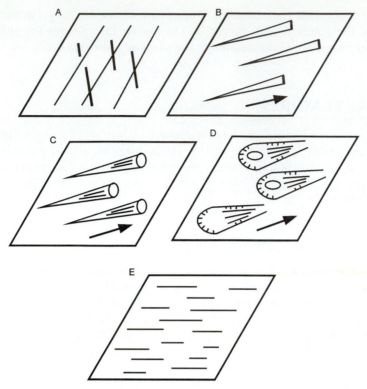

FIGURE 4.3 Diagrams of selected small-scale glacial-erosional features. *(A)* Two sets of striae. The heavier, deeper ones can persist when additional, but shallower, abrasion (striae) occur. *(B)* Wedge-shaped striae. *(C)* Nailhead striae. *(D)* Rat-tail striae are composite features showing convex upstream erosion in front of still-existing or re-moved obstructions, and tail furrows. *(E)* Polished surface made by numerous small scratches on the surface of hard rocks. These are generated by small clasts like silt grains.

Lunate fractures are similar to crescentic fractures but are concave upflow (Fig. 4.5C).

Chattermarks are chipped-out depressions, often occurring in grooves (Fig. 4.5D). They are formed by clasts that are cyclically retracted and re-exposed at the base of the glacier.

Crescentic marks form when localized diagonal pressure is exerted by the gla-cier on the bedrock through a clast. The localized force generates tensional stresses in bedrock in the upflow direction and compressional stresses in the downflow di-rection. When the strength of the substrate is exceeded, crescentic fractures and crescentic gouges, respectively, are formed (Fig. 4.6).

Grooves are linear erosional features formed in solid bedrock, and are up to 2 m deep and 50 to 100 m long. They may show overhangs and invariably show striae and other minor erosional features inside (Fig. 4.7). Grooves likely form partly by basal gouging by boulders or debris bands and partly by subglacial water erosion.

P-forms (plastically molded forms) are smooth, sinuous, streamlined erosion-al features cut into bedrock. They show different shapes, ranging from small (about

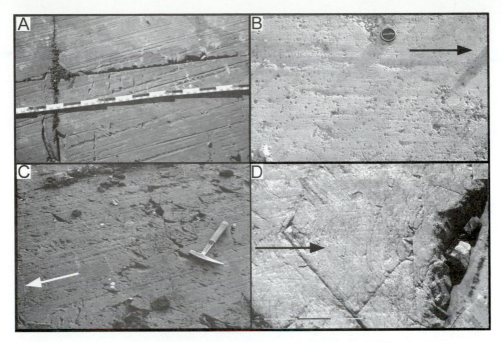

FIGURE 4.4   Photographs of selected small-scale glacial erosional features. *(A)* Simple striae; *(B)* wedge-shaped striae; *(C)* crescentic gouges; *(D)* crescentic fractures. (Arrows show direction of glacier flow.)

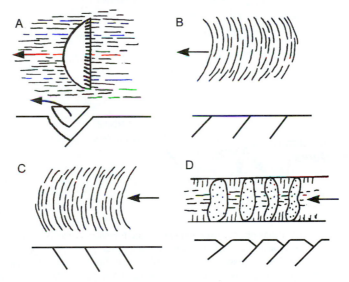

FIGURE 4.5   Diagrams of selected small-scale glacial erosional features in plan and section. *(A)* Crescentic gouges; *(B)* crescentic fractures; *(C)* lunate fractures; *(D)* chattermarks. (After West 1968.)

1 m wide, and 1 m deep) channels with streamlined ornamentation and at times overhanging sides, to sickle-shaped (*sichelwannen*) features open in the downflow direction, to subrounded features of various sizes (a few tens of centimeters to several meters in diameter, and from a few centimeters to tens of meters deep) (Fig. 4.8). Generally they are associated with other features that

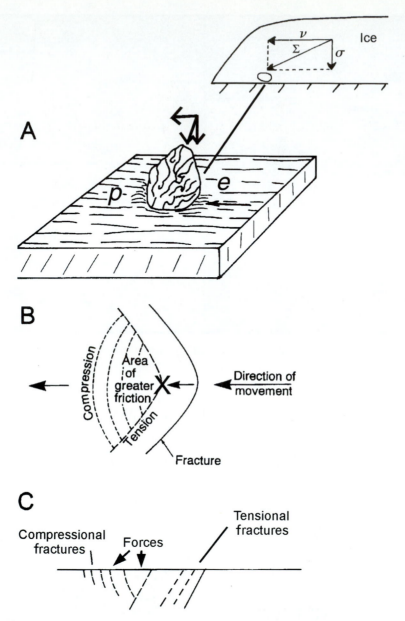

FIGURE 4.6 Illustrations of processes involved in the formation of selected crescentic marks. *(A)* Clasts pushed against bedrock obliquely, as at the base of a glacier. Downflow from the clast, a pressure ridge *(p)* develops. If the strength of the bedrock is exceeded and a piece is removed, a crescentic gouge develops. On the upflow side of the clast, tension *(e)* is developed in the bedrock. If the bedrock strength is exceeded but no material is removed, a set of crescentic fractures develops. *(B)* Plan view of stress distribution around a clast leading to the formation of crescentic gouges and crescentic fractures. *(C)* Cross-sectional view of stress distribution and fracture patterns around a clast. (After Harris 1943.)

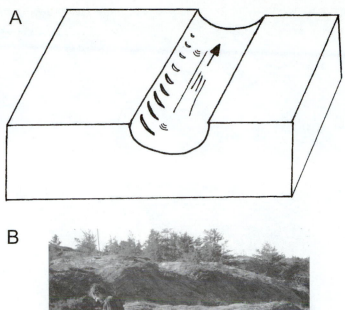

FIGURE 4.7   Illustrations of grooves. *(A)* Grooves cut by glacier ice and meltwater into bedrock, usually ornamented inside by chattermarks, striae, and other small-scale glacial erosional features. *(B)* Groove formed on Precambrian rocks in the Canadian Shield.

confirm the glacial setting. P-forms are generated by various processes working in combination or in sequence. The principal erosional process is most likely highly plastic warm ice and/or flowing water, charged to varying degrees with sediments. The mixture can range from dilute icy mudflows to highly concentrated turbulent water flows.

FIGURE 4.8 P-forms on bedrock. (Photograph from Phillip Kor, Ontario Ministry of Natural Resources.)

Channels. Subglacial meltwater flow can also cut channels downward into the bedrock (Nye, or **N-channels**), or, when under pressure, upward into the glacier itself (Rothlisberger, or **R-channels**), or both into bedrock below and glacier ice above (**C-channels**).

Potholes are round, at times very deep, bedrock scours. They almost invariably contain the rounded clasts that have drilled them (Fig. 4.9A,B). In glacial settings they may be partially filled by glacial sediments. Potholes are formed by streams moving quickly over bedrock, either under a glacier or in nonglacial areas. Their initiation has been recently associated with cavitation in very fast flowing, deep meltwater streams. **Cavitation** occurs when air bubbles formed in fast-moving water implode near the bed, generating sufficient tensile stress to pluck bedrock. The resulting incipient cavities may subsequently be enlarged and deepened by rock tools (clasts) caught in turbulent vortices. Doubts have been cast on whether cavitation is a significant erosive mechanism in glacial settings.

Large-Scale Erosional Features

Large-scale, glacial erosional features range in size from tens of meters to kilometers. They are formed primarily by glacial plucking but with some contributions from abrasion and flowing water.

Roches moutonnées (French term meaning sheepback-like rocks) are streamlined landforms with a smooth, gentle, upflow slope and a jagged, steeper, downflow

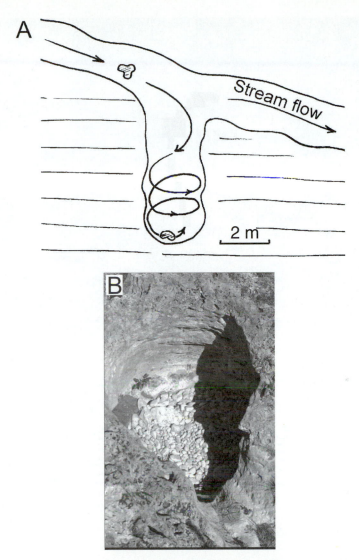

FIGURE 4.9   Potholes. *(A)* Diagrammatic representation of pothole formation. *(B)* Photograph of a pothole approximately 1 m in diameter showing rounded clasts (drilling tools) in place.

side. Typically, they vary from 1 to 50 m in height, a few meters to kilometers in length, and tens to hundreds of meters in width. A few exceptional roches moutonnées can reach heights of 250 m. They form on every type of hard rock substrate, and may be localized by pre-existing morphologic irregularities caused by variations in composition and structure of the rocks (such as igneous dikes, reefs). Roches moutonnées are formed both by ice sheets and valley glaciers (Fig. 4.10A).

Both quarrying and abrasion participate in the formation of roches moutonnées (Fig. 4.10B). As the glacier moves over an obstruction, it generates high stresses against any upstream bedrock slope. The basal ice may reach the pressure melting point and partially melt, causing the glacier to slide, while the embedded clasts abrade the bedrock. On the downflow side of the obstruction, the pressure drop causes the meltwater to freeze the glacier to its base, and fractured rock blocks

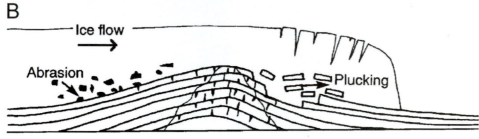

FIGURE 4.10   Roches moutonnées. *(A)* Photograph of a roche moutonnée in an alpine valley, Austria. *(B)* Diagram illustrating formation of a roche moutonnée, with abrasion on the upflow side and plucking in the downflow part.

may be removed by quarrying (plucking) as the glacier pulls away (Fig. 4.10B). Quarrying is more effective than abrasion, and the downflow side is transformed into a jagged step, forming a roche moutonnée.

In southern Ontario, Canada, roches moutonnées have developed on dolomitized Silurian reefs composed of a massive core overlain by beds of wave-reworked carbonates (Fig. 4.11). Through differential compaction, the capping beds formed a small anticline over the reef core, sloping away from the core (Fig. 4.11A). When the last Pleistocene ice sheet crossed the area, it abraded the beds sloping upflow, but froze to and removed (quarried out) the fractured beds that were sloping downflow (Fig. 4.11B–D).

Crag and tail has a resistant bedrock knob (crag) and a streamlined remnant of bedrock or sediments plastered onto the lee side (tail) (Fig. 4.12). Deep, semilunate troughs can form around the upflow part of the obstruction. Grooves may be cut into the crag and persist along the flanks of the tail.

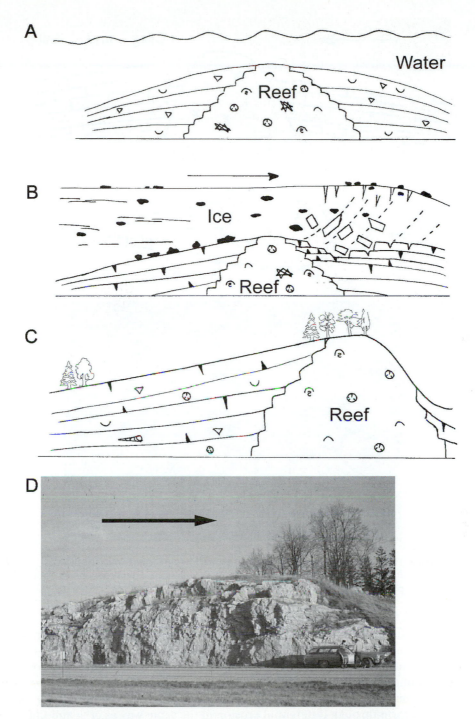

FIGURE 4.11 Illustrations of the complex history of a roche moutonnée formed in association with a small, dolomitized reef core in southern Ontario, Canada. *(A)* Deposition of a Silurian reef core (mostly algal material and brachiopods) covered and surrounded by water-reworked carbonate strata. A slight anticline was formed by differential compaction on the reef core and flanks. *(B)* Glaciers advanced over the area in the last 2 million years. The layers on the upflow part of the reef core were abraded, but pushed against the core and thus retained. The layers on the downflow side of the reef core were removed by plucking. This led to the formation of a roche moutonnée (arrow indicates direction of glacial flow). *(C,D)* The roche moutonnée as it is today after a face has been cut by highway construction.

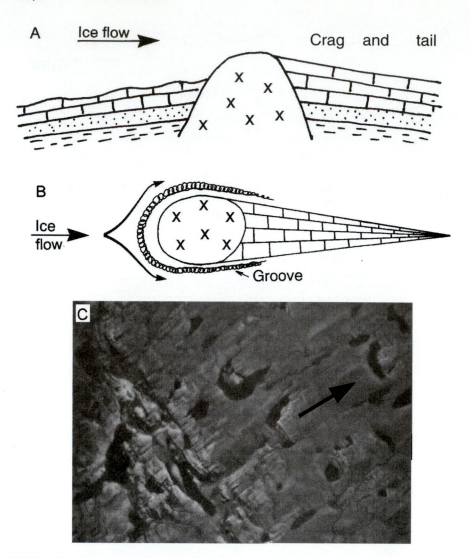

FIGURE 4.12 Illustrations of a crag and tail, an erosional feature formed by an ice sheet. *(A)* Cross-sectional view; *(B)* plan view; *(C)* crag and tail features from the Precambrian Canadian Shield (air photograph). (Arrow indicates direction of glacial flow.)

Edinburgh Castle in Scotland was built on a resistant igneous crag, and the rest of the old city of Edinburgh on its less resistant sedimentary tail (Fig. 4.13A,B).

Drumlins are streamlined hills commonly resembling inverted spoons, composed primarily of sediments. Their origin is controversial, as they may be formed by a combination of erosional and depositional processes. Those with a bedrock core may form through differential erosion in the same way as crag and tail, except that sediment is plastered around and over them. Those with layered sediments may have formed by differential erosion of pre-existent waterlaid deposits, or by subglacial floods in ice tunnels or cavities.

Flutes are subparallel grooves with ridges between them that vary greatly in size, ranging up to 200 km long, 100 m wide, and 25 m high (Fig. 4.14). They are aligned parallel to the direction of glacier movement, and form in flat areas, where they seem independent of minor variations in both lithology and morphology.

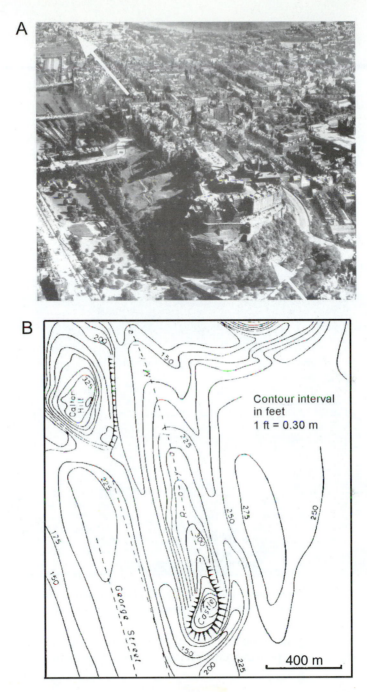

FIGURE 4.13   Edinburgh Castle, Scotland, built on a crag and tail. *(A)* General view (arrow indicates direction of glacial flow) (after Lorimer 1967). *(B)* Topographic map (after Sissons 1971).

They form on both bedrock and sediment-covered terrain and are mostly erosional, but they may also form when basal sediment is squeezed into fractures at the base of the glacier.

Cirques are hollows cut into a mountainside. They are generally semicircular in shape, with a steep, high headwall and an open side downvalley (Fig. 4.15A).

A

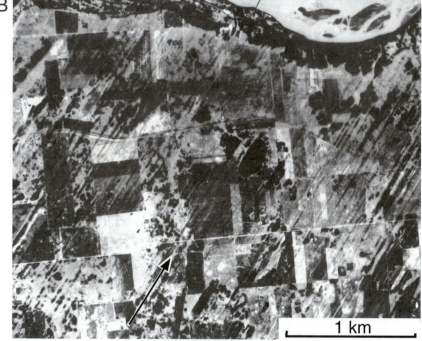

FIGURE 4.14   Flutes formed at the base of a glacier. *(A)* Diagrammatic representation. *(B)* Air photograph showing large-scale fluting in Saskatchewan, Canada.

Cirques vary greatly in shape and size, although in their simplest form the longitudinal section can be approximated by a logarithmic curve of the form

$$y = k(1 - x)e^{-x}$$

where

        x = distance from the headwall to lip
        y = depth of cirque at the upstream headwall
        e = constant (2.718)
        k = constant related to the shape of the hollows, generally between 0.5 and 2;
            the steeper the headwall, the greater the k value

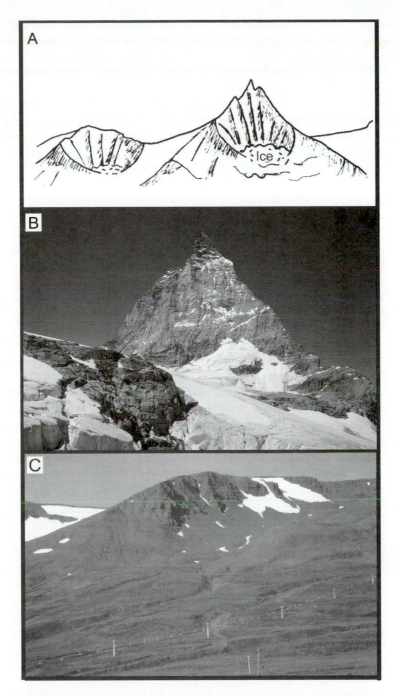

FIGURE 4.15   Mountain erosion. *(A)* Schematic representation of pyramidal mountains sculpted by cirques. *(B)* Photograph of the Matterhorn in the Swiss Alps. *(C)* Snow hollows, Iceland.

Cirques are formed by small glaciers (*névé*) near the snowline. Such glaciers are relatively thin and respond rapidly to changing climatic conditions. Rotational mass movements of the glacier carry ice and sediment to the downvalley lip of the hollow. Intense freezing-and-thawing on the valley walls and beneath the glacier make cirque erosion very effective, and mountains are readily sculpted into steep-sided **arêtes** (sharp-crested ridges between glaciated valleys) or **horns** (isolated pyramidal mountains) (Fig. 4.15B). K2, the second highest mountain in the world, is a horn, and so is the famous Matterhorn in the Swiss Alps. Nunataks emerging from ice sheets may be similarly sculptured.

Snow hollows are small niches (Fig. 4.15C) cut into the side of hills through **nivation.** Nivation is the sum of processes associated with numerous freeze-thaw cycles that break up the local rocks and slope processes that move such products downvalley. The snow serves mostly as a source of moisture for nivation to be effective. The snow patch should be relatively thin so as not to insulate the ground and prevent freeze-thaw cycles.

Glaciated valleys are scoured by streams and subsequently modified by glaciers, and they acquire characteristic morphologies.

Transverse profile. A glaciated valley tends to develop a U-shaped cross-sectional profile, due to the glacial modification of a pre-existing, V-shaped fluvial valley. The fundamental reason for the change in shape is to develop a channel that is most efficient for the transmission of a material like ice. In fact, a Newtonian (viscous) fluid, such as water, can flow efficiently through a V-shaped channel, whereas a plastic (Bingham substance) or a pseudoplastic material such as ice flows more efficiently (with less expenditure of energy) through a semicircular (U-shaped) trough. This can be inferred from the velocity distribution of a fluid over a surface: the Newtonian fluid displays a narrow, pointed, parabolic profile, whereas the plastic fluid displays a wide-fronted profile (Fig. 4.16A).

The glacier changes a V-shaped fluvial valley into a U-shaped valley through a combination of erosion and sedimentation. The cross-sectional velocity distribution of a pseudoplastic substance (like a moving ice mass) shows a steeper gradient near the walls of the V-shaped valley (Fig. 4.16B); hence the maximum stress develops against these walls (Fig. 4.16C). Erosion occurs along the flanks, until the channel shape is such that the velocity gradient is minimized: this leads to the formation of a U-shaped valley (Fig. 4.17A–D). Additional contributions to the shaping of U-valleys are "plugs" of entrained sediment that, like debris flows, move sluggishly and may stop along the narrower, center-bottom parts of the valleys, or, upon dewatering, may form levees along the valley sides (Fig. 4.17E).

Longitudinal profile. Complex glacial erosional processes are associated with the development of the longitudinal profiles of valley glaciers. In general, glaciers moving down river valleys tend to straighten them out, and alter the longitudinal profiles by overdeepening them. Several erosional features such as steps, rock basins, paternoster lakes, and hanging valleys are developed. These reflect the erodibility of the substrate as well as glacier behavior.

1. Erodibility of the substrate changes depending on its lithological and structural characteristics, that is, whether it is composed of more erodible shale, or cemented sandstones and resistant igneous rocks, and to what extent the rock has been weakened by fractures (joints), faults, and folds (Fig. 4.18A–C). Everything else being equal, **steps** develop and are maintained at the point where two substrate lithologies of different erodibility outcrop.

2. Differential erosion along the valley can also be related to changing slope and behavior of the glacier. On flatter reaches, the glacier develops compressive

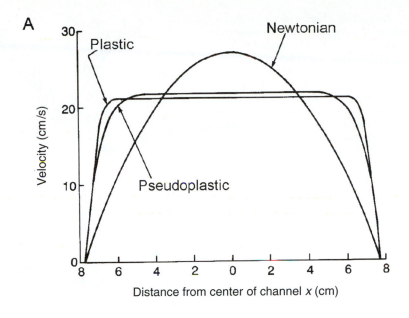

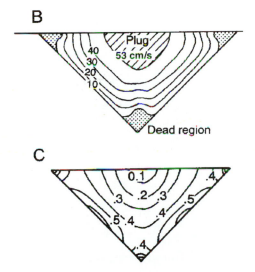

FIGURE 4.16   *Development of a U-shaped valley by a glacier. (A)* Plan-view pattern of different substances (Bingham = plastic material; pseudoplastic = akin to ice; Newtonian = akin to water). *(B)* Cross section of theoretical velocity gradients of a pseudoplastic material in a V-shaped valley — steep velocity gradients (contour lines closer together) toward the sidewalls. Note that no movement occurs at the bottom and levees are where fluid is lost. *(C)* Cross section showing pressure gradients — highest in correspondence with the steepest velocity gradients (units are dimensionless). (After unknown, please check website for update.)

flow, thickens, and basal ice can reach the pressure melting point. The glacier then slides and abrades instead of quarrying, and forms **rock basins** and carves transverse rock ridges locally called **riegels** (Fig. 4.18D). On steeper slopes, the glacier develops extending flow, crevasses form, and the glacier thins. There, basal ice may freeze to

FIGURE 4.17   Large-scale, glacial-erosional features —
transverse sections of a glaciated valley. The characteristic U-shape of the glaciated valley is acquired through
different steps. *(A)* V-shaped fluvial valley. *(B)* The
glacier funnels into the valley and modifies to adjust
it to movement of a quasi-plastic material, the ice.
The major forces (arrows) are against the flank of the
valleys, whereas, like a debris flow, frozen plugs form
at the bottom and as levees at the sides (black patches).
(C,D) Slowly the valley is modified by erosion. *(E)* Final
U-shaped glaciated valley. Note that although this is
primarily an erosional form, the valley bottom may
contain considerable glacial and fluvial deposits. (After
Davis 1906.)

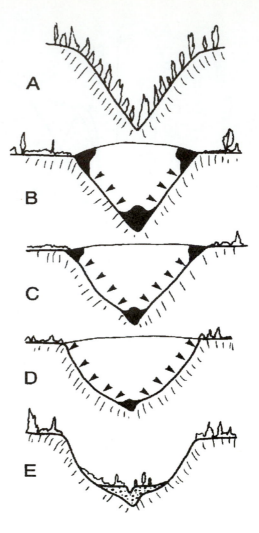

the bottom and pull out (quarry) fractured rocks, maintaining or even steepening
the slopes.

The result may be a fairly regular alternation of steep slopes and flat basins
down a glacial valley. The basins may fill with water during and after ice retreat,
forming a sequence of **paternoster lakes,** a reference to the large beads of a Chris-
tian rosary (Fig. 4.19A).

3. Sudden changes in glacier thickness may locally lead to greater erosional power and the development of bedrock steps. The increase in thickness occurs at the confluence of two glaciers. If confluent glaciers differ in size, the smaller
one is less powerful; hence its valley is smaller and shallower. After deglaciation, its
floor is at a higher elevation than that of the major glacier, forming a **hanging valley**
(Fig. 4.19B).

4. Glaciers can also spill over, and erode into the lower saddles, or cols, in
their valleys, forming *breached watersheds* that may be occupied and perpetuated
by streams.

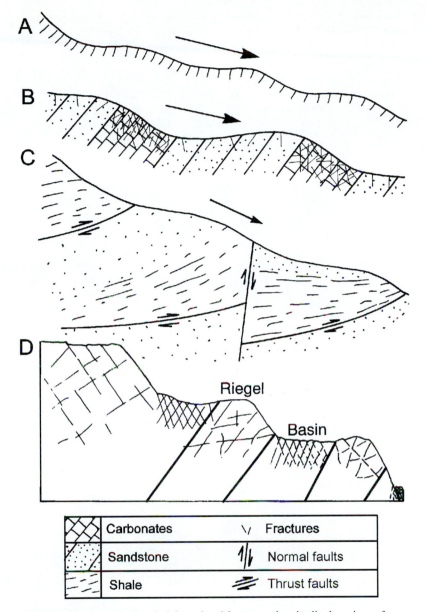

	Carbonates	\\/	Fractures
	Sandstone		Normal faults
	Shale		Thrust faults

FIGURE 4.18   Large-scale, glacial-erosional features — longitudinal sections of a glaciated valley. *(A)* Step morphology. *(B)* Variation in lithology and fracture pattern in relation to location of steps. *(C)* Faults can lead to variation of lithologies along the valleys, hence to formation of steps by differential erosion by glaciers. *(D)* Riegels are promontories formed by more resistant rocks; basins form where rocks are more readily eroded by glaciers.

Fjords (fiords). Glaciated valleys can be carved below sea level, and become flooded during and after ice retreat, forming fjords (Fig. 4.20A). The simplest fjords and lochs are deeper near their heads, giving them a down-at-the-heel profile. Underwater ridges often divide them into sub-basins and separate them from the open sea or main lake. These ridges are due to the buoyancy of a glacier. Where a

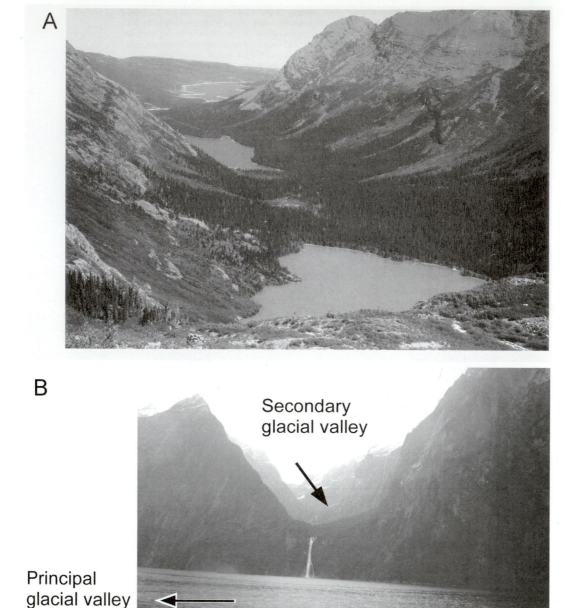

FIGURE 4.19   Photographs of glacial valleys. *(A)* Paternoster lakes (from Tarbuck and Lutgens 1999). *(B)* Hanging valley in background, on the flank of a large submerged (fjord) valley in foreground, New Zealand. (Arrows indicate direction of glacier movement.)

valley glacier ends in a body of water, its erosional power is progressively diminished by buoyancy (Fig. 4.20B). Thus, the valley may terminate against a bedrock ridge that could not be eroded. Often the glacier deposits sediments over the ridge, enlarging it. Other ridges form where icebergs preferentially ground and release sediments.

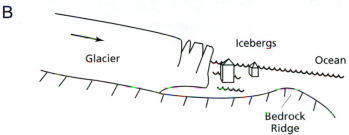

FIGURE 4.20   Illustration of large-scale, glacial erosional features. *(A)* Photograph of a fjord, New Zealand. *(B)* Longitudinal profile of a glaciated valley where the glacier terminates and floats into a large body of water (sea or lake). At the terminus, the glacier loses its erosional power, and the deepened glaciated valley can be separated from the open sea by a residual bedrock ridge.

EROSIONAL GLACIAL LANDSCAPES

Erosional glacial landscapes have unmistakable features, yet their classification is difficult because many processes have acted in concert over the time required for their development. One possible subdivision scheme is to contrast areas of low relief affected primarily by ice sheets with mountainous areas affected mostly by valley glaciers. Plateau areas are an intermediate case affected primarily by ice sheets, but also by valley glaciers upon deglaciation.

1. Low-relief glaciated areas may be uniformly scoured over extensive areas (Fig. 4.21A). Differential erosion of bedrock along fractures, faults, and igneous dikes form elongated basins both parallel to the ice flow and at various angles to it, depending on the structural grain of the region. These basins are usually filled with lakes that, together with low domes and ridges, impart a characteristic appearance to low-relief glaciated areas. Typical examples can be found in the vast Canadian Shield (Fig. 4.22A), Scandinavia, and other similar regions.

2. Mountainous areas can be affected by ice sheets, but, ultimately, they are sculpted by valley glaciers (alpine erosion). Glacially eroded mountains may have steep

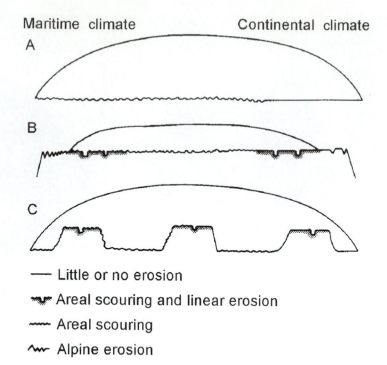

FIGURE 4.21   Models for glacial erosion as function of ice positioning, climate, and morphology. *(A)* Low-lying areas covered by ice sheets; linear erosion is minimal, areal scouring predominates toward the more humid termination of the ice sheets, and little or no erosion occurs under the cold-based continental part of the ice sheet. *(B)* Ice sheets cover highlands. Sculpting by valley glaciers (alpine erosion) is more effective in the humid maritime side of the ice sheet. *(C)* Plateau areas covered by ice sheets. The topography leads to an increase of linear and areal erosion on higher grounds. (After Sudgen and John 1976.)

slopes, rugged tops, and wide, relatively straight, U-shaped valleys (Fig. 4.21B). In many places, these glacially cut valleys radiate out from a divide (Fig. 4.22B). Wet-based alpine glaciers erode and sculpt the land rapidly, and their effectiveness is greatly enhanced by the intense frost shattering that occurs in high-altitude mountain regions.

3. Plateau areas are not only widely eroded by ice sheets but are also eroded along alignments (linear erosion) by valley glaciers at their edges and during retreat of the ice caps (Fig. 4.21B,C). The highlands are smoothed out, except for local, minor sculpting due to later cirque- and valley-glacier activity.

The differences between the smoothness of low-lying cratonic areas and parts of the plateaus and the rugged, steep terrain of glaciated mountain areas are due to the more uniform widespread erosion of large ice sheets compared to the localized erosion of smaller alpine glaciers. The differences may also be partly related to the response of Earth's crust to these surficial processes.

a. Cratons are generally stable areas. When loaded by ice sheets, they subside. When the ice sheets melt, cratons rebound differentially through time, but generally over wide areas.

b. Plateaus are found in tectonically more active zones, older mountain chains, and in parts of young mountain chains. The effect of ice sheets on plateaus may be dramatic, but for cratons the effect is spread out across the

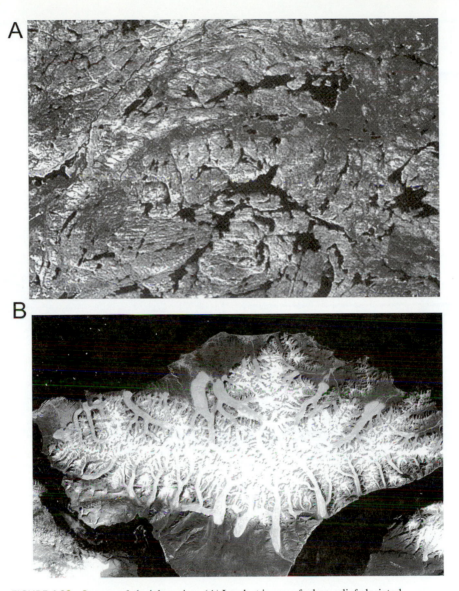

FIGURE 4.22   Images of glacial erosion. *(A)* Landsat image of a low-relief glaciated area on the Canadian Shield showing mostly areal erosion with some linear erosion along faults. *(B)* Valley glaciers, radially distributed along a divide, Bylot Island, Northwest Territories, Canada. (From Banks 1989.)

entire area. The erosion caused by cirque or valley glaciers during the last stages of glaciation at the margins may not be sufficient to differentiate one part of the plateau from another.

c. Mountainous areas may be affected by ice sheets and, at one time or another, by alpine glacial denudation (erosion). Formation and melting of large glaciers on tectonically active mountains may slightly modify their rate of uplift by adding or removing weight. Postglacial isostatic rebound may be contributing to the uplift of the European Alps, Himalayas, and Andes, among others, after the late Pleistocene glaciers that covered them melted away.

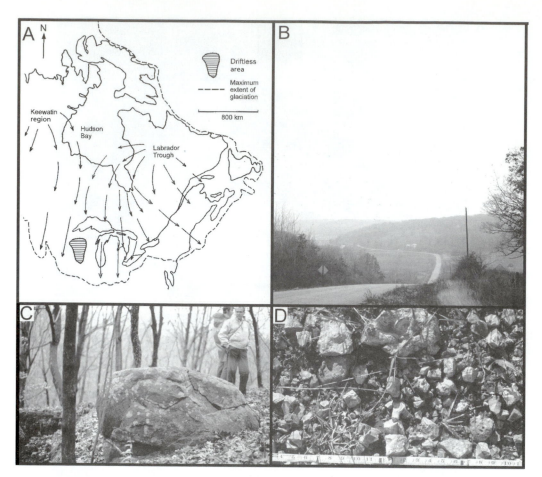

FIGURE 4.23   The driftless area of Wisconsin. *(A)* Map showing the maximum extent of Pleistocene glaciations in North America and the driftless area of Wisconsin, United States (arrows indicate ice paleoflow). *(B)* Landscape of the driftless area of Wisconsin. *(C)* Glacial erratic in the driftless area of Wisconsin. *(D)* Mature soil in the driftless area of Wisconsin composed of red residue from cherty carbonate parent material. The angular chert clasts are the only relatively unweathered portion of the original material.

In general, erosional glacial landscapes can vary a great deal as a function of:

1. the position of the area relative to the main body of the ice sheet;

2. the rate of deglaciation, especially during the transition from ice cap to valley glaciation; and

3. the climate of the region; for example, a humid maritime climate can support active glaciers and numerous freeze-thaw cycles, whereas a dry continental climate preferentially supports less active glaciers (Fig. 4.21).

Glacial erosion can be impressive in some glaciated areas, but in others is minimal or even nonexistent. The latter is indicated by glaciated areas that show fluvial rather than glacial valley morphology, or retain old, thick saprolites (surficial layer of weathered rock). These areas were undoubtedly glaciated, as indicated by erratics, striae, or other similar features, but not necessarily during the most recent glacial advance. For example, the driftless area of Wisconsin (Fig. 4.23A,B) contains large granitic erratics (Fig. 4.23C) associated with deep, mature soils. Figure 4.23D shows the surface of a highly weathered soil that formed on carbonate bedrock containing

FIGURE 4.24   Photographs of the glaciated Canadian Shield. *(A)* Preserved, thick, granitic saprolite. *(B)* Preserved mature soil on metamorphic rocks.

chert nodules. The high maturity indicates that this soil would have required much longer to form than the 15,000 years since the last deglaciation. Another example occurs in the Canadian Shield of northern Ontario, where thick granitic saprolites are present in an otherwise glaciated landscape (Fig. 4.24). It is possible, therefore, that during the last glaciation, in some areas, particularly where the glacier was thin, cold based, and diverging around obstructions, erosion was minimized.

ENDNOTES

1. Power is work per unit time and is measured in watts (joules/sec; J/s). Note that 1 cal = 4.1868 J.

2. Unless otherwise stated, we use the term *velocity* to indicate rate of movement (speed) in the direction of movement of the glacier.

Glacial Transportation and Deposition

GLACIAL TRANSPORTATION

A by-product of glacial erosion is sedimentary particles that are transported by glaciers and meltwater. Sediment is added to a glacier by (a) plucking and abrasion of the substrate, (b) falling from nunataks and the sidewalls and headwalls of valleys, and (c) wind transportation of material onto the glacier surface, sometimes from far away. Ice sheets get their sediment load mostly from their base, valley glaciers from the base and from the sides.

Glacial sediments may be transported at or near the top of the glacier (**supraglacial, or superglacial, drift**[1] **[debris]**), within the glacier (**englacial drift**), or at the bottom (**subglacial, or basal, drift**) (Fig. 5.1).[2] Glaciers can transport drift as isolated particles, but more often particles are in concentrated patches, called **moraines.**[3] **Lateral moraines** are formed by material derived from the valley walls. **Medial moraines** form from the joining of lateral moraines at the confluence of two glaciers (Fig. 5.2). **Basal moraines** form primarily from material eroded from the base of the glacier. **Internal moraines** form when sediment falls into crevasses, when lateral moraines coalesce at the confluence of glaciers, and when basal drift is thrust upward near the terminus.

Supraglacial (superglacial) drift (Fig. 5.1) is important in valley glaciers where the confining walls provide the material, generally in the form of angular particles. In ice sheets, supraglacial drift may derive from isolated nunataks, from up-thrusting of basal material, and, to a small extent, from windblown sediment and extraterrestrial bodies (meteorites). Indeed, many meteorite fragments have been recognized and collected from the surface of Antarctic Ice Sheets, where no rock outcrop sources occur above the glacier. Supraglacial sediments are affected by frost shattering, the clasts do not abrade one another, and much of the fine material is removed and locally sorted by meltwater. Thus, supraglacial drift mostly consists of angular clasts with minor lenses of finer, waterlaid sediments (Fig 5.3A). Supraglacial drift may be transported a long way. For example, Quaternary glaciers transported large boulders of Precambrian igneous and metamorphic rocks of the Canadian Shield thousands of kilometers to the southern part of Canada and the northern United States. Isolated clasts that differ in composition from the surrounding country rock are called **erratics.**

Subglacial (basal) drift is mostly composed of material eroded from the local substrate, although some exotic clasts may be added from other parts of the glacier or from reworking of previously deposited glacial sediments. Subglacial drift can form a water-saturated moving carpet in places, which greatly facilitates basal sliding of glaciers. Clasts abrade against bedrock and other drift particles, and can also be crushed. A by-product of abrasion is rounding, faceting, and scratching of the

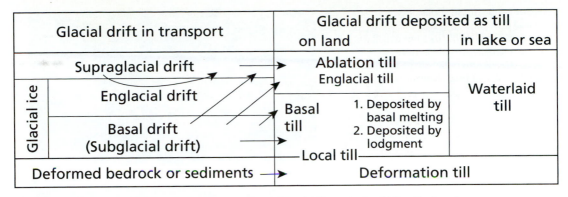

Glacial drift in transport		Glacial drift deposited as till	
		on land	in lake or sea
	Supraglacial drift	Ablation till Englacial till	Waterlaid till
Glacial ice	Englacial drift		
	Basal drift (Subglacial drift)	Basal till	1. Deposited by basal melting 2. Deposited by lodgment
		Local till	
Deformed bedrock or sediments →		Deformation till	

FIGURE 5.1   Chart of glacial materials transported and deposited by a glacier. (After Dreimanis 1969; Dreimanis 1988.)

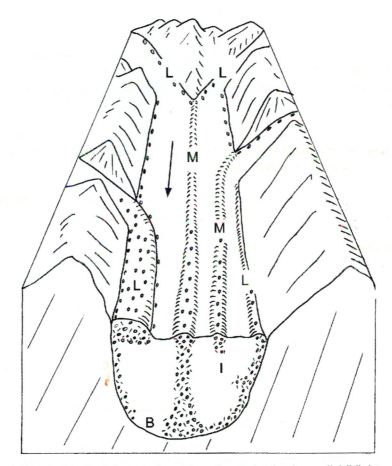

FIGURE 5.2   Idealized diagram of a moving valley glacier showing medial (M), lateral (L), internal (I), and basal (B) moraines. (After Sharp 1960.)

coarse clasts, which eventually acquire typical **flat-iron (bullet)** shapes (Fig. 5.3B). Some of these have an upward recurved abraded nose, and scratches (striae) on the principal flat surface, parallel to the long axis (Fig. 5.3C). The faceting may be related to the fact that clasts retract into the ice due to pressure melting, rotate, return to the base, and are abraded again for a time along a different facet (small face). Al-

FIGURE 5.3   Illustrations of typical glacial drift clasts. *(A)* Angular clasts of supraglacial drift. *(B)* Rounded, flat-iron, bullet-shaped cobble from basal drift. *(C)* Various bullet-shaped pebbles from basal drift (lines indicate striations).

ternatively the clasts may get stuck in the substrate and are abraded by ice and debris overriding them. Evidence of major abrasion on the principal, widest clast surface indicates that this is the preferred position of transportation. In that position, lower stresses are induced by the weight of the glacier on the larger cross-sectional area of the clasts. So there is less likelihood of the ice reaching the pressure melting point; hence retraction into the ice does not occur as frequently as for other positions. A corollary to this is that isolated particles may abrade and be abraded less than particles that are part of a sliding sediment mass and remain in contact with the substrate for longer periods. Another important by-product of abrasion is fine powder or silt, which leads to lower average size of basal-drift particles relative to other drift types of the same parent material.

   Englacial drift is composed of a mixture of basal and supraglacial drift, and its characteristics depend on the relative proportions of these. In temperate glaciers, englacial drift often has large particles coated with fine dust that develops as follows. Ice in temperate glaciers is at the pressure melting point and undergoes fre-

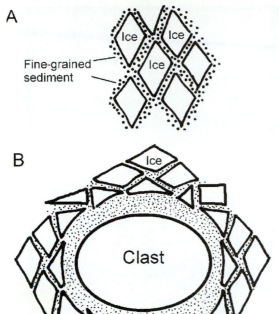

A

Fine-grained sediment

B

Clast

FIGURE 5.4   Diagrams of dust-coated clast of an englacial drift. *(A)* Fine particles extruded in front of developing ice crystals. *(B)* Fine extruded particles plastered around a large, unmovable clast.

quent freeze-thaw cycles. Because fine impurities cannot be accommodated within the crystalline structure of ice, they are extruded along the freezing front. Large, difficult-to-move pebbles or boulders in the path of the freezing fronts get coated with the fine material extruded from the ice (Fig. 5.4).

GLACIAL DEPOSITION

Transported debris is released from a glacier in ways that impart additional diagnostic characteristics to the deposited material. Material deposited directly by glacier ice is called **till.** If direct deposition from a glacier cannot be proved, the term **diamicton** is used. A diamicton may be a till, a till reworked by debris flows, or a nonglacial mass flow.

Till is commonly massive, ideally unsorted and unstratified sediment (Fig. 5.5). It is differentiated on particle-size distribution (particularly the type and amount of matrix relative to clasts), color, matrix and clast composition, clast shape, roundness, orientation and imbrication, and structure of the deposit (for example, whether it is massive, jointed, or pseudo-bedded). No single variable can discriminate among tills; rather, it is the combination of variables (multiple criteria) in the deposit that defines the tills. Various types of till are listed in Figure 5.1, and their idealized stratigraphic succession is illustrated in Figure 5.6.

Basal till is composed primarily of subglacial drift (Fig. 5.5A,B). Variations occur due to the mode of release of sediment from the glacier, either by lodgment or melt-out processes.

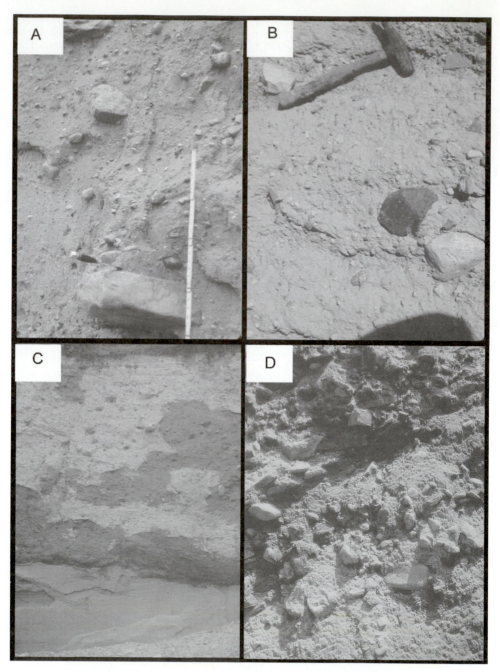

FIGURE 5.5   Photographs of till. *(A,B)* Massive basal till with abraded clasts predominantly of local origin. *(C)* Sandy, pebbly ablation till with several clasts from a distant origin. *(D)* Sandy, gravelly, matrix-poor ablation till with very little fine matrix. This till forms in areas where there is little clay in the substrate.

Deposition of basal drift by *lodgment* under a sliding glacier can occur by direct plastering of material on a rigid substrate (**lodgment till** *sensu stricto*), by shearing and deformation of a deformable drift layer (**deformation till**), or by releasing the drift in cavities (low-pressure areas) that develop downstream from substrate

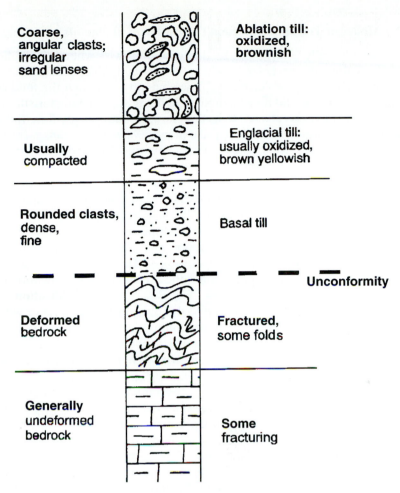

Coarse, angular clasts; irregular sand lenses — **Ablation till:** oxidized, brownish

Usually compacted — **Englacial till:** usually oxidized, brown yellowish

Rounded clasts, dense, fine — Basal till

— — — — — Unconformity

Deformed bedrock — **Fractured,** some folds

Generally undeformed bedrock — **Some fracturing**

FIGURE 5.6 Idealized vertical section (not to scale) of a till sequence over bedrock.

obstructions or clusters of immovable debris. These tills represent extreme cases of a continuum in which basal drift is released from the glacier under the effect of pressure and local melting. Basal tills may retain compacted, abundant, fine-grained matrix (where fines are available from the subsurface or are generated by subglacial abrasion) and may have numerous rounded, faceted, bullet-shaped clasts. The finer the sediment, the more compact and dense basal tills are, up to 2.12 g/cm^3 for siliciclastic matrices and up to 2.24 g/cm^3 for calcareous matrices.[4] The clasts may acquire preferred orientation in the direction of glacier flow or have bimodal orientation and, generally, an upflow imbrication (shingling). Some clasts are shoved into the water-saturated deposit and are tilted downstream. The high till density inhibits water percolation except along fractures (joints), which can develop once the glacier has retreated. Thus, basal tills formed by lodgment are generally less oxidized and may be darker than equivalent ablation tills.

Basal melt-out till is similar to lodgment till, but is less compacted. On land, basal melt-out till preferentially forms under stagnant ice toward the terminus of glaciers, where basal drift and some englacial drift are released through progressive, differential basal melting of the glacier. Underwater in lakes and seas, basal and

BOX 5.1   Modes of Transportation and Deposition of Sediments

Sediments are naturally transported by the force of gravity when it is applied to a fluid that drags along the particles, or when it is applied to the particles themselves. In the latter case the particles can move independently, although the fluid component of the sediment may aid movement and, in doing so, its characteristics may be changed. The first mode is referred to as **fluid flow,** or fluid gravity flow, the second as **sediment gravity flow,** or mass flow. Fluid flow prevails in the transportation and deposition of sediment in rivers, lakes, seas, and by wind. The sediment gravity flow prevails in mass movement along slopes in subaerial and subaqueous settings. These two modes are extreme cases of a continuum where mixtures of fluid and sediment particles exist. The intermediate case, where the mixtures have a solids concentration of 40–70% (by volume), is called hyperconcentrated flow.

The fundamental differences between these modes of transportation and deposition relate to the dominant sediment-support mechanism, that is, the mechanism keeping the particles loose and capable of movement. **Turbulence** of the moving fluid keeps particles in suspension, and **traction** is the process by which the moving fluid or other substance drags particles along (Fig. Box 5.1.1A). **Fluidization** of sediments occurs when fluids become entrapped and overpressured under or within the sedimentary body and escape upward, sustaining particle movement (Fig. Box 5.1.1B). **Dispersive pressure** is related to the elastic bouncing of the particles as they impact one another (Fig. Box 5.1.1C), and **matrix strength** is related to the buoyancy of a mixture of fine-grained sediment and fluid (Fig. Box 5.1.1D). Usually a combination of these mechanisms participates in sediment transportation, and their relative importance may change in space and time during any transport event. For example, as in an avalanche of ice and snow, a mass of material starting movement as a slump may become a sediment gravity flow sustained primarily by matrix strength, that is, a debris flow. As it acquires fluid during travel, it may change into a hyperconcentrated flow where both matrix strength and fluid turbulence contribute, and finally, into a fluid flow where the particles are primarily dispersed by turbulence.

	A	B	C	D
Sediment-gravity flow	Turbidity current	Fluidized sediment flow	Grain flow	Debris flow
Principal sediment-support medium	Turbulence	Upward intergranular flow	Grain interaction	Matrix strength
Deposit	Ripple cross-laminations / Graded bed (F, C)	Fluid-escape and dish structures	No grading / Inverse grading (C, I, F)	Massive or poor grading

FIGURE BOX 5.1.1   Diagrammatic representation of sediment gravity flows and their main deposits. (Circles = particles in general, usually sand and coarser; large dots = pebbles or larger particles; small dots = sand; dashes = clay and silt; arrows = relative movement of fluid or of impacting particles; F = finer particle size; C = coarser particle size; grading = vertical variation in particle size, fining upward.) (After Middleton and Southard 1978.)

BOX 5.1   (*continued*)

Where one mechanism prevails, specific processes and deposits can be recognized, as shown for sediment gravity flows in Figure Box 5.1.1. Each category can be further subdivided depending on the environmental settings—for instance, whether subaqueous or subaerial—and on the material involved, such as prevalent particle size and whether matrix-poor or matrix-rich. Extreme cases of mass moments are **landslides,** where the whole mass of material starts movement along a detachment surface, and **solifluction** and **soil creep,** whereby surficial materials move slowly downslope, in the case of cold-climate settings aided also by meltwater percolating and refreezing. In relation to all this, the term *flow till* is restricted to debris flows and hyperconcentrated flows involving till material. By analogy, a glacier can also be considered a debris flow, with ice as the matrix, although such a matrix is close to the melting-recrystallization point, such that fluid and ice crystals may become interchangeable under certain conditions.

englacial drift released from floating glaciers settles to form a slurry, and the deposit acquires some sorting. If there are currents or waves, crude stratification develops.

Deformation till is material deposited from mobile deforming layers at the base of the glaciers. It can derive from intense reworking of other tills or other sediments, and from poorly cemented sedimentary rocks.

Ablation till (or supraglacial melt-out till) is composed primarily of supraglacial drift, with some subglacial and englacial drift material thrust to the surface. So, it shows characteristics derived mainly from supraglacial drift, such as coarse, angular pebbles, some sorted material in stratified lenses, and numerous coarse erratics as well as some bullet-shaped clasts derived from basal drift (Fig. 5.5C,D). The drift is released by differential melting of stagnant ice and is not greatly compacted (generally less than 1.84 g/cm^3);[5] and because particles move and slump during release, it may not develop good directional fabric (preferred particle orientation). Ablation tills are more porous and permeable than basal tills, and percolating waters oxidize them, imparting a brownish tinge both along joints and throughout the matrix.

Sublimation till is formed under very cold conditions where ice is vaporized without going through the liquid phase and releases the supraglacial drift. Sublimation till is rare except in parts of Antarctica.

Flow till develops where water-saturated debris undergoes significant downslope mass movement on the stagnant glacier or in front and away from it. Flow till develops in both subaerial and subaqueous conditions, although the term is primarily used for subaerial settings. Flow till is often nothing but supraglacial debris reworked by sediment gravity flows (Box 5.1) and, locally, by flowing meltwater. The sedimentological characteristics of flow tills vary greatly, depending on the materials involved, ranging from matrix-rich deposits with disseminated clasts of various sizes, to matrix-poor, coarse gravelly sediments with various degrees of clast orientation and imbrication. Apart from their bullet-shaped clasts, excess fine silt formed by abrasion, and overall geological setting, ancient flow tills are indistinguishable from sediment gravity-flow deposits formed in nonglacial settings. This has led to major controversies on the existence of ancient glaciations in some areas where lithified diamicton (diamictite) has been interpreted as lithified till (tillite) or non-glacial deposit of sediment gravity flows.

Englacial till theoretically forms from englacial drift, but is almost impossible to distinguish from ablation till, except from its stratigraphic position in an orderly till succession (Fig. 5.6). Englacial till should lack lenses of sorted, stratified material, and have boulders with powdery coatings, if not washed by meltwater.

On the whole, an ideal succession of terrestrial glacial deposits may form: showing fractured bedrock at the base, in part broken due to addition and removal of various glaciers, overlain by deformed softer substrate or older sediments, overlain by tills of various thickness from basal tills below to partially water-washed ablation (supraglacial) till at the top (Fig. 5.6).

Waterlaid (or waterlain) till forms from glaciers ending in seas or lakes. The sediment is released into the water through rockfalls and debris flows of supraglacial and englacial drift, and by bottom melting of the basal drift. The resulting chaotic deposits have a high water content and thus can easily flow downslope, be pushed or squeezed in front of the glacier, or squeeze into basal cavities. Waterlaid tills can be readily reworked by waves caused by falling ice blocks, and can be mobilized into sediment gravity flows (slumps, debris flows, or even turbidity currents). Another type of diamicton is formed offshore by iceberg-rafted material.

DISPERSAL FANS

Dispersal fans are outlined by glacial redistribution of distinctive material. When a glacier advances over an area, it may pick up diagnostic rock clasts, fine debris, or even submicroscopic particles and disperse them over a wide region. Through mapping of the distribution of these materials and appropriate geochemical markers in the glacial deposits, the dispersal pattern can be reconstructed and the source area of the material pinpointed. Dispersal fans can extend over large areas of continents and can be used to reconstruct the paleoflow of past glaciers, and locate the source area of useful mineral deposits. Once target areas have been identified by analyzing the materials dispersed by the glacier, conventional, more expensive mining exploration methods, such as local detailed bedrock mapping and drilling, can be applied to assess the value of the deposits.

Dispersal fans outline the flow of the last ice sheets that crossed Norway and fanned southward into the Baltic Sea and northern Europe (Fig. 5.7A). In Canada, the analysis of dispersal fans has been successfully used to determine the paleoflow of the Quaternary ice sheets and for ore exploration. For example, a portion of the Laurentide Ice Sheet originated and picked up Precambrian iron-ore material from the Labrador Trough, and it picked up iron-rich conglomerate as it crossed the eastern part of Hudson Bay, and then moved southward to deposit these materials in southern Ontario and the northern United States (Fig. 5.7B). This dispersal fan outlines a paleoflow very different from that part of the Laurentide Ice Sheet originating to the west of Hudson Bay. This part of the Laurentide Ice Sheet picked up iron-ore-free Precambrian metamorphic and igneous rocks from the Keewatin area, advanced along the western shores of Hudson Bay, and continued southward. The reconstruction of these dispersal fans has helped to locate economic minerals, such as large gold deposits at Hemlo in northern Ontario. Similarly, traces of diamonds in the dispersal fans have indicated the existence of diamond-bearing diatremes in the

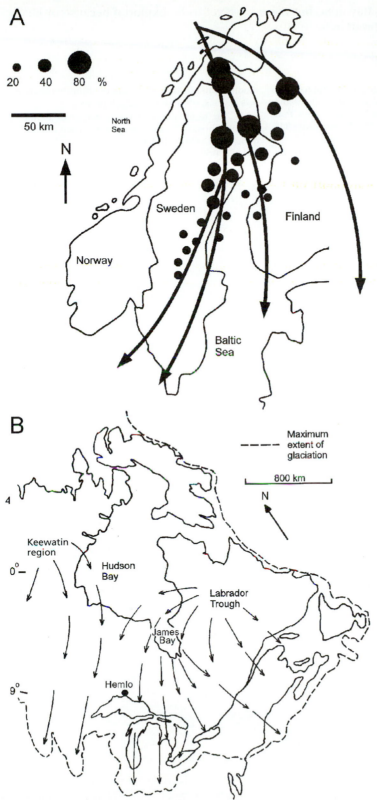

FIGURE 5.7 Generalized maps showing paleoflow directions of Pleistocene ice sheets. *(A)* Glacial dispersal fan in southeastern Sweden defined by percentage of relative particle size (coarse, medium, and fine as indicated by relative dot size), lithologic type, and glacial striae. *(B)* Extent and paleoflow directions (arrows) of the Pleistocene Laurentide Ice Sheet of east-central North America. (After unknown, see website for update.)

James Bay/Hudson Bay area, but these have yet to be explored because of heavy cover by forests and wetlands.

ENDNOTES

1. Drift is an old, but still often used term for generally bouldery, unsorted glacial sediment. The name comes from the early-nineteenth-century idea that such sediment was deposited in marine environments by drifting icebergs.

2. Many classifications of glacial debris in transport and deposited (tills) exist, all useful and, at the same time, unsatisfactory because they are based on interpreted genetic factors. We are here presenting one of the oldest, simplest schemes, very slightly modified, to illustrate the basic relations between debris in transport and resultant deposits.

3. The term *moraine* refers to an accumulation of glacial debris. In North America it is used mostly for deposited material; in Europe it is also used for material in transport.

4. $2.12 \text{ g/cm}^3 \sim 132 \text{ lb/ft}^3$; $2.24 \text{ g/cm}^3 \sim 140 \text{ lb/ft}^3$

5. $1.84 \text{ g/cm}^3 \sim 115 \text{ lb/ft}^3$

C H A P T E R   6

Glacial Landforms Formed by Glacial Sediments

Sediments deposited directly by, or in proximity to, glaciers can form several landforms. Some landforms develop under actively moving glaciers, others under stagnant ice at the glacier terminus, and still others just in front of the glacier itself. The widely known features are those formed on land, where they can be readily observed. Others develop underwater in lakes or seas. The origin of many landforms is still debated; here, a few of the most typical ones, composed primarily, but not exclusively, of till are presented to explain a few processes.

DRUMLINS

Drumlins are oval, streamlined, teardrop-shaped hills with blunt, rounded heads and long, pointed tails, generally with a long, straight axis (Fig. 6.1). They can be simple or composite in form, and their sizes vary from 1 to 2 km in length, 0.4 to 0.6 km in width, and 15 to 30 m in height. Drumlins often occur *en echelon* (in a staggered pattern; Fig. 6.1B) in fields behind end moraines (Fig. 6.2). They may be rather uniform within a field, but vary in size and shape in different fields. They may also be associated with small end moraines, crevasse fillings, and small eskers.

Many drumlins are composed of till (Fig. 6.3), but some contain much stratified drift, such as in New York State and in various parts of Canada. Their shapes can be described and quantified by using ideal reference models, such as an ellipsoid or a lemniscate loop (Fig. 6.4A). In the first instance, each contour line drawn for a map representation of a drumlin approximates an ellipse. The ellipsoid model has been found reasonably adequate for the base of the hill, but not for higher levels. The teardrop shape of a drumlin, instead, can be better quantified by approximating it to the cross section of the upper side of an aircraft wing (Fig. 6.4B). This shape resembles a lemniscate loop, which is defined by the equation

$$\rho = L \cos K\theta \text{ (in polar coordinates);}$$

where

L = length of long axis (when $\theta = 0$)
K = dimensionless number expressing the elongation of the loop
($K = L^2 A\, \pi/4$). $K = 1$ when the form is circular, and increases in value with the elongation of the loop (Fig. 6.4B).
A = area of the loop

Note that $\rho = 1$ when $\theta = 0$

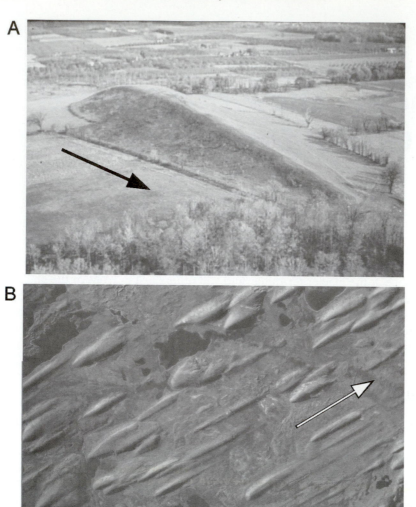

FIGURE 6.1 Photographs of drumlins. *(A)* In New York State: The glacier paleoflow is from the upper left to the lower right. (After Wards Natural Science Establishment Inc., Rochester, N.Y.) *(B)* Drumlin field with hills positioned en echelon (air photograph, Canada. (Arrows indicate direction of ice movement.)

The maximum width of this wing shape is about three-tenths of its length from the round leading edge.

Like wings, slender, more elongated drumlins (as opposed to shorter, stubby ones) may be adjustments to higher glacier speeds (Fig. 6.5A). However, the resistance of the sediment to deformation can influence drumlin shape as well; that is, higher concentrations of cohesive clay lead to higher, bolder outlines, whereas high concentrations of noncohesive sand lead to flatter, more subdued profiles. Deviance from these ideal forms occurs (Fig. 6.5B–D).

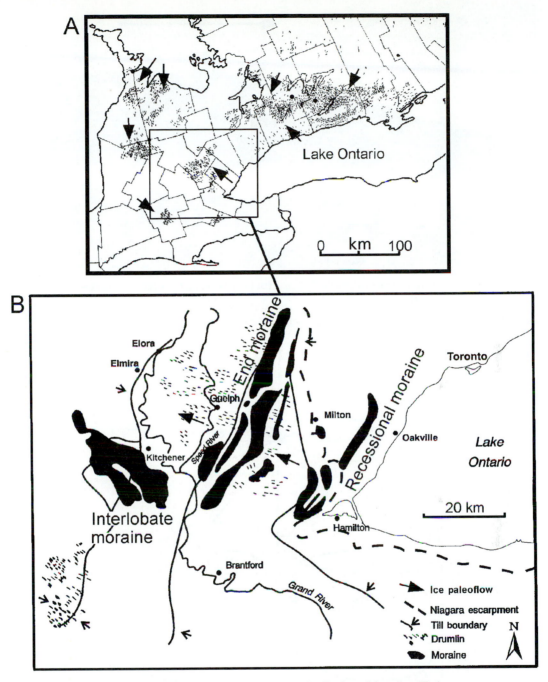

FIGURE 6.2   Maps of drumlin fields of Ontario. *(A)* Regional distribution of drumlins. Their formation is associated with the various ice lobes flowing from the lake basins. They occur behind end or interlobate moraines (arrows indicate paleoflow directions). *(B)* Drumlins between Hamilton and Guelph, Ontario. (After Delorme, Thomas, and Karrow 1990.)

FIGURE 6.3 Cross section of a drumlin composed of till, Wisconsin, USA (arrow indicates paleoflow direction).

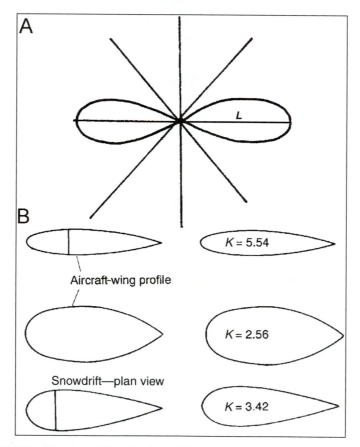

FIGURE 6.4 Plan view of drumlin shape. *(A)* Diagram of a lemniscate loop (*L* = length of loop). *(B)* Characteristic natural (snowdrift) and artificial (aircraft wing) shapes compared with calculated shapes (diagrams to the right) with specific form factors *K*. (After Chorley 1959.)

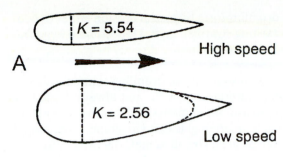

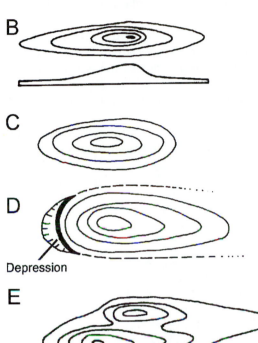

Depression

FIGURE 6.5 Plan-view shape of drumlins. *(A)* Given a certain constituent material, narrower and longer drumlins may indicate higher glacier speed than wider and shorter ones (after Chorley 1959). Drumlins may deviate from the ideal shape. *(B)* Hill proceeded by a long, low ridge. *(C)* Drumlin with centrally located maximum height and width. *(D)* Drumlin with a depression at the base of the upstream slope. *(E)* Serrated or twin-crested drumlin.

   The origin of drumlins has been hotly debated; yet no single hypothesis explains every feature or occurrence. Drumlins probably form in more than one way. Any hypothesis of drumlin formation must explain the following features:

1. Drumlins show a streamlined shape with a blunt end pointing upflow.
2. Drumlins may contain a variety of materials; though many are composed of till.
3. Some drumlins have rock cores, but many do not.
4. Rock-cored and sediment-cored drumlins have the same shape and occur in the same drumlin fields.
5. Stratified drift may be present inside drumlins.
6. Drumlins may have faulted and folded layers.
7. Drumlins often occur behind moraines and are aligned parallel to the ice-flow direction.
8. Drumlins occur in fields that are wider than most moraines, and they rarely occur in isolation.
9. Many glaciated areas do not contain drumlins.

Erosional hypotheses state that drumlins form by glacial erosion either of bedrock (rock-cored drumlins) or of pre-existing stratified drift. **Depositional hypotheses** state that drumlins form by sediment deposition either directly by glaciers, or by meltwater under the glaciers.

The **erosional hypotheses** note that, in some areas, drumlins contain a significant amount of stratified material, often interlayered with, and always draped by till. One possible explanation is that pre-existing sediments can become water saturated and easily reworked by advancing temperate glaciers. Erosional, streamlined landforms are developed with part of the original stratified material little deformed or not at all. Everything is eventually draped by till that is plastered onto the landform by the moving glacier, or let down later during melting of the ice.

These hypotheses appear reasonable in places, but they do not fully explain why drumlins form in fields, frequently behind end moraines. Is there a stress pattern within the glacier that leads to this? If so, what leads to the development of such stress patterns?

The **depositional hypotheses** note that drumlins need unusual conditions to form because they do not occur everywhere.

The direct glacial deposition hypothesis requires a continuously deformable deposit under temperate glaciers, and some obstruction in the substrate, or clusters of coarser sediment that resist movement.

An early, elegant analysis of the reasons sediments can resist movement at the base of a glacier starts by considering that some may be granular dilatant material, which must change its packing, thus expand to move (Fig. 6.6) (Smalley and Unwin

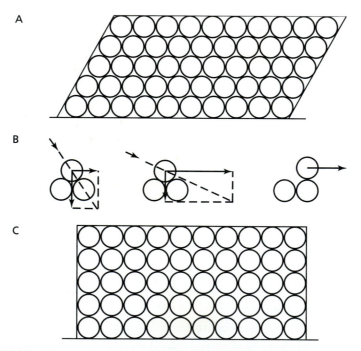

FIGURE 6.6 Diagrams showing ideal packing arrangements of spheres. *(A)* Maximum-density packing (close packing)—the spheres are positioned *en echelon. (B)* Diagrams showing the sum (diagonal of parallelogram) of the forces exerted on packed grains (vertical arrows indicate weight of the glacier and horizontal arrows the force associated with the forward movement of the ice), and the expansion needed between particles for movement to occur. This means that the applied stress must exceed the confining pressure (weight) of the system. *(C)* Minimum-density packing (open packing)—the spheres are positioned in straight lines.

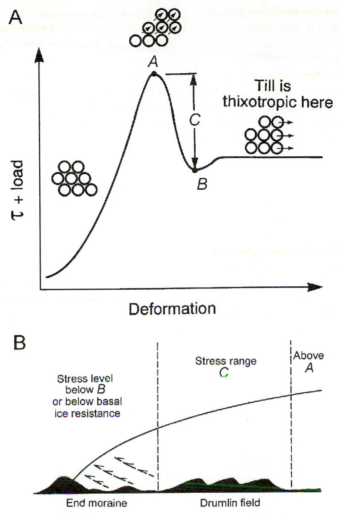

FIGURE 6.7 Diagrams of deformation-stress relations in a natural granular system. *(A)* Experimental relation between τ (shear stress) + load and deformation in a dilatant material. Little deformation occurs until a stress value of *A* is reached; that is, the dilatant material resists deformation. After the stress *A* is reached, the fabric of the dilatant material is destroyed and deformation can be maintained and increased at lower stresses (zone *C* and beyond). *(B)* Idealized diagram showing zones of the glacier where the various stress ranges could be found. Where the glacier is thick, the stress may be greater than a stress value of *A;* hence the erosional power is high, everything is eroded, and no deposition occurs. Where the glacier has intermediate thickness, the stress range may be in the *C* zone, where some dilatant material may still resist movement, but other material may be moved and plastered around obstructions. Theoretically, drumlins may form in this zone. Toward the thin terminus of the glacier, stress level is insufficient to move much material at the base of the glacier. In addition, the ice may be frozen to the bottom, and the stresses could be released along concave-upward thrust faults (shear zones). Here, end moraines may develop. (After Smalley and Unwin 1968.)

1968). This is not possible until a limiting stress level is reached (Fig. 6.7A). Once the strength of the material has been overcome, however, it becomes **thixotropic (liquid-like),** and it can flow and plaster around nuclei of more resistant, drier or coarser clusters (Fig. 6.7A). This triggers the formation of streamlined deposits that can eventually grow into drumlins. If the stress exceeds the strength of all material at the base of a temperate glacier (τ + load $> A$; Fig. 6.7A), the entire basal drift flows and a sheet of lodgment till can be deposited; or, alternatively, everything can be eroded if the effective glacier power is sufficiently high. For a τ + load value less than B (Fig. 6.7A), there is no deformation of the basal drift. In this hypothesis, drumlins form within a restricted pressure range (τ + load value in the C range; Fig. 6.7A), which, in a glacier, may develop behind end moraines (Fig. 6.7B). The sizes and shapes of drumlins depend on the amount of material the glacier carries, its speed, and the amount of water at its base.

Recently, a more complex, all-embracing hypothesis considers that under the glacier there are sediments with pressured interstitial waters moving at different speeds in a deformation carpet (Boulton 1987). The finer materials can move faster in a sort of pressurized slurry. The fabric of coarser materials is less easily disrupted, and static or slow-moving obstructions develop, around which the finer material may form streamlined mounds; drumlins form if there is enough material. Material forming the drumlins can consist of basal drift or previously deposited material, such as glaciofluvial sediment, reworked as the glacier advances.

The **meltwater hypothesis** states that drumlins form when catastrophic meltwater floods flow under temperate glaciers (Shaw, Kvill, and Rains 1989). At peak discharge, these floods may float the glacier itself for a short time, and melt (or abrade) streamlined caverns in it. The flow, upon expanding into these caverns, loses competence and deposits stratified sediments into streamlined mounds. Later as it melts, the glacier releases till, draping the mounds. This hypothesis requires sudden releases of large quantities of meltwater as subglacial megafloods. Large quantities of meltwater are generated in places like Iceland, where volcanoes occasionally erupt under glaciers. This causes the rapid filling of subglacial lakes. Large subglacial lakes have also been detected by seismic investigations under the Antarctic Ice Sheet. The meltwater flood hypothesis can possibly explain some drumlin fields, such as in central-eastern Canada, where outbursts from large subglacial lakes may be inferred. However, this hypothesis does not explain drumlin fields formed primarily of till. It further requires ice-stagnant conditions for the preservation of the hills.

MORAINES

Moraines are accumulations of glacial sediments that can form both under actively moving parts of glaciers and under stagnant ice at the glacier margins.

Ground Moraines

Ground moraines consist of basal lodgment till and may be draped by ablation till. They vary in thickness, frequently being thin, on the order of 0.5 to 3 m. They can have a corrugated surface with irregular ridges transverse to the ice flow, or have fluting and drumlinized features parallel to ice flow.

Ground moraines are deposited by constantly, relatively rapidly retreating glaciers so that only the material in transport at any one time can be deposited, without the possibility of thickening the layer through protracted glacial transport.

End (Terminal, Recessional) Moraines

End moraines are ridgelike accumulations of glacial drift at a glacier front (Fig. 6.7B). They are generally formed of till and water-reworked, stratified material. The word **terminal** is used for moraines formed at the maximum glacier extent.

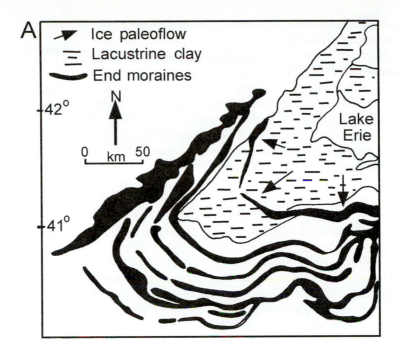

FIGURE 6.8 Illustrations of end moraines. *(A)* Lobate end moraines around the western tip of Lake Erie in Ohio and Indiana, USA. *(B)* Hummocky morphology of an end moraine characterized by alternating knobs and kettles (Paris moraine near Guelph, Ontario, Canada). This morphology develops from differential melting during the letdown of supraglacial drift into ablation till.

The word **recessional** is commonly used for moraines formed during short-lived interruptions in glacier retreat.

End moraines consist of one or more subparallel ridges. They are generally discontinuous, either because some parts of the glacier carried only small amounts of drift, or because the ridges were cut by meltwater channels. End moraines mimic the shape of the glacier terminus, and can be curvilinear or lobate (Fig. 6.8A). They vary depending on the regime of the glacier; its location and direction of flow; the amount, type, and mode of transport of glacial drift; and the amount and competence of the meltwater flow.

Subaerial end moraines may have hummocky topography (Fig. 6.8B), characterized by local ridges (knobs) and lows (kettles) formed by differential ice melting and release of sediments from the glaciers. Moraines with such a hummocky topography are formed or draped by ablation till, but some consist primarily of stratified drift interstratified with till. In every case, end moraines form where the rate of ice wastage equals the rate of transport; that is, the glacier terminus remains stationary while the internal glacial conveyor-belt carries sediment and builds up the landform. If instead, the glacier continuously advances and retreats across a wide area, it leaves only a thin till sheet. For example, rapid retreat of a glacier allows only a thin ground moraine to be deposited.

Two slightly different models have been derived from the study of moraines forming at the terminus of modern glaciers in Baffin Island and in Greenland.

1. Baffin Island has a small (150 km long, 72 km wide) ice cap at its center, called the Barnes Ice Cap (Fig. 6.9). Much of the ice is cold, with surficial temperatures averaging $-8°C$, and internal temperatures averaging $-9°C$. The outer 150 m surface of the glacier is smeared with drift thrust upward along shear zones at the ice margin. Several zones can be recognized on the ice surface, grading outward to the terminus:

  a. White ice free of debris

  b. Black ice with thin (a few millimeters) debris cover

  c. Ice-cored zone with much debris

These zones react differently to solar radiation. White ice has a high albedo (it is highly reflective), absorbs little energy, and thus melts slowly. Black ice has a lower albedo because rock particles on the surface can absorb much energy. This energy can be transmitted to the ice below the thin sediment cover, and more melting occurs here than in the white ice zone. The ice-cored zone also has low albedo, but the absorbed energy cannot be transmitted through the thick debris cover to the ice below; so melting is limited. Faster melting in the black-ice zone creates a depression. Sediment slumps from the ice-cored zone into the depression and accumulates, generating a locally thicker deposit. The resultant thicker sediment blanket changes the black-ice zone into an ice-cored stagnant zone. The moving glacier thrusts up behind the debris-laden ice-cored zone, and a new pattern of white-ice, black-ice, and ice-cored debris zones forms. Eventually a series of ridges develops as the material is released by total melting of the glacier ice.

2. Moraines at the termini of glaciers in the Thule area of western Greenland are similar to those of Baffin Island. However, the overall environment is more dynamic, and the various moraine ridges may form entirely by back shifting of the thrust zones behind the stagnant terminus of the glacier (Fig. 6.10). The rate of glacier movement and the rate of melting at the terminus are usually not steady. Periods of more active glacier movement bring greater amounts of ice to the terminus, and generate thrusting of ice and basal debris to the surface. During periods of lower activity, the glacier thins by melting, and previously active, debris-laden thrust zones become stagnant. Renewed glacier movement develops a new thrust zone upslope from the previous one, and this in turn stagnates during quiescent periods. As the process repeats, debris-rich and debris-poor bands alternate at the glacier terminus. As the material is released through basal melting, a series of ridges (end moraines) develops.

Other types of end moraines occur in fields of small (maximum height of 5–15 m) ridges in a swell-and-swale pattern, perpendicular to glacier flow. It has

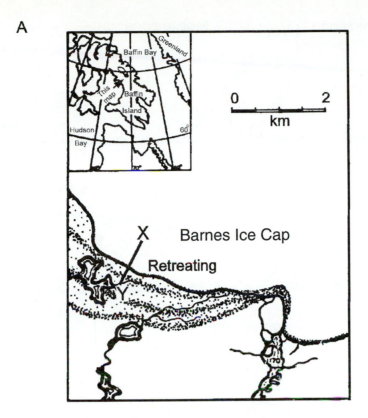

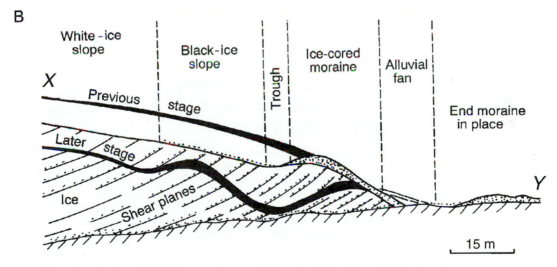

FIGURE 6.9 Diagrams illustrating end moraines of the Barnes Ice Cap, Baffin Island, Canada. *(A)* Succession of end-moraine ridges at the south side of the Barnes Ice Cap. A cross section of the transect *X–Y* is shown in *(B)*. *(B)* During a starting stage ("previous stage"), the terminal part of the glacier is subdivided into zones depending on the amount of debris on the surface. The "black-ice slope" is prone to melting; hence, in time, it develops a trough (shown in the ideal surface during a "later stage"). Much of the material of the ice-cored moraine slumps into this trough, generating a thick accumulation that will become a morainal ridge once all the ice has melted. (After Goldwait 1951.)

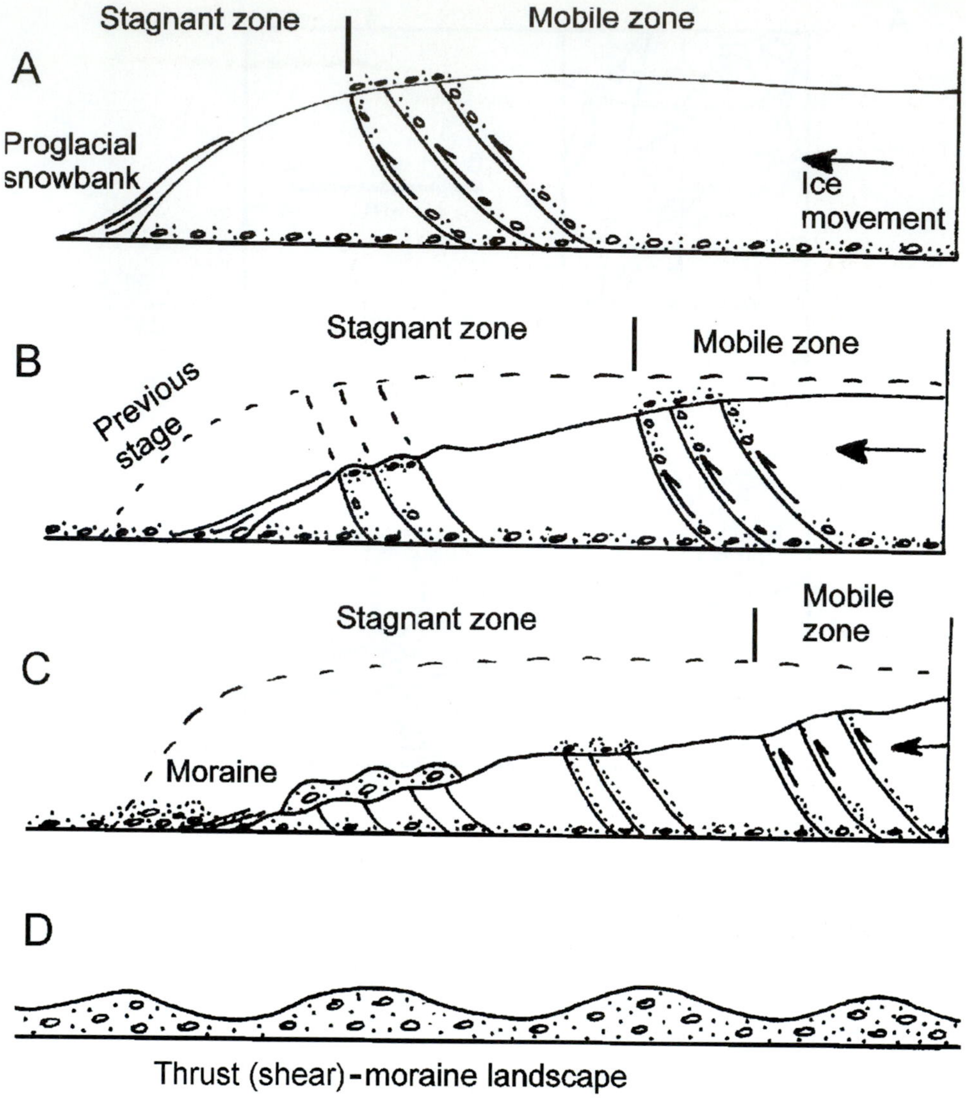

FIGURE 6.10   Diagrams showing inferred origin of moraines formed toward the terminus of part of the Greenland Ice Sheet near Thule, Greenland. The ridges form in correspondence to active thrust (shear) zones where much debris has been carried to the terminus. Periods of active thrusting alternate with periods of melting and thinning of the glacier; and once released, the debris forms separated subparallel ridges. *(A–D):* progressive melting of terminal ice and shift of active (mobile) shear zone (thick arrows indicate ice flow direction; half arrows indicate relative movements along active thrusts). (After unknown; see website for update.)

been suggested that such ridges can form under an active glacier, either at the base of the thrust zone, or at the grounding line of glaciers terminating in lakes or the sea.

Ribbed, or **Rogen,** end moraines often are partially fluted in a direction parallel to the glacier flow, and grade into drumlins (Fig. 6.11A). Several hypotheses have been proposed for their origin, but they are usually interpreted as reworking of subglacial sediment ridges originally perpendicular to ice-flow direction.

Subaqueous moraines form at or near the grounding line of glaciers that terminate underwater. These moraines resemble terrestrial end and push moraines, al-

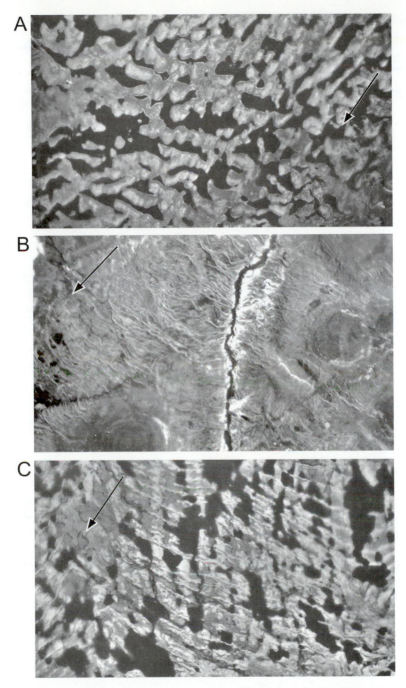

FIGURE 6.11   Photographs of small end moraines that develop during ice recession (recessional moraines). *(A)* Ribbed or Rogen moraines; *(B)* cross-valley moraines; *(C)* washboard, or De Geer, moraines. (Air photographs, Canada; arrows indicate paleoflow direction.)

though they may have more chaotic deposits, including water and mass-flow re-worked sediments. Pushed or squeezed materials in front of the glaciers form small ridges called **cross-valley ridges** (Fig. 6.11B), **De Geer moraines,** or **washboard moraines** (Fig. 6.11C).

Interlobate Moraines

Interlobate moraines are high (up to about 50 m), long (on the order of tens to hundreds of kilometers) ridges, which consist primarily of stratified sand and gravel, locally interstratified with thin, discontinuous diamicton layers. They form when a large volume of sediment-laden meltwater is funneled between two adjacent receding glacier lobes (Fig. 6.12A). The Oak Ridges moraine north of Toronto (Ontario, Canada) is a good example of a large interlobate moraine (Fig. 6.12B).

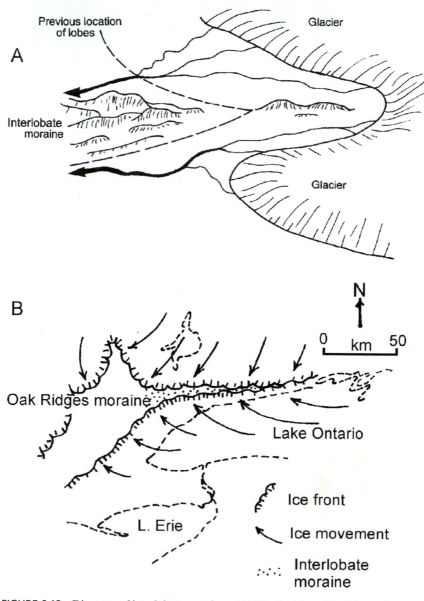

FIGURE 6.12   Diagrams of interlobate moraines. *(A)* Idealized location of formation of interlobate moraines between two ice lobes. This is the area where meltwater and sediment from both lobes is funneled (arrows indicate meltwater paleoflow directions (After West 1968). *(B)* The Oak Ridges is a typical interlobate moraine of southern Ontario, Canada (arrows indicate paleoflow directions of ice lobes; the outline of present lakes is given as geographic reference).

Push Moraines and Ice-Thrust Ridges

Push moraines and ice-thrust ridges look like end moraines, but their structure shows pervasive deformation indicating compression rather than simple slumping. They form perpendicular to ice flow.

Push moraines form when a glacier bulldozes and deforms glacial drift. This occurs toward the margin of the glacier where ice is thin, debris laden, and frozen to the bottom. Push moraines are rarely more than 10 m high.

Ice-thrust ridges are composed of deformed bedrock with folds and faults involving basal till, the whole usually capped by ablation till (Fig. 6.13A). They occur

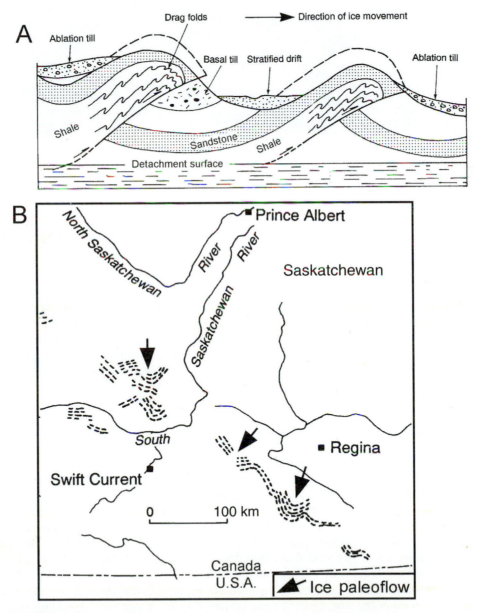

FIGURE 6.13 Schematic illustration of ice-push ridges in Saskatchewan, Canada. *(A)* Cross section of folds and thrusts with basal till overridden by bedrock. *(B)* Distribution map of ridges formed by glaciers bulldozing till and weak bedrock (arrows indicate ice paleoflow directions). (After Kupsch 1962.)

FIGURE 6.14   Photograph of ice-cored, lateral moraine, Iceland.

in fields of ridges that can be on the order of 30 m high, spaced 200 to 300 m apart, and can be traced laterally for up to 500 km (Fig. 6.13B). The ridges form preferentially in areas where bedrock has layers of varying competence, such as in Saskatchewan, Canada, where the bedrock is composed of Mesozoic sandstone and relatively soft (incompetent) shale.

Lateral Moraines

Lateral moraines are long ridges of diamicton aligned along the flanks of glaciated valleys (Fig. 6.14). They accumulate ice and meltwater debris transported by the glacier and rocks fallen from the valley walls. Meltwater streams may in part rework the deposits, and some terracing may develop. The moraines are ice-cored during formation, and some deformation and slumping of the material occurs during let-down by total melting of the ice.

STAGNANT-ICE TILL FEATURES

Stagnant-ice till features develop on stagnant ice mainly near the glacier terminus. They may exist on their own or as ornamentation to previously described features. They may be affected to various degrees by meltwater.

    Hummocky disintegration moraines show a variable topography with numerous knobs and kettles (Fig. 6.15). They look like end moraines mantled by ablation till, but do not necessarily parallel the ice lobe terminus. Disintegration moraines usually have a round, broad shape that does not stand out in the landscape, and they may grade imperceptibly into ground moraines.

    Prairie mounds are small hills with a regular donut-shaped top characterized by a smooth, low, central area surrounded by a rim or ridge (Fig. 6.16A). Most of these mounds consist of till, although some deformed lenses of stratified drift may occur.

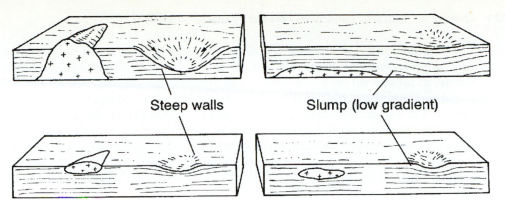

Steep walls

Slump (low gradient)

FIGURE 6.15 Diagrams showing the development of a kettle (sag) in till and ice-stratified drift. The size of the kettle depends on the size of the ice block involved. The steepness of the walls may depend on whether the ice blocks protrude above the sediment and on the depth at which they are buried. (After Flint 1957; based on Fuller 1914.)

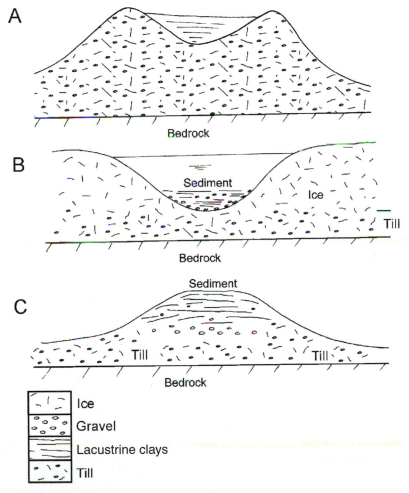

FIGURE 6.16 Diagrams showing landforms derived from stagnant-ice melt-out. *(A)* Cross-sectional area of a prairie mound. They are donut-shaped hills, common in Saskatchewan, Canada. These hills are formed by a central flat area covered by fine sediments surrounded by raised rims of diamicton. *(B,C)* Cross sections showing the development of a moraine plateau due to topographic inversion, with a shallow lake on stagnant ice *(B)* changing into a mound of sediment upon melting of the ice *(C)*. The mound may develop a raised rim if slumps of diamicton occurred along its borders, and thus become similar to a prairie mound.

Moraine plateaus are relatively flat areas of hummocky moraines. Their surfaces are level with, or slightly higher than, the top of the surrounding knobs. They consist of till capped by a few centimeters of lacustrine silt and clay, or are composed predominantly of lacustrine silt and clay. They form through a topographic inversion of a supraglacial lake as the glacier melts down (Fig. 6.16B,C).

Till ridges are small (approximately 30 m long) ridges, some of which cross and drape one another. They are often associated with moraine plateaus and consist mostly of till. They may form when crevasse fillings are let down onto the land surface as a glacier melts. Alternatively, they may form when water-saturated till is squeezed into basal cracks by the weight of the overlying ice.

CHAPTER 7

Fluvial Sediments and Landforms

Meltwater reworks, transports, and deposits debris from glaciers to form a variety of sediments and landforms. In order to understand these, the following need to be considered: first, how the dynamics of water flow in open channels determines sediment transport; second, how various flow regimes affect deposition and sediment character and structure; third, how sediment character and structure, conversely, can be used to reconstruct past hydrological and environmental conditions; and fourth, how meltwater streams form characteristic deposits and landforms. Basic concepts of fluid flow are essential for understanding sediment transport and deposition; so they are presented in the first part of this chapter.

DYNAMICS OF WATER FLOW AND SEDIMENT TRANSPORT: BASIC PRINCIPLES

Though all streams have similar dynamics, glacial and cold-climate ones are affected by ice cover and floating ice blocks (floes) for part of the year. The increasing water viscosity near freezing also has measurable effects on flow.

Flow velocity is a measure of how fast a stream moves downslope and is given in meters per second (m/s). It depends on the slope, size, and shape of the channel and the amount of water available at a given time (discharge). Stream flow can be **steady,** when flow velocity remains the same through time at one locality, and **uniform,** when flow velocity is the same for all points along part (reach) of the stream at a given time (Fig. 7.1). A natural flow can be uniform and nonsteady or steady and non-uniform, but it is rarely both steady and uniform.

Discharge is the amount of water flowing through a given cross section of the stream in a given time and is usually reported in m³/s. Glacial meltwater discharge is typically greatest during spring and summer melting; but also has daily fluctuations, being greater during daytime melting than at night.

Capacity is the amount of sediment the stream can carry, expressed in either volume (m³/s) or weight (kg/s) and is related to the discharge.

Competence is the maximum particle size the stream can carry, and depends mainly on flow velocity.

Material is transported by water as discrete particles (sediment) and in solution.

The sedimentary load is the total amount of sediment carried by a stream, and is determined by the amount and type of material available for erosion as well as by the capacity and competence of flow. The solution load is the total amount of dissolved ions transported by a stream and is determined by the source material and

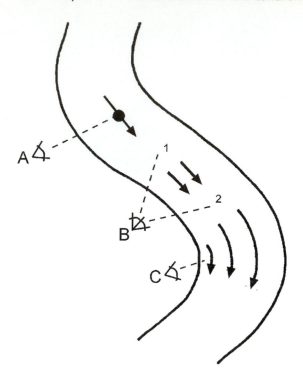

FIGURE 7.1   Diagram of stream-flow conditions. Steady flow occurs when the current velocity does not change at one point through time *(A);* uniform flow occurs when current velocities at various points (1, 2) are the same at the same time *(B);* non-uniform flow occurs where the current has diverse velocities across the stream *(C).* (Arrows indicate direction and strength of flow.)

the physico-chemical characteristics of the transporting stream. Natural or artificially dammed lakes along a stream may not affect water discharge much, but they can drastically change the sediment and solution loads, because coarse particles can be physically trapped in the lake (Fig. 7.2) and ions in solution can be precipitated.

The sedimentary load is transported as bedload (generally coarse particles moving along the bed), suspended load, and washload (generally the fine particles moving within the water column) (Fig. 7.3).

The **bedload** is composed of sand and coarser particles that have high settling velocity with respect to the water flow. These particles are transported near the streambed where they are kept loose by water turbulence and particle interaction (bouncing against one another = dispersive pressure).

Particles with relatively low settling velocity with respect to the water flow, and that are carried in the body of the flow, form either the suspended load or the washload.

The **suspended load** is composed of particles that are sorted by weight in the water column, with the larger and heavier ones toward the bottom. Like bedload, the suspended load is influenced by water discharge, and generally the higher the discharge, the greater the suspended load.

The **washload** is composed of fine particles with very low settling velocity, which travel at the same speed as the flow. This load is almost independent of discharge, and the particles are dispersed throughout the water column.

The various loads vary both in space and time not only among streams but also within an individual stream because of differences in source material, capacity, and competence in the various reaches. For example, during very high discharges (high capacity and competence), coarse particles that usually move as bedload at lower flow velocities may be transported in suspension, and particles usually transported as suspended load may move as washload. A high washload increases flow

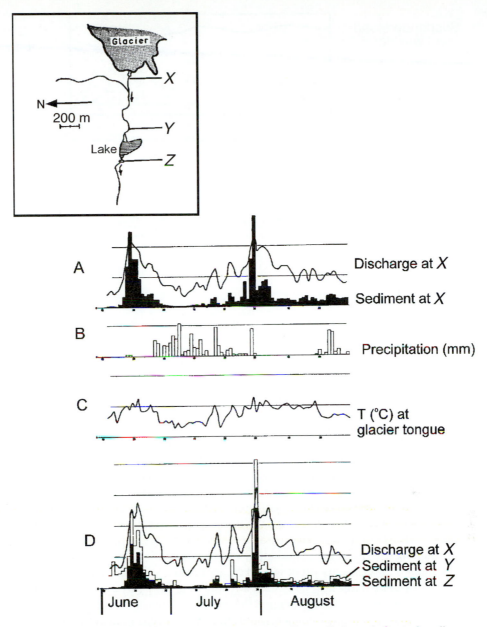

FIGURE 7.2   Map and diagrams showing that water and sediment discharge in front of a valley glacier in Norway is cyclical *(A)*, due to seasonal and daily variation in rain events *(B)*, and melting (temperature) *(C)*. The sediment load closely follows the discharge everywhere *(A,D)*. The principal effect of the lake is to trap much of the coarser fraction of the sediment load, as can be seen by comparing the load upstream and downstream of the lake (points *Y* and *Z* in *(D)*) (arrows indicate flow directions). (After Østrem, Ziegler, and Ekman 1970; see translation in Østrem, Ziegler, and Ekman 1973.)

density, which in turn reduces the settling velocity of particles; so under these conditions, particles may be transported as suspended load, whereas, under smaller washload, they would have been transported as bedload.

Cold climate streams, including glacial meltwater streams, have one more special type of suspended load: the **ice-rafted load** (Fig. 7.4A,B). This load is picked up

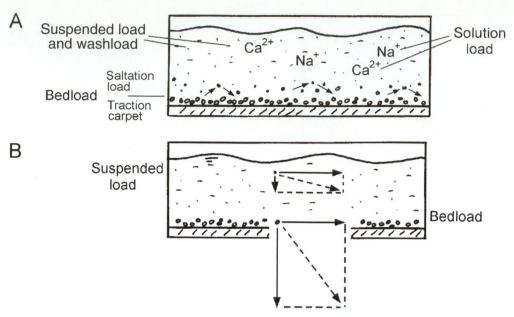

FIGURE 7.3   Diagram showing sediment transport in a fluid flow. *(A)* Modes of transport. The particles are free to move according their size and density and sort themselves out accordingly in the water column. *(B)* Relation between current velocity (horizontal arrows) and particle-settling velocity (vertical arrows). The vector sum of the forces (diagonal dashed arrows) indicates the path of the particles, and thus the ability of the finer, lighter ones to be transported farther downstream than the coarser or heavier ones.

and transported by floating ice blocks. It may contain coarse to very coarse particles mixed with fine ones. Coarse ice-rafted particles may be deposited along the steams in areas where ice jams occur (Fig. 7.4C), and also in areas where fine-grained suspended-sediment deposits generally prevail (Fig. 7.4D).

Sediment bulking refers to the addition of sediment to a stream in excess of its capacity. That is, under normal conditions, the sedimentary load of a fully turbulent stream cannot theoretically exceed the flow capacity, because at that critical point a dynamic equilibrium between erosion and deposition is established. However, sediment bulking can occur by a rapid input of excess material through slumps, slides, or other sudden mass gravity process, and by an increase in washload. Under these circumstances the stream flow can change from being fully turbulent, where fluid turbulence is the major force keeping the particles separated and in motion, to a hyperconcentrated (highly concentrated) flow. In this type of flow the force separating particles is in part due to turbulence, in part to the strength (buoyancy) of the matrix (mixture of fine particles and water) and the interaction between particles (dispersive pressure). Finally, if enough sediment is incorporated, a debris flow is generated in which turbulence is virtually absent and particles are kept separated (thus able to move) by the matrix strength (buoyancy). Fully turbulent flows, hyperconcentrated flows, and debris flows thus form a transport continuum, and they can transform into one another depending on the local environmental conditions.

Flow Regimes

The flow conditions of natural streams can be further described by considering the flow regimes that reflect the combined effect of gravity and viscosity. Regardless of the scale of the system, these regimes can be defined by dimensionless (pure) num-

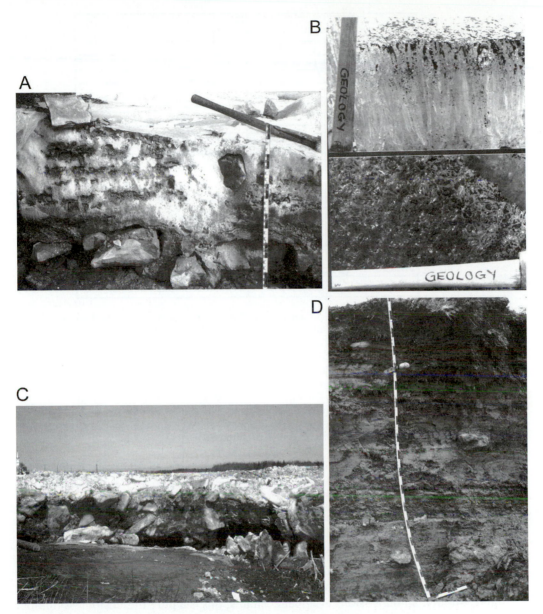

FIGURE 7.4 Photographs of ice-rafted material in north-central Canada. *(A)* Clasts transported by an ice block (ice floe). *(B)* Fine sediment carried on and within the ice among crystals (upper half = cross-sectional view, lower half = surface view). *(C)* Ice floes piled up in an ice jam, Albany River, Canada. *(D)* Coarse, ice-rafted clasts deposited on river levees together with water-transported, fine suspended-load material, Albany River, Canada.

bers, such as the Reynolds (R) and Froude (F) numbers, that are derived by comparing the relative importance of various forces affecting the flow.

The **Reynolds number (R)** is the ratio of inertial forces (which keep the fluid moving) to frictional forces internal to the fluid (which resist fluid deformation and motion). For an open channel, like a stream:

$$R = \frac{\text{inertial forces}}{\text{internal frictional forces}} = \frac{\rho v^2 / l}{\mu v / l^2} = \frac{\rho l v}{\mu}$$

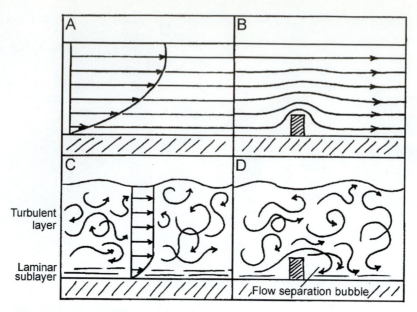

FIGURE 7.5   Cross-sectional representations of flow under different regimes, with su-perimposed vertical velocity gradients. *(A)* Laminar flow (horizontal lines represent flow tubes, arrows indicate flow speed and directions). *(B)* Quasi-laminar flow hugging an obstruction, with change in separation between flowlines but no major breakage, that is, without significant development of turbulent eddies. *(C)* Turbulent flow, where the flowlines are broken into turbulent eddies that have random direction and speed of instantaneous local currents. The whole mass flows downstream as indicated by the average, box-shaped velocity profile. Friction with the bed reduces the velocity, and the flow becomes quasi-laminar near the bed (laminar sublayer). *(D)* Significant flow separation downstream from an obstruction in a turbulent current.

where

     ρ = fluid density
     l = flow depth
     v = overall mean velocity
     μ = viscosity of the fluid

To demonstrate that this is a dimensionless number, substitute in the above equation the dimensions (usually indicated in square brackets) of the various units and simplify: all dimensions disappear. That is, this is a pure dimensionless number and applies equally well to very small systems (such as a laboratory flume) and to very large systems (such as the Mississippi or Amazon rivers). A consequence of this is that the flow regime of any stream can be simulated in a laboratory flume.

$$[R] = \frac{[\rho l v]}{[\mu]} = \frac{[ML^{-3}][L][LT^{-1}]}{[ML^{-1}T^{-1}]} = \frac{[L^{-1}]}{[L^{-1}]} = \text{no dimensions}$$

where the dimensions are of

     M = mass
     L = distance
     T = time

When internal frictional forces are greater than inertial forces, the Reynolds number is small (order of 500), and the flow movement is said to be laminar or quasi-laminar (Fig. 7.5A,B). This flow can be visualized as parallel, unidirectional flowlines or flow tubes that are independent of each other. That is, in a perfect laminar flow, there is no transfer of momentum or matter from one flow tube to the next. Theoretically there cannot be any slippage from one flowline to the next, nor from the static bottom substrate to the slightly (infinitesimal distance) higher flowline. Yet, the velocity of a stream with this type of flow shows a progressive increase from the bottom up (Fig. 7.5A). This implies that water molecules in the flow do not actually move (slip), but deform to infinity. Furthermore, the parallelism among flowlines implies that the flow is not affected by any roughness elements. The flow hugs any obstruction, and no separation of flow occurs. Perfectly laminar flow is difficult to achieve in streams, but quasi-laminar (almost laminar) flow occurs when water flows slowly on a smooth bottom. Under quasi-laminar flow conditions, a slight transfer of momentum and matter, and some separation of flow around obstructions occurs; that is, the flowlines are not perfectly parallel to each other (Fig. 7.5B).

When inertial forces prevail over internal frictional viscous forces, the Reynolds number is large (order of 2000 or more) and the flow becomes turbulent. Turbulence develops from laminar flow as flow velocity and/or bed roughness increase. Flowlines are disrupted, and turbulent eddies move en masse downflow. In turbulent flow, there is no direct relation between flow depth and instantaneous, local velocities, except in the basal laminar sublayer, where the flow is slowed by friction against the bottom. Therefore the average downslope velocity of the whole fluid mass is used (Fig. 7.5C). Transfer of momentum and mass occurs throughout the flow, and flow separation bubbles develop downstream from obstructions or steps (Fig. 7.5D). In open-channel water flows, the transition from laminar to turbulent flow takes place at Reynolds numbers between 500 and 2000.

The **Froude number (F)** is the ratio of inertial to gravitational forces in flows with a free surface, that is, not constricted in tubes or pipes.

$$F = \frac{\text{inertial force}}{\text{gravitational force}} = \frac{\rho v^2/l}{\rho g} = \frac{v^2}{gl} = \frac{v}{\sqrt{gl}}$$

where

g = force of gravity
l = flow depth
v = overall mean velocity

$$[F] = \frac{[v]}{[\sqrt{gl}]} = \frac{[LT^{-1}]^2}{[LT^{-2}][L]} = \text{dimensionless number}$$

The fraction $v/\sqrt{gl}$ can be interpreted as the ratio between the average velocity (v, rate of downslope transfer) of the whole liquid mass, and the rate of downflow migration (called celerity = $\sqrt{gl}$) of the surface waveform.

When inertial forces are less than gravitational forces, the Froude number is less than 1, and the flow is said to be subcritical. In this case, the celerity of surface waves is greater than the stream velocity; that is, the surficial waveforms move faster and overtake the main body of water, and the waves break downstream (Fig. 7.6A). When inertial forces equal gravitational forces, the Froude number equals 1, and the flow is called **critical.** In this case, the main body of water and the surficial waves move at the same speed, and there are no waves breaking (no white water) (Fig. 7.6B). When gravitational forces exceed inertial forces, the Froude number is greater than 1 and the flow is said to be supercritical or shooting. In this case, the main body of water moves faster than the celerity of surficial waves; the main flow

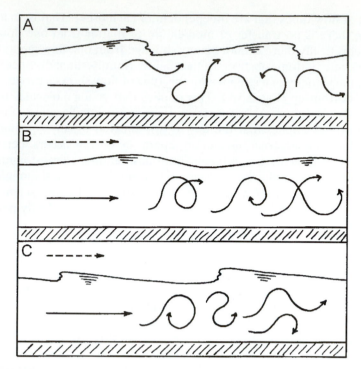

FIGURE 7.6   Cross-sectional diagrams representing flow regimes. *(A)* Turbulent, subcritical flow with surficial waves breaking downstream. *(B)* Turbulent, critical flow with standing surficial waves. *(C)* Turbulent, supercritical flow with surficial water waves breaking upstream. (Dashed arrow = celerity of surface wave form; full arrow = average current velocity; lengths of arrows suggest relation between waveform celerity and average flow velocity.)

shoots through and undermines the surface waves, causing them to break upstream (Fig. 7.6C).

SEDIMENT STRUCTURES AND FACIES

Sediment Bedforms and Structures

Each flow regime imprints granular sediments with diagnostic bedforms that may be preserved as sedimentary structures, and can then be used to accurately reconstruct the flow conditions.

Experiments and field observations have shown that when loose sand is present in a stream, a variety of sedimentary structures are formed. Sand ripples, showing ripple cross-laminations in cross section, form under quasi-laminar flow. Sand dunes showing cross-beds in section form under a turbulent subcritical flow regime. Horizontal beds, showing plane laminations or plane beds (if the layers are thicker than 2 mm) in section, and minute strings of particles aligned parallel to the flow direction at the surface (parting lineations), form when the flow is critical (F about 1). Massive beds or poorly developed dunes with the steepest side facing upstream (called antidunes) form when the flow is supercritical (Fig. 7.7A). Sand dunes and antidunes move, respectively, downstream and upstream in unison with the direction of breaking surficial water waves in the subcritical and supercritical flow regimes.

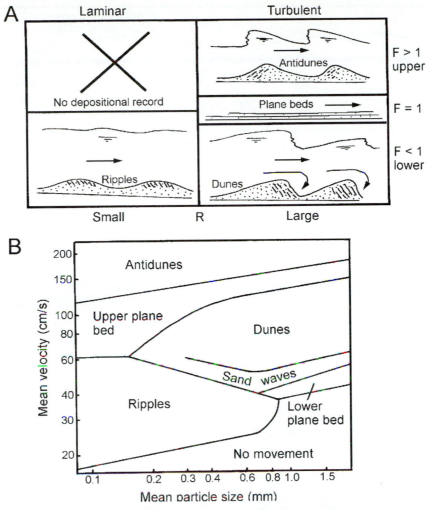

FIGURE 7.7   Diagrams of sedimentary structures and fluid flow. *(A)* Flow regimes as defined by the bedforms and sedimentary structures of sandy deposits (R = Reynolds number; F = Froude number). *(B)* Stability fields of various sedimentary structures as a function of water-flow velocity and particle size, as determined in flume experiments. (After Middleton and Southard 1978.)

To distinguish between the flow regimes defined by hydraulic measurements and calculation of the Reynolds (laminar, turbulent) and Froude (subcritical, supercritical) numbers, and flow regimes inferred from the deposits, the latter are separated into four fields. These are identified on the basis of the sedimentary structures developed in sands under quasi-laminar and turbulent flows, and lower and upper flow regimes (Fig. 7.7A).

The relation between flow regimes and sedimentary structures is imperfect, due to the varying influence of the amount and size of sediment in transport and the changing topography (bedform roughness) of the streambed. Thus, ripples form under quasi-laminar flow, but only if the sand is finer than about 0.9 mm (very coarse) (Fig. 7.7B). Also, sand dunes form under turbulent subcritical conditions, but change the flow as they develop. These variable conditions mold the dunes and their internal cross-beds (Fig. 7.8). On the upstream side, variable flow conditions, from turbulent subcritical to turbulent supercritical, develop toward the dune crest where coarse-grained sediment is released at the brink. Downstream, a flow bubble develops where mass flows (modified grain flow) carry the sediment downward

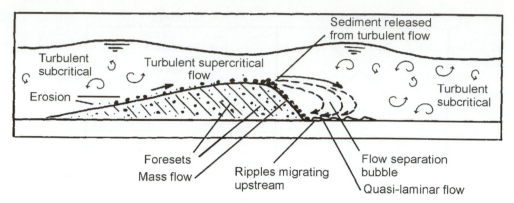

FIGURE 7.8 Diagram of flow conditions and sediment transport on a subaqueous sand dune (thick arrows = water and sediment flow directions; thin arrows = turbulent eddies).

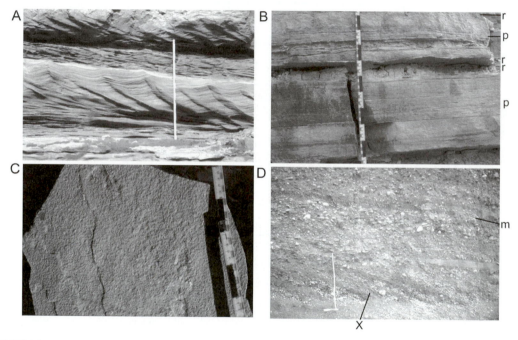

FIGURE 7.9 Sedimentary structures of sandy deposits. *(A)* Cross section of climbing ripple cross-laminations; *(B)* cross section of plane beds (p) and ripple cross-laminations (r); *(C)* parting lineations as seen on a bedding surface; *(D)* cross section of massive (m) and cross-bedded (X) sandy gravel of an outwash deposit.

along the frontal slope of the dune (precipitation slope). At times, a reversed, upstream-directed, quasi-laminar flow develops at the toe of the slope, generating sand ripples that migrate upstream.

Both ripples and dunes are lower flow regime structures, differing in size due to the dimensions of their flow-separation bubble; but they are similar in that they both form perpendicular to flow and migrate downstream. Where the sediment load is large, the trains of ripples and dunes migrating downstream climb relative to the horizontal surface (bedding plane), forming so-called climbing ripple cross-laminations visible in cross sections. The rate of deposition relative to bedform celerity determines the rate of climb of the bedforms (Fig. 7.9A).

Plane beds develop only when rapidly moving flows are charged to capacity with sediment; otherwise erosion occurs. During the formation of plane beds, the particles are transported very rapidly in trains of grains parallel to each other and to the current direction, forming parallel laminations (Fig. 7.9B). The parting lineations (Fig. 7.9C) form at the surface as trains of grains get quickly buried.

Antidunes are similar to subcritical dunes, except that they migrate slightly upflow in response to the upstream-breaking surface water waves. They form only when the flow is charged to capacity. The power of the flow is capable of moving all the sediment present; hence there is little opportunity for size or density sorting to occur, and the backsets (cross-beds directed upstream) of the antidunes are ill defined and not readily recognizable.

Laminar, upper flow regime conditions rarely develop in natural streams, except when a current moves very fast over a perfectly smooth hard bottom. So, this type of flow does not produce a sedimentary record.

Coarse-grained gravels usually show massive beds because a high flow velocity is needed to move the clasts, and usually the flow is supercritical during their transport (Fig. 7.9D, top part). Cross-bedded gravels generally form along frontal slopes of large, relatively stable transversal bars (Fig. 7.9D, lower part). However, some gravelly cross-beds are formed by true dune migration during rare, extremely large floods (megafloods) when all the sand and most pebbles travel in suspension and the coarse clasts travel as bedload.

Sedimentary Facies, Facies Associations, and Fluvial Models

The term **sedimentary facies** indicates the sum of characters of a deposit. Lithofacies refers to the physical and/or chemical characteristics. Biofacies refers to its biological (fossil) characteristics. Codes may be used to rapidly record the most common clastic lithofacies. These are defined according to a combination of grain size and structure (Table 7.1). Though such codes are useful, they are not standard nomenclature, and it is common practice to devise similar codes to best fit particular cases.

Sedimentary facies provide a good record of changing conditions through time and space. For example, hydrological conditions change in a predictable manner at any one locality due to climatic and geomorphological constraints.

1. In a meandering stream, water flows faster along the outside of the bend, and progressively slower toward the inside. From the outside to the inside of the bend the current changes from erosional, supercritical flows; to flows depositing massive, coarse-grained deposits or parallel laminated sand deposits; to subcritical flows depositing cross-bedded sand from dunes and, in the inner parts of the bend, fine-grained, cross-laminated sand ripples (Fig. 7.10). As erosion occurs and the meander bend migrates, the lateral facies also stack vertically: coarse-grained, massive beds rest on erosional surfaces and are in turn overlain by progressively finer-grained sediments occurring in plane beds overlain by cross-beds and ripple cross-laminations. This relation between sedimentary characters (facies) is the sedimentological signature (facies association) of a channel fill (called a point bar) of a regularly migrating meander. The signature remains the same, whether it is from a large river, small creek, or tidal channel. Other criteria, such as dimensions, biofacies, or geochemistry, may be required to differentiate among these alternatives.

2. A decelerating flow may also deposit a characteristic stack of textures and structures. There may be a massive unit at the base grading up into a plane-bedded

TABLE 7.1   Facies Codes for Fluvial and Related Sedimentary Facies

Facies Code	Lithofacies	Sedimentary Structures
Gms	Massive, matrix-supported gravel	None
Gm	Massive or crudely bedded gravel	Horizontal bedding, imbrication
Gt	Gravel, stratified	Trough cross-beds
Gp	Gravel, stratified	Planar cross-beds
St	Sand, medium to very coarse; may be pebbly	Solitary or grouped trough cross-beds
Sp	Sand, medium to very coarse; may be pebbly	Solitary or grouped planar cross-beds
Sr	Sand, very fine to coarse	Ripple marks of all types
Sh	Sand, very fine to very coarse; may be pebbly	Horizontal laminations, parting lineations
Sl	Sand, fine	Low-angle (<10°) cross-beds
Se	Sand, erosional scours with intraclasts	Poorly defined cross-beds
Ss	Sand, fine to coarse; may be pebbly	May show broad, shallow scours including cross-stratifications
Fl	Sand, silt, mud	Fine lamination, very small ripples
Fsc	Silt, mud	Laminated to massive
Fcf	Mud	Massive, with freshwater mollusks
Fm	Mud, silt	Massive
Fr	Silt, mud	Rootlets
C	Coal, carbonaceous mud	Plants, mud films
P	Carbonate	Pedogenic features

After Miall 1978.

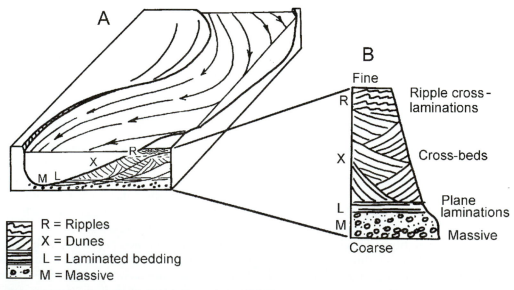

FIGURE 7.10   Sedimentary model of stream bend deposits. *(A)* Diagrams showing water flow (arrows), particle size, and sedimentary structure distributions. *(B)* Correspondent vertical variations in sedimentary deposits, resulting from migration and lateral filling of the channel. (After Blatt, Middleton, and Murray 1980.)

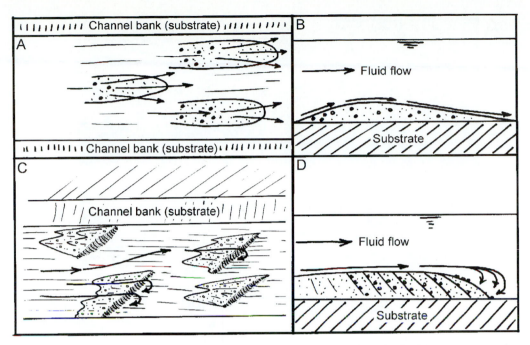

FIGURE 7.11   Diagrams of major bars of braided streams. *(A)* Plan view of longitudinal linguoid bars. *(B)* Cross-sectional view of fluid flow and deposits of a longitudinal linguoid bar. *(C)* Oblique plan view of transverse bars. *(D)* Cross-sectional view of flow and sediments of a transverse bar. (Arrows indicate fluid and sediment flow direction.)

unit, ripple cross-laminations, and finally into fine deposits such as clay, but all within a single bed, that is, during a single depositional event. This is the signature (facies) of a waning storm (tempestite), or of a decelerating subaqueous turbidity flow.

3. In shallow streams with highly variable water and sediment discharge where deposition alternates with erosion, several large-scale bed features, called bars, develop. High-competence flows may move sheets of pebble- or boulder-sized material to form sheet bars a few particles thick across the whole stream, leaving the fine sand and silt to be transported farther downstream in suspension. Large quantities of moving, coarse-grained gravel may form longitudinal bars parallel to the current. These bars may have several shapes, ranging from linguoid (tonguelike) to rhomboid and others (Fig. 7.11A). They are commonly characterized by having coarser particles at the upstream end and a tapering of the finer downstream end without development of a flow separation bubble (Fig. 7.11B). Therefore the resulting deposits (facies) are coarse grained and massive. Fine-grained gravel and sand may instead form bars at an oblique angle to the current, called transverse bars, like sand dunes except that they are less mobile, being somewhat constricted by the geomorphology of the stream. Transverse bars have a variety of shapes, but their defining characteristic is a pool downstream, allowing a flow separation bubble and inclined depositional surfaces to develop (Fig. 7.11C). The bars migrate downstream, forming coarse-grained cross-beds (Fig. 7.11D). Changing hydraulic conditions during floods can erode parts of the bars to form scours and subsequently redeposit material into them. Such fluvial deposits often show numerous cut-and-fills (Fig. 7.12). These are typically filled with sandy gravelly layers formed during peak flows, capped by ripple cross-laminated sandy layers deposited during receding flow and slack water.

Remnants of various types (facies) of bar deposits and cut-and-fill features indicate rapidly changing flow conditions. An assemblage or stacking of these facies

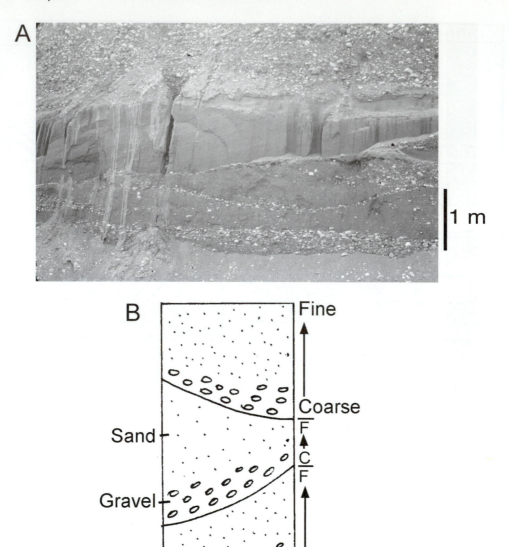

FIGURE 7.12   Cut-and-fill structures of a braided stream deposit. *(A)* Photograph of an outwash deposit, southern Canada. *(B)* Diagrammatic representation, showing fining upward sediments (C = coarse grained; F = fine grained) in each fill.

forms distinctive patterns (facies associations, facies succession) typical of various types of braided streams (Fig. 7.13). These facies successions can be useful in interpreting the origin of ancient sedimentary deposits. Where good outcrops, closely spaced wells, or seismic profiles are available, the three-dimensional relations (facies architecture) of the facies can be mapped, paying particular attention to bounding surfaces separating component facies or facies associations. Orderly changes in facies associations and facies architectures can be used to reconstruct paleogeography, because they reflect the regularly succeeding environments of sedimentation present at Earth's surface. An alluvial system, for instance, has an orderly lateral sequence of environments, albeit not equally developed everywhere, from alluvial fans, to braided streams, meandering streams, alluvial plains, coastal areas, and offshore areas.

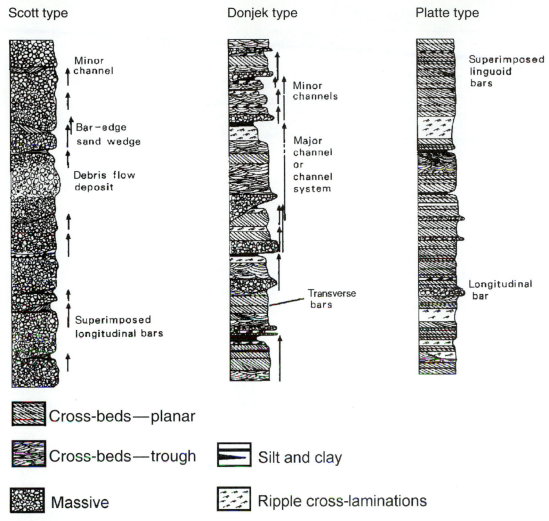

Scott type

Donjek type

Platte type

Cross-beds—planar

Cross-beds—trough   Silt and clay

Massive   Ripple cross-laminations

FIGURE 7.13   Diagrams of braided-stream sedimentary successions derived from and named according to typical rivers. Scott type: gravelly, predominance of longitudinal bar deposits; Donjek type: sand and gravel, predominance of transverse bar deposits; Platte type: sandy, mostly transverse bar and dune deposits. (After Miall 1978.)

MELTWATER STREAM CHARACTERISTICS AND DEPOSITS

The briefly introduced, basic concepts of fluid flow and sediment transport can be used to analyze meltwater streams. Water occurs everywhere in temperate glaciers, increasing in importance from the firn line toward the terminus and beyond. On the glacier itself, meltwater flows in supraglacial streams, accumulates in temporary supraglacial lakes, and plunges inside the glacier through crevasses and cylindrical holes (**moulins**). Within and under the glacier, water flows as sheets or through interconnected cavities, enlarging them and developing tunnels. When such conduits are full, the water is subjected to the hydrostatic pressure of the water column above, even if the vertical conduits are small and tortuous. When the connection to the surface is cut by freezing, the internal water can become overpressured as the

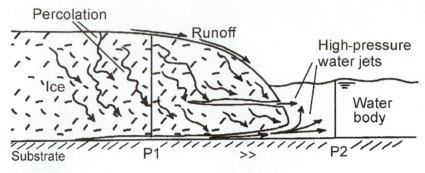

FIGURE 7.14 Diagram showing water flow on, within, and below a glacier. The extrusion of meltwater in front of a subaqueous glacier terminus is possible because the tunnel water is at pressure P1, greater than P2 in the basin. P1 is generated by a continuous column of water reaching the top of the glacier and, at times, pressured by the weight of the ice itself.

weight of the ice squeezes the water-filled cavities and pores; that is, part of the weight of the glacier is carried by the trapped water. The pressure may be strong enough to locally force water to flow uphill or to exit the glacier as jets at both terrestrial and subaqueous termini (Fig. 7.14).

The meltwater discharge changes seasonally or even daily depending on melting rates, on the conditions within the glacier (whether the ice is at the pressure melting point throughout or locally frozen at the base), and on whether continuously open conduits are present. The water can sometimes pond in lakes under ice sheets, or in tributary valleys behind valley glaciers, until it is deep enough to float the glacier or generates enough pressure to break through it. The water then forces its way out under pressure as a sheet or channelized flood. As the pressure is released, the glacier settles down again and the flood is terminated. In places, such as southern Iceland, very large glacier outburst floods (called *jökulhlaups,* literally "river leaps") have a quasi-regular multiyear recurrence time.

Supraglacial and Subglacial Deposits

For the most part, supraglacial, englacial, and subglacial water flows follow the same processes as those of open channel flows (only occasionally as flows in full pipes); so they generally produce deposits similar in structure and overall composition to those of other streams. However, sediments deposited in contact with the glacier ice (**ice-contact stratified drift**) have the following characteristics. (1) They show extreme range and abrupt changes in particle size; the composition of the clasts is similar to that of nearby till. (2) They contain fragments of till (Fig. 7.15A). (3) They show deformations because they were deposited against or over ice, and ice push or slumps occur during melting (Fig. 7.15B). (4) Whether deformed or not, they form landforms with steep ice-contact slopes, indicating that they once had confining ice banks.

Ice-contact stratified drift generates a variety of sediment accumulations on, within, and on the flanks of the glacier (Fig. 7.16A). Sediments deposited on stagnant glaciers are let down on the land surface when the ice melts, topographic inversions occur, and characteristic landforms develop (Fig. 7.16B). The type of landforms generated depends on the amount of debris present, the position of debris in the glacier, the amount and rate of meltwater generation, and the amount of

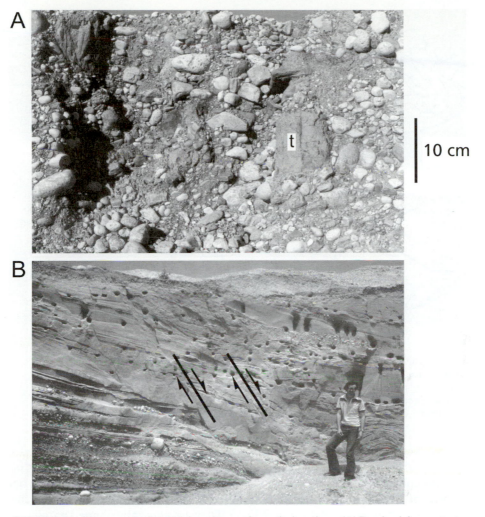

FIGURE 7.15 Photograph of glacial deposits seen in vertical sections. *(A)* Proximal, ice-contact stratified drift with clumps of till (t). *(B)* Faults (arrows) and folds of ice-contact stratified drift due to differential melting of a buried ice block.

erosion and deposition by meltwater. These landforms and sediments can often be preserved because they develop on dead (stagnant) ice and are neither modified nor destroyed by glacier readvances.

Kames are moundlike hills composed of gravel and sand originally deposited in depressions on stagnant ice or in small deltas or fans built at the terminus or sides of the glacier. Melting of the ice leaves them isolated and with steep slopes that often slump. Kame deposits generally lack fines due to washing by meltwater, and their pebbles and cobbles vary greatly in roundness.

Kame terraces are linear features also composed mostly of gravel and sand, though fine sediments may also be present. They form along the edges of valley glaciers mainly by streams running between the ice and the valley walls, or, occasionally, by lakes in similar position. Their sediments are derived from both glacial drift and scree from the valley walls. Kame terraces commonly develop kettles due to differential melting, and often slump along the steep, ice-contact slope as the glacier melts.

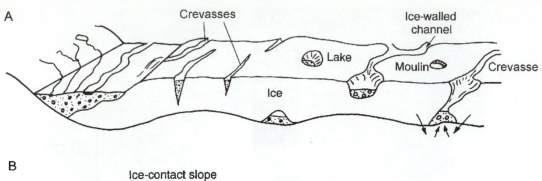

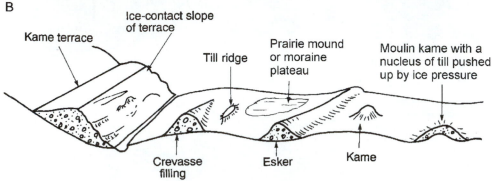

FIGURE 7.16 Diagrams showing ice-contact sediments and landforms. *(A)* Various features formed on stagnant ice. *(B)* Sedimentary accumulation after ice melting. During sediment release from the glacier, a topographic inversion occurs; that is, what were low areas in the stagnant ice receive much sediment and form topographic highs once the supporting ice is removed. The deposits maintain steep, ice-contact slopes. (After West 1968.)

Eskers are long, narrow ridges often composed of well-sorted gravel and sand, generally deposited in subglacial or englacial tunnels. Eskers may be straight or sinuous, simple or braided, and vary in size. They range between 2 and 50 m in height, between a few meters and couple of hundred meters in breadth, and can reach hundreds of kilometers in length (Fig. 7.17A). They are generally aligned parallel to the direction of ice movement, as determined by other directional landforms, and frequently follow valleys in the substrate if these do not diverge much from the direction of glacier flow. But they do not need to be located at the valley bottom, since they form perpendicular to the steepest pressure gradients in glaciers. Water under such gradients can locally flow uphill; so esker alignment can be locally independent of topography. Esker sediments may also be deposited in englacial tunnels and then let down upon melting of the ice. Thus, eskers may be found draping drumlins. Esker deposits and landforms are preserved unmodified only in the stagnant, terminal portion of a glacier, since active ice would destroy them.

Esker deposits show marked changes in particle size, but also good stratification of gravel and sand (Fig. 7.17B). Occasionally, boulders and lumps of till are present. The sand can be well sorted and well structured, for instance in sequences of cross-beds. Their ice-contact slopes can slump and disturb and reorient the sediment fabric.

Simple linear eskers form in tunnels below stagnant ice and end up as deposits in meandering or braided patterns, standing high on the landscape once the glacier has melted away (Fig. 7.18). During extreme floods, the tunnels fill with water and behave as pipes, allowing water under pressure to move great quantities of material as a massive carpet, resulting in massive deposits.

FIGURE 7.17   Photographs of esker landform and sediments. *(A)* Small esker in the Great Lakes area, Canada. *(B)* Good stratification of well-sorted sediments in eskers. Abrupt variation in particle size from bed to bed attests to the strong variation in meltwater discharge.

Compound eskers show discontinuous beadlike features joined by connected ridges. The narrow ridges are formed by tunnel deposits, whereas the beads are fan-delta sediments that form where channelized flows expand either into larger subglacial cavities or at the glacier terminus. Strings of beads indicate the punctuated retreat of the glacier terminus. This can occur on land or underwater. Underwater, large **subaqueous outwash** fans can form at the mouths of glacial tunnels.

Some eskers are not continuous but segmented, although not clearly beaded. They were formed in relatively short-lived tunnels, or may have been partially disrupted by renewed glacier movement.

Deposition in subglacial or englacial tunnels can reasonably well explain relatively short eskers formed near the terminus of glaciers. However, the extreme length (several hundred kilometers) of some eskers implies that tunnels maintained continuity and overall orientation for hundreds of years as the ice sheet stagnated

FIGURE 7.18   Photograph of a glacial landscape revealed as a glacier has melted out and
the terminus has retreated (1 = glacier; 2 = ice-cored hummocky end moraine; 3 = fluted till;
4 = esker; 5 = fan; 6 = braided stream). (Photograph by B. Washburn, from Friedman, Sanders,
and Kopaska-Merkel 1992.)

and retreated. This is difficult to accept. An alternative hypothesis is that some
eskers were formed almost instantaneously (perhaps in weeks or months) by megafloods draining very large subglacial meltwater reservoirs. However, this catastrophic flood hypothesis still faces difficulties in explaining the source of the necessary
large meltwater reservoirs, and the preservation of the landforms under non-
stagnant ice conditions.

Eskers seem also to be controlled by the type of material beneath the glacier.
Most large eskers in North America (Fig. 7.19) and Eurasia occur on hard rock.
Fewer and smaller eskers occur on softer substrate and thicker glacial deposits. One
explanation is that, on hard rocks, ice tunnels at the base of the glacier (R-channels)
form more readily because meltwater cannot penetrate the impermeable substrate.
Instead, the meltwater flows through connected ice cavities to eventually form tunnels in which esker sediments are deposited. On softer, more readily deformable
rock substrates or glacial deposits, meltwater can percolate into, erode, and flow
through substrate channels, so that eskers tend to be short and discontinuous.

Crevasse fillings are linear, relatively short (a few tens of meters long), low (up
to 10 m high) ridges formed by the accumulation of debris in crevasses of stagnant
ice, which are then let down by ablation. Crevasse fillings invariably overlie landforms composed of till, indicating their origin within the ice itself. The sediments
of crevasse fillings are generally coarse and irregularly bedded, with some degree
of pebble sorting and roundness. They may contain layers of silt and clay alternating with sand. Mapping ancient crevasse fillings can help to establish the stress

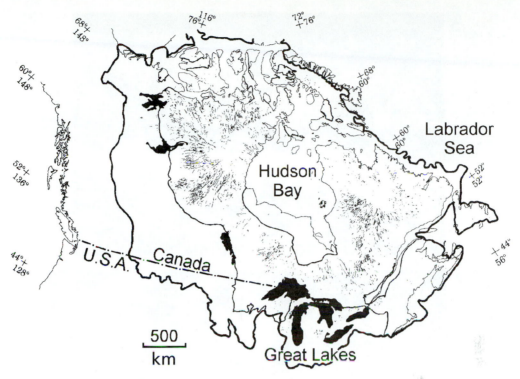

FIGURE 7.19   Distribution map of eskers (numerous short lines) in central North America (thick outline = limit of the Laurentide Ice Sheet; thin outline = boundary between solid crystalline rocks of the Precambrian Canadian Shield and softer sedimentary rocks to the south and west; black = glacial lakes). (After Clark 1997.)

distribution that existed in the glacier, and perhaps even the former existence of ice streams.

Proglacial Stream Deposits

Streams and rivers fed by glacial meltwater have large and rapid fluctuations in discharge. They move large amounts of sediments, and may carry, or are affected by, ice blocks from the glacier or from seasonal ice cover. There is normally little vegetation on floodplains near the glacier; so sediments are unprotected and easily eroded by water and wind. Such conditions lead to instability in watercourses and to development of braided streams whose channels are cut and partially refilled during each flood. Proglacial rivers form a complex network of changing braided channel patterns, which is called a valley train (Fig. 7.20A) where confined in a mountain valley, or outwash[1] where spread out on a proglacial surface of various dimensions (Fig. 7.20B). Valley trains may be long and narrow and outwash plains have various shapes and dimensions, but both eventually terminate in lacustrine or marine deltas of various steepness (Fig. 7.21A,B). Subaqueous outwash fans form where glaciers terminate directly into lakes or seas (Fig. 7.21C).

The clast composition of the proglacial streams is variable, depending on the composition of the source material, and may consist of mixtures of igneous, meta-

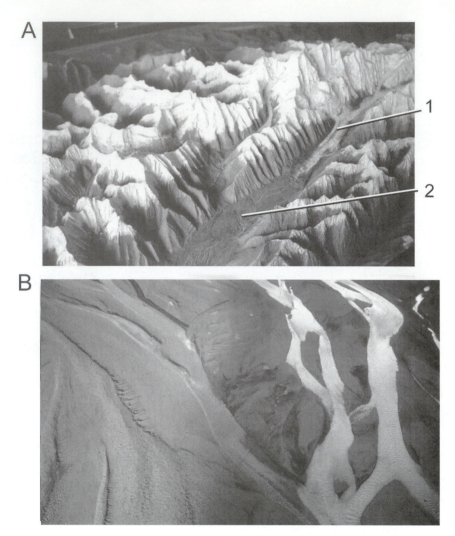

FIGURE 7.20   Photographs of proglacial stream systems in the southern island of New Zealand. *(A)* Valley train (1 = glaciers; 2 = braided stream); *(B)* outwash.

morphic, and sedimentary materials. The larger clasts are generally fairly well rounded. This roundness is partly inherited from the glacial material, and is partly produced by abrasion during fluvial transport. Some clasts may split along fractures by impact during transport or by frost shattering if exposed at the surface. These form characteristic broken-rounds (split rounded pebbles, with sharp corners at one or more sides). Postdepositional weathering of the less stable minerals of igneous and metamorphic debris may lead to formation of isolated sandy ghosts of the original clasts in outwash deposits (Fig. 7.22A). Generally, no glacial striations are preserved on clasts transported more than a few hundred meters downriver.

The outwash deposits have a great variety of particle sizes, but generally they contain little clay (Fig. 7.22B). Their sorting varies as well, and poorly sorted layers may alternate with well-sorted ones (Fig. 7.22C,D). Strong variations in grain size occur both between beds and vertically within thick beds, reflecting strong changes in flow strength, respectively, of different floods and during a single flood. In areas proximal to the glacier, vertical particle-size cyclicities caused by daily or seasonal variations in meltwater discharge may occur. Such cyclicities disappear farther from

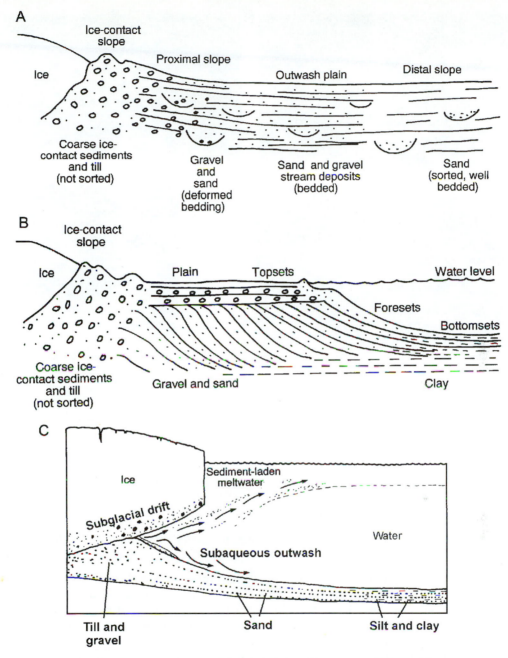

FIGURE 7.21 Diagrams of outwash. *(A)* Wide/long outwash where deposits are well differentiated, graded, and structured. *(B)* Narrow outwash terminating into a delta. Usually the deltas are well differentiated into bottomsets, forests, and topsets when developed in lakes (after West 1968). *(C)* Subaqueous outwash formed in front of a submerged glacier terminus (after Rust and Romanelli 1975).

the glacier as incoming tributaries dampen and steady discharge fluctuations. In long valley trains or outwashes radiating away from the glacier, the particle size diminishes in a quasi-regular, logarithmic fashion with distance away from glacial point sources. Associated sedimentary structures, facies, and facies associations

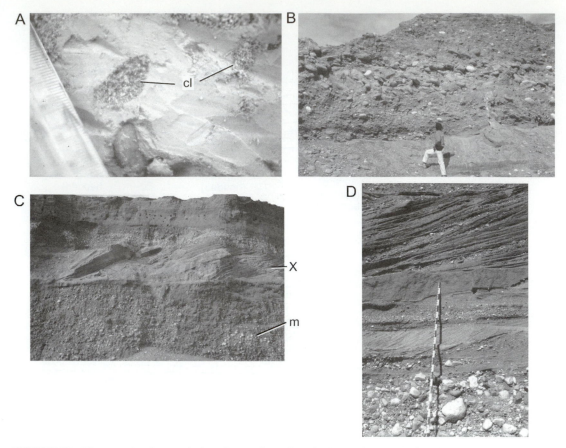

FIGURE 7.22   Photographs of outwash deposits, southern Canada. *(A)* In situ weathering of transported crystalline clasts. *(B)* Coarse-grained bouldery deposits of an ice-proximal setting. *(C)* Massive sandy gravel (m) of longitudinal bars alternating with large cross-beds (X) of transversal bars in more distal parts of the outwash. *(D)* Superimposition of cross-beds formed by migrating sandy bars.

also change with distance from the source to reflect changing flow conditions and sediment type.

1. Close to the source, coarse gravel predominates and there is development of longitudinal bars (Scott model, Figs. 7.13, 7.23). Close to the glacier, the outwash (pitted outwash) may have many kettles, caused by melting of buried, isolated ice blocks left behind during the retreat of the glacier or transported in during strong floods (Fig. 7.24).

2. At intermediate distances from the source (Donjek model, Fig. 7.13), the outwash is split into numerous channels where there is mixed sedimentation of gravel and sand, and foresetted transverse bars become prevalent.

3. Farthest from the source (Platte model, Fig. 7.13), sand prevails in cross-bedded layers. In places, such as in distal parts of the southern outwash of Iceland, the channels are more poorly defined and vast floodplains with shallow overland sheet flows and shallow channels develop. There, the sediments are sandy with sparse pebbles, and subaqueous ripple cross-laminations are common.

4. Fine (silt and clay) particles, which are originally formed in large quantities by glacial abrasion, are usually washed farther downstream. As the outwash surface dries out during low flow stages, the silt is readily lifted by strong winds and

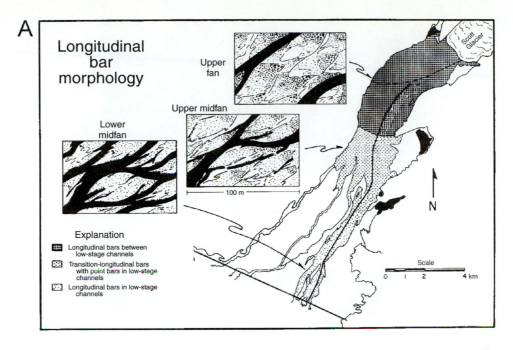

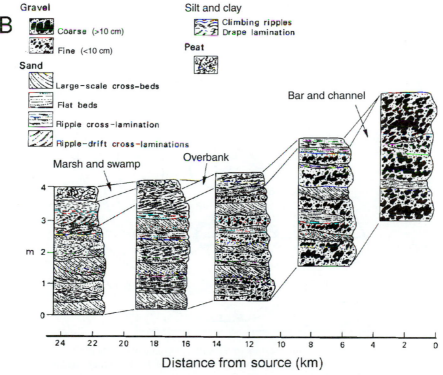

FIGURE 7.23   Diagrams showing variations in particle size and bedforms along the Scott Glacier outwash fan, Alaska. *(A)* Variation in bar morphology with distance from the glacier. *(B)* Downstream variation in particle size and sedimentary structures. (After Boothroyd and Ashley 1975.)

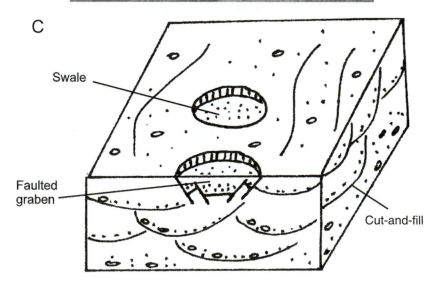

FIGURE 7.24   Photographs and diagram of outwash environments (Iceland).
(A) Large blocks of ice strewn over the outwash after a large flood (southern Iceland).
(B) Kettles developed on the outwash after melting of strewn ice blocks (southern Iceland). *(C)* General model of a sandy-gravelly, pitted outwash.

redistributed over vast areas, at times continent wide, to form loess inland and fine muddy deposits in lakes and seas.

Deltas formed by proglacial streams vary, but most tend to be river dominated. If they develop on shallow coasts with multiple sources from braided streams, braided deltas develop. If they form instead in deep basins, the sediment-charged river waters plunge, generating turbid flows (hyperpycnal flows, turbidity currents) that transport the material to the lake bottom.

Proglacial, braided-outwash sediments can be deposited during glacial advance or retreat; however, those formed during glacial advance may be reworked or remolded in some fashion. Deposits formed when the glacier is in a **steady state** may be characterized by thick, stacked cut-and-fills. Under these conditions, sediment is continually conveyed by the ice to the quasi-static terminus; it accumulates in moraines, and is in part reworked by meltwater streams and carried onto the outwash. For deposits formed when the glacier terminus slowly retreats, any point in the outwash gets increasingly more distal from the sediment source, receives increasing finer sediments, and develops a fining-upward braided stream succession generally characterized by cut-and-fills, cross-beds, and ripple cross-laminations.

Meltwater Megafloods

Floods are common in proglacial environments, and, at times, they may reach enormous to catastrophic dimensions (megafloods). Megafloods usually develop where ice-dammed lakes burst. Examples of large floods are the *jökulhlaups* of Iceland, some of which have been estimated to exceed flows of 100,000 m³/s. Possibly the largest floods ever experienced on Earth occurred in Washington State (USA) during the late Pleistocene, with peak discharges estimated to 21×10^6 m³/s. These megafloods had a relative short duration and were caused by the failure of large lake dams formed by glaciers themselves or by glacial deposits.

1. In Iceland, the large floods are related to bursts of subglacial lakes fed by meltwater generated by the high geothermal heat flux, and at times by subglacial volcanic eruptions (Fig. 7.25). Under normal circumstances, the lake level rises until it can float part of the glacier and then subglacial floods burst out onto the outwash plain. In places, these events can be predicted, because the size of the lake and the thickness of the ice above are known; at times though, subglacial eruptions melt the ice very rapidly and generate more water than usual. The large floods can generate tunnels and canyons within the glacier, and transport large quantities of sediments and ice onto the outwash. The great stream power of these outburst floods is usually dispersed over large outwash areas; but where the flow remains partially constricted, it causes large erosion. In 1996 one such gigantic flood transported and deposited enough material to cause the southern coast of Iceland in front of the outwash (*sandur*) to advance 12 m into the sea in a few weeks. Numerous similar floods have occurred in Iceland and were responsible for the development of the many outwashes, for partial filling of the fjords, and for the erosion of large canyons excavated into the lava bedrock.

2. Most Washington floods were probably due to the sudden failure of a glacier dam that contained the upper Pleistocene large glacial Lake Missoula, in the western United States (Fig. 7.26). The enormous floods are thought to be responsible for the regional erosion of the Washington Scablands and deposition of bouldery dunelike features.

3. Very large floods have also probably occurred along other marginal parts of the ice sheets and other glaciers in North America (Laurentide Ice Sheet), in Europe, and in Asia, particularly in the Altai region of western Siberia. They are recorded both in erosional and in depositional features, such as gravelly sidebars, several tens of meters thick, formed at the mouth of tributaries of the main flood

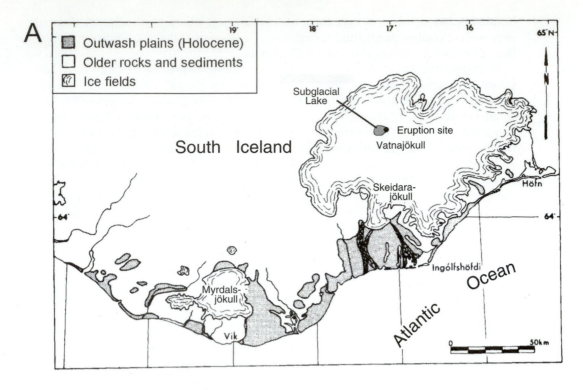

FIGURE 7.25   Wide southern Iceland outwash (*sandur*) subjected to frequent large floods.
(A) Site map; *(B)* satellite image showing details of the outwash in front of Skeidarajökull.
(After Einarsson 1994.)

valley, and meters-thick, single-event, massive waterlaid beds in lakes. Large, ancient subglacial floods may have participated in the erosion of large channels and
valleys (tunnel valleys) that have been subsequently filled with till, glaciofluvial, and
other deposits (Fig. 7.27).

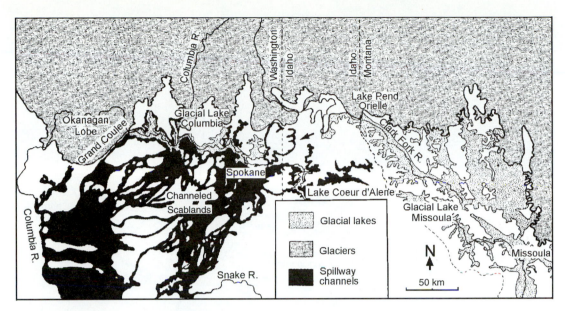

FIGURE 7.26   Map of the Washington Channeled Scablands: a legacy of megafloods in the northwestern United States. (After Baker 1983.)

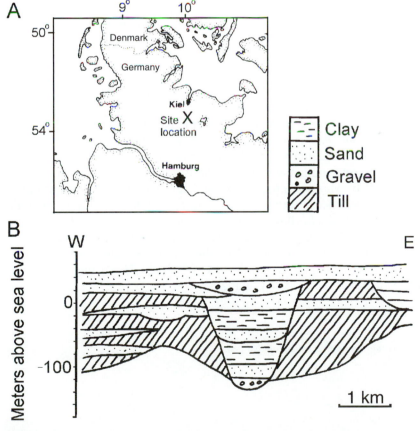

FIGURE 7.27   Tunnel valley filled with till and waterlaid deposits, Germany. *(A)* Location map; *(B)* generalized cross section. (After Piotrowski 1993.)

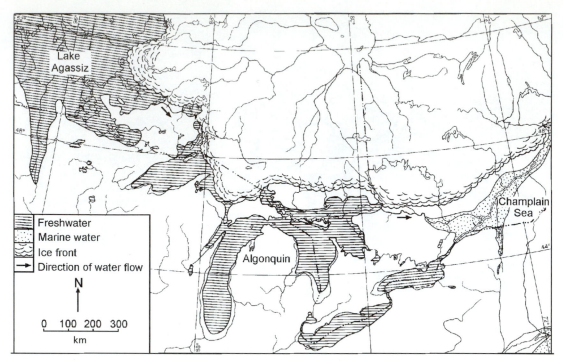

FIGURE 7.28 Map showing reconstructed terminus of the Laurentide Ice Sheet of approximately 11,200 years BP, at a time when an ice dam burst may have occurred and the waters of Lake Agassiz could have flooded eastward to the Gulf of St. Lawrence. Arrows indicate water outlets. (After Prest 1970.)

4. Megafloods may have influenced the climate of entire regions. One example may be the catastrophic freshwater floods that probably occurred along the southern edge of the Laurentide Ice Sheet in North America (Fig. 7.28). The waters of the vast glacial lake (Lake Agassiz) that covered large parts of central Canada on occasion were suddenly released southward along the Mississippi River channel, temporarily freshening and changing the isotopic composition of waters of the ancestral Gulf of Mexico. Similarly it has been hypothesized, admittedly with little supporting evidence, that as a glacier barrier broke, the waters of Lake Agassiz (south-central Canada) flowed rapidly across the precursors of the Great Lakes to the Gulf of St. Lawrence. These waters may have been responsible for a sudden local rise (about 0.5 m) in sea level, cooling of the seawater, and causing the onset of short-lived colder conditions of about 11,000 years BP (Younger Dryas equivalent) in the Maritime Provinces of Canada.

ENDNOTE

1. In Iceland the outwash is called *sandur* (sand plain; pl., *sandar*).

CHAPTER 8

Glaciomarine and Glaciolacustrine Environments and Deposits

INTRODUCTION

Seas and lakes have many similar physical, chemical, and biological processes. For example, waves and currents act similarly everywhere, and deltas, beaches, and basinal turbidite deposits are similar whether they form in large lakes or small seas. But there are also many differences between seas and lakes. Seas have heavier, saline waters, are larger, and most are globally interconnected. In contrast to lakes, they respond to glaciation by quasi-synchronous changes in water level and isotopic composition. Large seas have tides; landlocked seas and lakes do not. The biota is also very different. For example, many life forms, such as corals and starfish, can live only in seawater. Since these organisms affect sedimentation and provide sediment particles, many marine sediments differ from lake sediments. The processes and deposits that are similar in seas and lakes are presented together here, but any peculiarities of specific settings will be pointed out (Table 8.1).

Marine and lacustrine environments range from coastal plains to deep offshore areas (Fig. 8.1). Glaciers strongly influence all these but particularly the shallow environments from the coast to the edge of the continental shelf. Glaciers trap a lot of water inland, reducing the volume of water in the ocean basins; so sea level falls. Coasts and shelves may become exposed to a depth of about 150 m during global glacial periods. Glaciers also depress Earth's crust isostatically under the weight of ice. During deglaciation, coastal areas cannot rebound to their original elevation as fast as the glacier melts. So, much land is first inundated by the postglacial rise in sea level, and later part of it re-emerges through postglacial isostatic rebound. This leads to the disruption and re-establishment of shelf ecosystems, and to changes in water currents and related changes in climate and sediment distribution. Furthermore, glacial effects on seas and large lakes are felt for thousands of years after the glaciers have disappeared. For example, although the ice sheets of Europe and North America disappeared thousands of years ago, northern lands are still uplifting due to residual, postglacial isostatic rebound, affecting areas like Hudson Bay and Baffin Island.

Glaciolacustrine and glaciomarine (glacimarine, glacial marine) environments are those affected by lacustrine (lake) and marine processes (hereafter all referred to as oceanographic processes) modified by glacial ice and meltwater flows. Seasonal or multiyear ice cover influences many cold lakes and seas, but not all glaciomarine environments have ice cover, and not all lakes and seas with ice cover are glacial. For example, the southwestern Alaskan shelf is partly affected by glaciers and is considered to be a glaciomarine environment even though the sea never freezes. Conversely, Lake Baikal in Russia has a thick ice cover in winter, but is not

TABLE 8.1   Similarities (>>>) and Differences (>><) Between Marine
and Lacustrine Settings

Sea		Large Lakes
Denser saltwater, with many ions in solution.	>><	Lighter freshwater, few ions in solution. (Some small lakes may have saltwater.)
Generally larger basins, in most cases globally intercommunicating, with approximately uniform water level.	>><	Smaller, isolated basins, with independent water levels controlled by the elevation of outlets.
Restricted marine basins occur, such as landlocked seas and fjords, which are strongly affected by fluvial and glacial processes.	>>>	Large lakes may be hydrographically similar to small, restricted marine basins. (Small lakes are overwhelmed by glacial and fluvial processes.)
Wave action similar to large lakes.	>>>	Wave action similar to seas.
Tides are active in large intercommunicating seas.	>><	Tides do not exist. Recurrent changes in water level are due to wind action (seiche).
Currents are caused by wave refraction, tides, density differences (some due to freezing of salt water), and formation of cold, deep waters in polar seas. Some currents move globally.	>><	Currents are caused by refraction of waves and density differences. Currents are restricted to single basins.
Water stratification in cold seas with ice or glacial cover may be caused primarily by — freshening of the surface layer during melting — increase in salinity during formation of ice cover due to extrusion of salts — development of polar deep waters and related deep-sea currents — injection of fresh meltwater at different levels within the marine water column.	>><	Water stratification is related to — presence of ice cover, which reduces potential water mixing — temperature variations in the water column.
Biota is more diverse and abundant than in lakes.	>><	Biota is present, but not as abundant or diverse as in seas.
Glacial sediment transport and release mechanisms similar to large lakes.	>>>	Glacial sediment and transport release mechanisms similar to seas.
Calving, formation of icebergs, ice-rafted material, generation of dropstones, similar to large lakes.	>>>	Calving, formation of icebergs, ice-rafted material, generation of dropstones, similar to seas.
Suspended sediment plumes from meltwater streams may be lifted by the buoyancy of seawater, come up to the surface, and spread out.	>><	Suspended sediment plumes from meltwater streams may be dispersed along thermal discontinuities in the water column.

(continues)

156

TABLE 8.1   (*continued*)

Sea		Large Lakes
Cold freshwater flows must be sediment laden in order to sink into the saline water and develop density (hyperpycnal) flows along the bottom.	>>>	Cold freshwater flows do not need to be as heavily sediment laden as in the sea to be able to develop turbid (hyperpycnal) bottom flows.
Ice welding and pelletization of suspended particles similar to lakes.	>>>	Ice welding and pelletization of suspended particles similar to seas.
Flocculation of clays occurs except in embayments receiving much freshwater. No true varves can form in open seas.	>><	Flocculation does not occur, and true varves can form.
Rhythmic deposits (rhythmites) can form through turbidity flows (cyclopsams and cyclopels).	>>>	Rhythmic deposits (rhythmites) can form through turbidity flows.

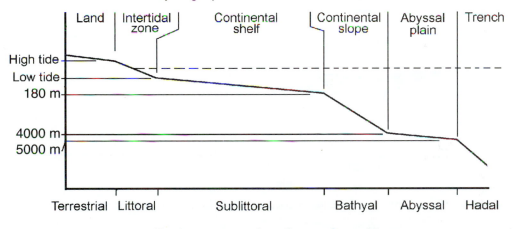

FIGURE 8.1   Diagrams showing marine environments in cross section.

now a glacial lake. Nonglacial sediments and features affected by ice in fluvial, lacustrine, and marine environments characterize the so-called *glaciel* (Dionne 1976).

Glaciers terminate in seas or large lakes in two main ways. They may have a floating terminus, as in ice shelves (nowadays essentially restricted to Antarctica), fringing ice shelves, and floating ice tongues, or a grounded terminus where the glacier mostly remains in contact with the bottom, typical of tidewater glaciers. The **grounding line** is the line (area) from which the glacier starts floating. It is located far inland from the terminus in ice shelves and is at or near the terminus in tidewater glaciers (Fig. 8.2). The terminal parts of some tidewater glaciers may experience recurring floating due to rising tides.

Floating ice shelves are now marine extensions of ice sheets; but during the Pleistocene, some may have been associated with thickening and grounding of sea

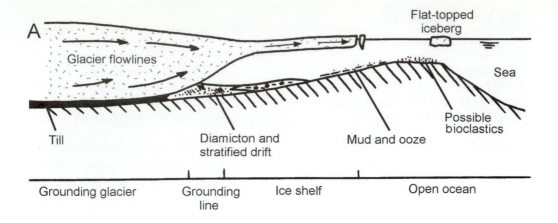

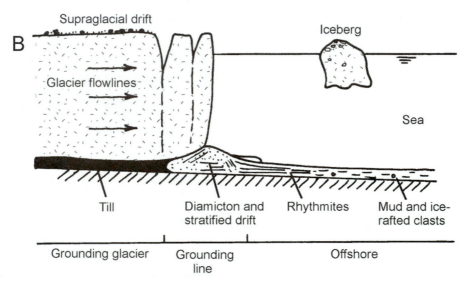

FIGURE 8.2 Diagrammatic representations of glacial terminations. *(A)* Floating terminus; *(B)* tidewater terminus.

ice on the shelves themselves, such as in the Kara Sea shelf in the Arctic. Conversely, large, thick ice sheets may ground in deep seas (order of hundreds to thousands of meters), as in Antarctica.

Tidewater glaciers are now most common in embayments and fjords. They are usually formed by temperate and sometimes by subpolar glaciers. Surges may expand temperate glaciers over large lakes or sea shelves, but they are quickly broken up and removed. This may have occurred frequently during the late Pleistocene in the Great Lakes and northeastern seaboard of North America.

Generally, a subaqueous terminus of an active glacier forms a cliff because ice is removed by calving into deep water faster than it is by surface melting (Figs. 8.2, 8.3). As the glacier becomes less active and its terminus partly stagnant, calving causes retreat of the terminus to the grounding line, surface melting rounds the profile, and the cliff changes into a glacial ramp (Fig. 8.4A–C). Eventually the terminus becomes landlocked (Fig. 8.4D).

FIGURE 8.3   Photograph of calving ice blocks from a tidewater glacier in southern Iceland.

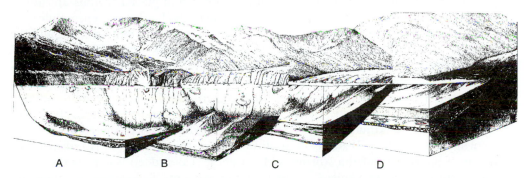

A   B   C   D

FIGURE 8.4   Idealized representation of terminations of tidewater glaciers. *(A)* Thick advancing glacier, grounding and bulldozing sediments. *(B)* Melting and retreating glacier with abundant meltwater developing ground-line fans. *(C)* Thinned-out glacier covered by debris, prone to subaqueous sediment slumping. *(D)* Thinned terminus has retreated inland, developing a sub-aerial outwash. (After Powell 1981.)

OCEANOGRAPHIC PROCESSES: BASIC PRINCIPLES

Oceanographic processes are those that affect large bodies of water, from oceans (oceanographic proper) to lakes (limnic). The effects of ice on these processes are analyzed here; glaciers and ice cover influence fluid movements, and thus sediment dispersal and deposition.

Surface **waves** are primarily generated by winds; so they are not effective under ice cover. In front of a calving glacier, waves are also formed by the impact of falling ice blocks. These waves and the induced longshore currents are particularly important in semi-enclosed basins like large lakes and fjords, where they may re-work glacial grounding-line sediment and sweep shallow coasts. Waves also affect the glacier terminus and icebergs by abrading them and aiding in melting and thus weakening the ice.

Tides are caused in oceans and interconnected seas as local waters are raised and lowered by the attraction of the Moon and the Sun as the Earth rotates around its axis. Tides entering a restricted area can generate strong currents capable of re-working bottom sediments. Coastal sediments can also be scratched by ice blocks (ice floes) moved by tides and currents. One interesting effect, called **ice shelf pumping,** is hypothesized to occur at the grounding line of ice shelves. That is, it has been observed that the large shelves of Antarctica experience significant vertical movement, presumably associated with tides. These movements may be translated into a bellowing effect in cavities at the grounding line, expelling water and sediment, and disturbing bottom deposits each time the ice is lowered. Landlocked seas and large lakes are not affected by tides, but recurrent water-level variations (seiches) occur, caused by wind pushing the water mass to one side of the basin or the other.

Currents are generated by wave refraction along coasts, by funneling of tidal water prisms, by local or global temperature or salinity variations, and by the rotation of the planet (Coriolis force, which has the effect of moving the fluid clockwise in the Northern Hemisphere and counter-clockwise in the Southern Hemisphere). Glaciers and perennial sea ice cover may act as a physical barrier, or modify salinity and temperature to change the water density gradients, and can thus turn off, trigger, or modify currents.

During interglacial periods, like the present one, glaciers terminating in the sea and perennial sea ice cover are restricted to high latitudes. This means that temperature and salinity gradients between the tropics and glacial regions are not as steep as they were during Pleistocene glacial periods. Under present conditions, wide-ranging currents sweep the world's oceans, governing global climate and the marine biological productivity of certain areas. Noticeable among these are warm tropical currents moving northward, such as the Gulf Stream and the North Atlantic current. These are partly balanced by cold southward-moving currents, such as the Greenland and Labrador coast currents, by local deep currents generated in cold shelves, and, mostly, by the deep ocean current associated with the polar-generated **North Atlantic Deep Water** (Fig. 8.5). Over 90% of the colder, denser, North Atlantic Deep Water exits southward from the Arctic and is redistributed over all the oceans of the Northern Hemisphere. Cold, deep water also forms in Antarctica, but its movement, unimpeded by continents, is governed by planetary rotation and is restricted to circumantarctic regions.

Other more localized currents are generated on shelves. Some develop in Antarctica, where sub-ice shelf waters become supercooled to about −2°C, and being denser, spill over the continental slope. There they can reach velocities in excess of 1 m/s and are capable of eroding and redistributing sediments. Similar local, cold currents develop in the Arctic, where shallow shelf waters have their salinity and therefore their density increased by the formation of sea ice cover. Formation of ice removes water from the system but not salt; hence the residual water is proportionally more saline. The flows of these more saline, denser waters have been recorded in several places, for instance from Foxe Basin into the deeper Hudson Bay, and from the shallow Barents Sea into adjacent deeper basins.

These currents could not have existed during glacial times when shelves were exposed or occupied by grounding glaciers that extended into deep, off-shelf waters. Furthermore, during deglaciation, large quantities of freshwater were added to the seas, freshening them and probably preventing the formation of the North Atlantic Deep Water, thus reducing or cutting off the associated oceanic currents. Conversely, the expansion of glaciers and sea ice cover during glacial periods may have triggered other types of currents and water mixing. In fact, the temperature gradients from

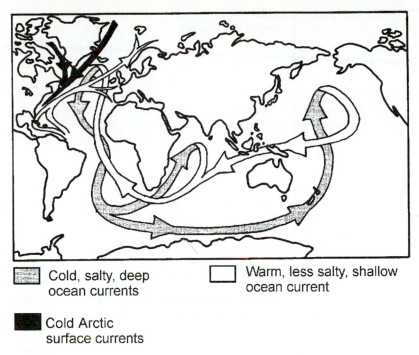

Cold, salty, deep
ocean currents

Warm, less salty, shallow
ocean current

Cold Arctic
surface currents

FIGURE 8.5   Schematic map showing selected large oceanic currents (cold deep
ocean current = North Atlantic Deep Water). (After Andersen and Borns 1994.)

the icy regions to warmer equatorial areas were steep; thus strong winds were generated that forced stronger wind-driven currents and turbulence.

Other marine and lacustrine density currents are generated by high suspended-sediment concentrations. These are derived from debris-laden stream flows or from dilution of sediment gravity flows such as slumps or debris flows that develop into turbidity currents farther from the source. Turbidity currents are almost frictionless, riding on a pressured basal fluid sublayer, and can travel widely, at times thousands of kilometers across entire large lakes and restricted marine basins. Depending on the size of sediment transported and its concentration, density currents may hug the bottom or become interstratified within the water column along thermal, density discontinuities. The sediments are later released from these plumes by settling.

Vertical density stratification of the water column as a function of temperature and salinity is an important property of the basinal environments and affects sedimentation and biota. These stratifications are particularly important in lakes but also occur in restricted seas, fjords, and estuaries.

In lakes, such stratification may not exist at all when full mixing (polymictic conditions: *poly* = multiple, *mictic* = mixing) occurs in shallow basins due to wind- or current-generated turbulence. In other cases, stratification may be short lived, such as when there is a sudden input of freshwater into a sea. In many well-stratified lakes, mixing occurs once a year during the summer (monomictic lakes) or twice a year during the spring and fall (dimictic lakes). In other cases, mixing rarely, if ever, occurs (amictic = nonmixing) during the life span of a lake. Amictic conditions occur only in polar regions where lake waters perennially covered by ice can be heated by solar radiation and conduction, and temperature stratification can develop and persist, such as in the ice-free areas of Victoria Land, Antarctica.

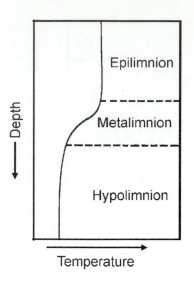

FIGURE 8.6   Generalized vertical section showing temperature stratification in lakes.

Stratified lakes are separated into two main temperature zones by a thermocline or thin zone (metalimnion) where there is a steep temperature gradient. The epilimnion is the upper few meters of the water column, and the hypolimnion is the layer below the thermocline (Fig. 8.6). A few degrees' difference between the various zones is sufficient to inhibit water circulation (mixing). In glacial settings, water stratification is greatly complicated by the presence of ice cover and the injection of cold glacial water with differing amounts of sediment, hence different densities, entering as tongues at different levels in the water column.

In seas, stratification is caused by differences in temperature, salinity, and suspended sediment concentration. The temperature effect is frequently associated with changes in salinity due to formation and melting of ice and to inputs of fresh meltwater from inland areas. Here, a number of points are relevant.

1. The surface salinity of polar seas changes due to formation and melting of sea ice. During freeze-up, salt is progressively concentrated in the seawater beneath the thickening ice cover because no salt can be accommodated in the structure of ice crystals. Conversely, the freshening of the surficial layer is caused by meltwater during warmer periods.

2. Marine stratification occurs in fjords that receive much fresh meltwater from glaciers and from the melting of trapped icebergs. Some fjords act as stratified estuaries with a fresh- to brackish-water upper layer and denser saltwater below. The lower layer of a partially restricted (barred) fjord that doesn't receive many turbid bottom flows seldom turns over (mixes with overlying water), and its waters may become anoxic (oxygen deprived), inhibiting colonization by aerobic organisms. This type of environment is characterized by the preservation of settled plankton, only part of which is consumed by anaerobic microorganisms, and by the formation of minerals such as pyrite (Fe_2S).

3. Stratification may also occur in some deepwater bodies, such as Hudson Bay, which may have a very slow turnover, with bottom waters having a residence time of several tens of years.

4. Upwelling of deep oceanic water may be localized in some areas and recur yearly or in longer cycles associated with global climatic changes, giving rise to the so-called *el niño* effect. This effect has repeatedly changed weather patterns and is responsible for anomalous high-frequency flooding and windstorms in certain areas and prolonged aridity in others.

Biological Factors

Biological activity varies greatly in cold basins depending on organic matter input from inland areas, abundance of terrigenous material, ice cover, and upwelling of nutrient-rich bottom waters. Open waters of polar continental seas are generally very productive. In certain arctic water bodies where strong currents and/or upwelling occur, the sea is open year-round, forming so-called **polynyas.** These have high organic productivity, attract fishes and sea mammals, and, accordingly, Inuit and other hunters. Antarctic waters are also highly productive and, for example, support huge populations of krill, which directly and indirectly feed numerous other organisms, including whales, penguins, and seals. Bottom fauna consists of echinoids, bryozoans, foraminifers, mollusks, occasional brachiopods, and worms. Some organisms encrust the hard bottom, and others burrow into sediments. Flora is present both on open coasts and offshore. Of particular importance are the annual or biannual blooms of silica-secreting diatoms.

Minerals

Mineral distribution in seas reflects the physico-chemical and biological conditions of surrounding lands and of the marine settings themselves, as indicated by the following few examples.

Clay Minerals and Feldspar   The relative distribution of clay minerals in modern oceans and ancient marine deposits can be used as a proxy (indirect) indicator of cold conditions. The clay minerals present depend primarily on substrate composition in the source area and the prevalent type and intensity of weathering, thus on the type of clay minerals existing inland. Areas with predominantly volcanic rock undergoing strong chemical weathering under warm climate produce a lot of montmorillonite-type clays. These are relatively abundant in the South Pacific Ocean. Similarly, kaolinite develops preferentially in highly weathered tropical areas, and abounds in tropical seas. Illite is the dominant clay in sedimentary bedrock and is common throughout seas that have a strong influx of terrestrial material at mid-latitudes. Cold northern seas are relatively abundant in chlorite derived from the surrounding less intensely chemically weathered crystalline rocks of the Precambrian shields of North America and Europe. Similarly, it may be possible to discriminate between deposits containing relatively fresh feldspar, eroded, transported, and deposited quickly or in cold-climate regions (therefore less weathered); and deposits containing highly weathered feldspar, which has undergone lengthy transport under warm climatic conditions.

Silica and Carbonate   A considerable amount of silica and calcium carbonate is produced in seas and some lakes through biogenic processes. The distribution of oceanic bottom sediments reveals that calcium carbonate prevails in tropical to temperate regions, whereas silica is more abundant in polar regions. This is partly due to the prevalent production of silica skeletons in polar regions and calcareous skeletons in warmer areas, and partly due to the fact that calcite is more soluble in cold waters and under high pressure. So, for example, once the small calcite skeletons of planktonic organisms lose the organic protection of the living organisms, they are readily dissolved in cold polar waters or everywhere they settle down into cold, deep, high-pressure abyssal zones. These areas are deprived of calcium carbonate minerals and, by default, enriched in silica. Particularly high concentrations of primary silica can develop where annual or biannual diatom blooms occur.

Calcium carbonate accumulations derived from benthic organisms occur on cold shelves that receive little or no terrigenous material. These calcareous deposits generally consist of reworked accumulations (coquina) of mollusk shells. In some areas, buildups of bryozoans and worm tube reefs develop as well. Below the high-latitude thermocline, coral reefs with an associated diverse fauna can form. These can occasionally be mistaken for tropical reefs in ancient deposits.

Glendonite   Newly formed minerals may be diagnostic of cold-climate conditions as well. Glauberite ($Na_2Ca(SO_4)_2$), for example, forms under present freezing conditions, and its pseudomorph glendonite is commonly found in ancient glaciomarine rocks.

GLACIAL SEDIMENT TRANSPORT, RELEASE, AND DEPOSITION

Sediment transport and release from glaciers terminating in large bodies of water are similar to that of any other glacier, except that much sediment is released by bottom melting and by removal in icebergs (calving). Another fundamental difference is that stagnant ice does not exist at sea or deep-lake termini; thus ablation till does not form except locally during the final wasting of tidewater glaciers, when some subaqueous ice may get buried under debris (Fig. 8.4C).

Unlike landlocked glaciers, the flowlines of a floating terminus are not concave upward, but are essentially horizontal with a slightly downward component (Fig. 8.2A). This reflects the replenishment of ice lost at the snout by calving and from bottom melting. In fact, under ice shelves, salty seawater is at or just below the freezing point, and this, together with abrasion by currents, causes melting and release of bottom sediments. Large ice shelves have vast surfaces and could potentially lose a lot of surficial ice as well. Most ice shelves, however, receive heavy snowfall and are invariably formed by cold ice in cold settings. Although sublimation occurs, there is little melting, and relatively little ice is lost from the surface. Most of the sediment of ice shelves is indeed released within a few kilometers of the grounding line by bottom melting, unless sealed into the glacier by bottom freezing of oceanic waters. By contrast, surficial melting is an important process in tidewater glaciers of cold-temperate regions like Alaska. There, sediments released by melting may slump and fall out directly into the sea or wash into crevasses. In fact, the terminal parts of these glaciers are intensely fractured, and meltwater carries a lot of supraglacial drift into englacial or subglacial positions. From there, it exits through englacial or subglacial tunnels near the water line. Furthermore, whereas some englacial drift may be exposed on top of the glacier by surface melting, essentially no subglacial drift reaches the surface because no thrusts exist. That is, the snouts of calving glaciers are primarily extensional rather than compressive settings.

Much sediment carried into the basin by the glacier and released at the grounding line of subaqueous termini, or sediment carried by glacial meltwater streams to the shore, is locally redistributed by sediment gravity flows, waves, and currents. Some sediment is carried farther offshore by ice streams, and some is dispersed farther still by icebergs, seasonal ice rafts, and marine currents.

Ice Rafting

Ice-rafted debris reflects cold, but not necessarily glacial conditions, because sediment of any size can be rafted by both icebergs and seasonal ice rafts (floes). Fine-grained material, as well as boulders up to tens of meters in diameter, is transported to shelf and offshore areas where little coarse-grained, clastic material would otherwise be deposited. In deep oceanic deposits, any sediment coarser than 63 microns (μ) is considered to be rafted. However, details of sorting and lamination must be analyzed, so as not to mistake better-structured turbidite deposits for poorly sorted and structured ice-rafted deposits. The principal evidence for ice-rafted sediment is (1) the textural discontinuity whereby particles of strongly different size and shape may be found together in the same deposit, indicating different processes of transport; and (2) the presence of disrupted laminae caused by clasts (dropstones, lonestones) falling into finer, water-saturated bottom deposits (Fig. 8.7). The mixture of poorly sorted, coarser-grained, ice-rafted material and local fine deposits may also form diamicton in oceanic settings.

The amount of material rafted by icebergs varies depending on the origin of the iceberg and whether the glacial or interglacial periods are considered. (1) During glacial periods, many glaciers terminated in the sea, and many icebergs were generated. However, icebergs derived from ice shelves carry very little sediment; most is carried by icebergs derived from ice streams or ice lobes. During these periods, ice-rafted material could be distributed to low latitudes. (2) During interglacial periods, fewer glaciers terminate in the open sea, but many reach the sea in fjords. Less material is rafted and generally only at high to mid-latitudes. Rafting by seasonal ice, however, is common during these times.

Although interpretation of the data requires caution, deep-sea sediment cores from certain areas, such as between Greenland and Iceland in the North Atlantic, show a cyclical vertical variation in grain size. This variation (the so-called **Heinrich events**) occurs apparently in unison with glacial and interglacial periods. The coarser-

FIGURE 8.7   Photograph of large dropstone in Paleozoic glaciomarine deposits, Brazil.

grained layers are interpreted to represent glaciers extending over the shelves and producing more ice-rafted material, and the finer-grained layers to represent inter-glacial periods with little iceberg rafting.

Discriminating between iceberg and seasonally ice-rafted material is not always simple, either in Pleistocene or more ancient deposits. But the composition and texture of the deposits may help. Iceberg-rafted material mostly comes from glacial drift and should, when compared to seasonal ice-rafted material, (1) have a more variable composition, (2) contain distal erratics, (3) be poorly sorted, and (4) have numerous angular or glacially imprinted pebbles. Although shore-ice rafts and icebergs both gouge their substrate, icebergs may leave deep diagnostic scour marks. In the northern Labrador Sea, for instance, iceberg-generated, linear to curvilinear gouges and craterlike forms have been observed with average widths of 30 m, a mean scour depth of 5 m, and lengths in excess of 3 km. Iceberg scours are commonly found to depths of 200 m, but even exist to depths of 600 m, indicating formation by very large icebergs during periods of low sea level.

Sediment Plumes

Cold meltwater entering warmer lakes, or sediment-laden flows entering lakes or seas, may hug the bottom as density currents, disperse along stratifications, thermal or otherwise, within the water body, or rise to the surface (Fig. 8.8). Density currents lose velocity and competence as they travel away from their injection point and release sediments as a result. The sediments of the plumes eventually settle according to particle size and density, partly affected by chemical and biochemical processes in the water column (Fig. 8.8A). Sand and silt particles settle faster and closer to the sediment injection point. Fine clay-sized particles have low settling velocity and cannot settle as long as there is water stratification, waves, and currents. Settling of these small particles occurs when they cluster together and behave as larger single particles. This can happen through ice welding (formation of ice-welded pellets), flocculation, agglomeration, and pelletization.

Ice welding of fine particles to form pellets is a common occurrence in glaciers, meltwater, and surficial basinal waters. Pellets of till, clay, and silt can develop and behave as coarse particles.

Flocculation involves clay minerals. The flat clay mineral particles have unsatisfied electrical (ionic) charges at their edges. These negative charges cause the particles to repel one another, thus keeping them dispersed as long as they are in a freshwater environment where there are no sufficient ions to attach to the clay particles and neutralize them. Dispersed clay particles are found in lakes and in the surface freshwater layers of seas, or near meltwater injection points. However, as soon as the clay particles enter saline waters containing positive ions like Ca^{++} and Na^{+}, the negative charges are neutralized, the clay particles no longer repel one another, and they can cluster together. This process of flocculation forms coarser pellets that can settle rapidly to the bottom.

Agglomeration occurs when fine particles are covered by algae or other organic matter. The organic matter coats charged surfaces, thus impeding repulsion, and particles can bind together. As with flocculation, agglomeration forms coarser clasts that can settle to the bottom.

Pelletization occurs when fine particles and their organic coatings are ingested by planktonic organisms and re-expelled as fecal pellets coarse enough to be able to settle to the bottom.

All these processes occur to various extents in both lakes and seas, except for flocculation, which requires salty water. So in lakes, coarse-grained sediment entering

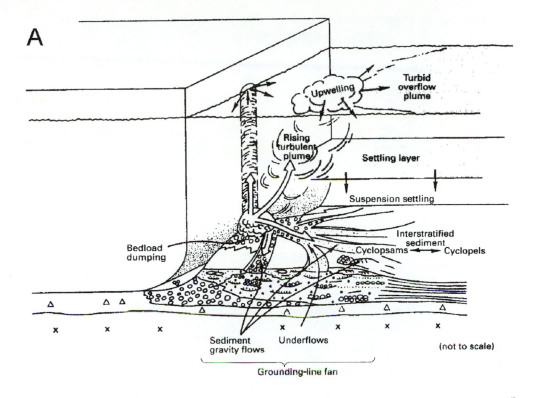

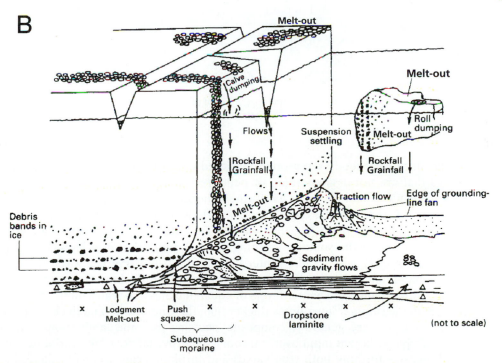

FIGURE 8.8   Schematic diagrams representing processes and deposits at the grounding line of a tidewater glacier. *(A)* Glacier with much meltwater developing a grounding-line fan. Depending on the suspended sediment load, the power of the meltwater jet exiting glacial cavities, and the thermal stratification in the basin, the plumes may rise vertically or diagonally through the basinal water column. *(B)* Slightly advancing glacier without much meltwater developing a morainal bank. (From Powell and Domack 1995; modified after Powell and Molnia 1989.)

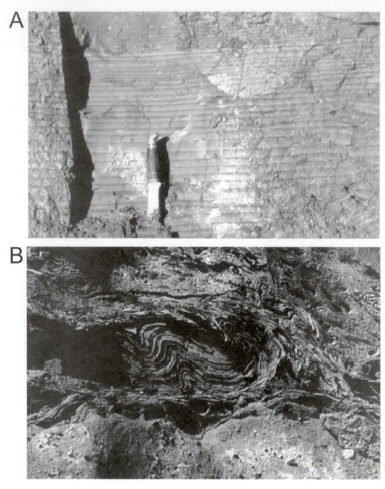

FIGURE 8.9 Photographs of varves and rhythmites of the Great Lakes of North America. *(A)* Regular glacial varves. *(B)* Deformed packets of rhythmites over basal till.

the basin during the ice-free season is deposited immediately, but fine-grained clay particles that are not agglomerated or pelletized remain in suspension. It is only during the winter when ice cover has formed, thermal stratification disappears, and water motion is reduced that these fine clay particles can settle. This process generates annual cold-lake deposits called **varves.** These are characterized by a coarse layer formed during the summer (melting season) and a thin clay lamina formed during the winter (Fig. 8.9). The winter lamina of a true varve is always very fine clay, and the summer layer is silty or sandy. Varves vary in thickness from a few centimeters, with summer and winter layers of equal thickness, to scales on the order of meters, with the summer layer much thicker than the winter one. The thicker, coarser, summer layers are generally composite, showing multiple laminations, which reflect multiple sediment input events through hyperpycnal flows and turbidite currents. Processes forming both true varves and turbidites can occur in a lake, but varves do not occur in the sea, so true varves are strictly an indication of a cold lake.

Two other types of laminated deposits form in front of and near (within 1–2 km) marine tidewater glaciers. These may be partly related to the above processes, but are also related to tide effects during settling of material from suspended sediment plumes. During low tides, the coarser-grained, sandier type (**cyclopsams**) is formed in more proximal areas, and during high tides the finer, siltier type (**cyclopels**) develops everywhere, but prevalently in more distal areas. Mistaking turbidites for exclusively marine tidal laminites, or exclusively lacustrine true varves, may lead to (and has in the past led to) incorrect paleogeographic reconstructions. Microscopic, mineralogical analyses of the fine-grained laminae and recognition of other characteristics, including fossils, can aid in correct identification of the processes involved and hence differentiate lacustrine from marine environments of sedimentation.

GLACIOMARINE AND GLACIOLACUSTRINE DEPOSITIONAL SYSTEMS

Glaciomarine and glaciolacustrine deposits may be related directly or only indirectly to glaciers. One can speak of **glacier-related depositional systems,** where the sediments derive directly from a glacier, and **land-related depositional systems,** where the sediment derives from overland drainage, possibly tied to meltwater, but still indirectly affected by glaciers (Fig. 8.10).

Glacier-Related Depositional Systems

The most typical subaqueous glacial sediment accumulations develop at the grounding line (grounding-line fans, morainal banks, and grounding-line wedges), and some occur under sea ice shelves. Each deposit is characterized by diagnostic associations of lithofacies (sediment type) imprinted by the prevalent environmental processes acting at those localities. Good examples of ancient deposits of this difficult-to-observe environment occur in southeastern Ontario, southwestern Canada (former glacial Champlain Sea).

Grounding-line fans develop as a jet of meltwater and sediment exits a glacial tunnel to form a fan-shaped, outwash-type deposit (Fig. 8.8A). The deposit is characterized by coarse-grained, proximal material deposited from the main jet and locally reworked by gravity flows, rapidly grading laterally into finer-grained materials deposited from turbid underflows hugging the bed and settling from suspended

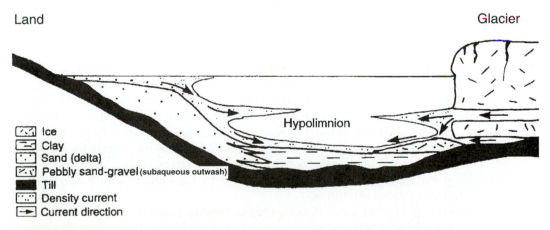

Land   Glacier

Ice
Clay
Sand (delta)
Pebbly sand-gravel (subaqueous outwash)
Till
Density current
Current direction

Hypolimnion

FIGURE 8.10   Model of land-related and glacier-related systems in a large lake or fjord; turbid plumes may become interstratified with basinal water along density discontinuities. (After Gustavson 1975.)

plumes. The most proximal area may have backward-facing slopes with layers inclined toward the glacier (backsets). The rest of the proximal zone is characterized by massive to plane-bedded and cross-bedded gravel, sandy gravel, and sand. Local synsedimentary deformation structures and slumps may occur. These proximal deposits may laterally interfinger with ripple cross-laminated sand as well as sand, silt, and clay laminites of various types, formed by underflow or, in seas, from settling from suspended plumes modulated by tides (cyclopsams and cyclopels). The grounding-line fan grades farther offshore into fine-grained silt and clay laminites partly generated by turbid underflow and partly by sedimentation from suspended sediment plumes. Some of these fans may be deformed and pushed by local glacier readvances. Although different in detail, these fans develop both in lakes and seas, and they have been previously variously named deltas, eskers, beaded eskers, subaqueous outwash, and glacier-contact fans.

Morainal banks are ridges similar in origin to terrestrial end moraines and push moraines (Fig. 8.8B), but they are formed subaqueously and may have more chaotic deposits (diamicton), including water and mass-flow reworked gravel, sand, and mud. Pushed or squeezed materials in front of the glaciers form small ridges, which can be related to some types of cross-valley moraines (Fig. 6.11B), De Geer moraines (Fig. 6.11C), and washboard moraines.

Grounding-line wedges (also called till deltas) form in front of the grounding line. Deformed subglacial till is reworked by mass flow into a gently sloping sediment accumulation, and this apron may be overlain by rain-out and basal till complexes when overridden and partially eroded by a glacier whose grounding line is prograding seaward. These deposits are still poorly known, but there is seismic evidence that grounding-line wedges exist under the present Ross Ice Shelf (Antarctica), and they have been tentatively recognized in Pleistocene lacustrine deposits of North America and in glaciomarine successions where they were named **till tongues.**

Subglacial shelf. Relatively little is known about the subglacial shelf. It is assumed to contain mostly mud, and, in the first one or two kilometers seaward from the grounding line, material released from the bottom of the ice shelf as it melts from below. Where this melt-out material is relatively important, a diamicton may develop on the shelf, partly laminated due to settling of the material through the water column and partial reworking by marine currents. Marine currents may further rework parts of these deposits to the extent of generating local lags of coarser-grained terrigenous or bioclastic materials.

Land-Related Depositional Systems

Land-related depositional systems include coastal and shelf deposits. The cold-climate deposits are similar to those of other settings, except for the influence of ice. The peculiarities of cold-climate environments and deposits are emphasized here.

Deltas. Large lakes and seas are the ultimate basins of deposition. When a stream enters a large body of water, it loses competence and releases its sedimentary load, the coarser fraction closer to the entrance and the finer particles farther away. These sediments may be further redistributed by waves, tides, and currents, or may slump and be transported into deeper areas by subaqueous sediment gravity flows, primarily dense underflows.

Deltas vary in form, depending on the size (discharge) of the river, the amount and type of material it carries, the slope of the basin, and the intensity of waves, currents, and tides that rework the incoming sediments (Fig. 8.11). A large river carry-

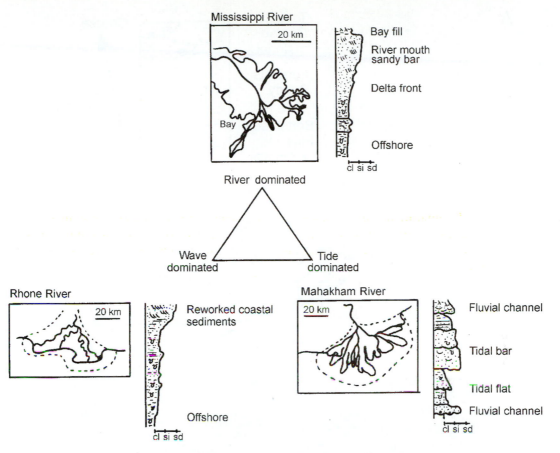

FIGURE 8.11 Schematic diagrams of principal types of delta. Plan view and idealized vertical section of end member deltas: river-, tide-, and wave-dominated. (After unknown; see website for update.)

ing a lot of sediment in a restricted basin deposits the sediments at the river mouth, and rapidly progrades onto its own deposits, generating a typical birdfoot delta, like that of the Mississippi River in North America. A river flowing into a large basin where strong waves actively redistribute sediments generates an arcuate delta, like that of the Niger River in Africa. A river that flows into a sea with large tides capable of entering distributaries and removing sediments forms an estuarine delta, like that of the Brahmaputra River in Asia. Deltas differ in detail, but all show a fining in particle size from the river mouth outward and, in marine environments, a change from fresh to brackish water near the river mouth grading to normal salinity offshore, with a related distribution of fauna and flora. By basinward progradation of the system, this lateral variation is translated into a vertical sedimentary succession. A typical vertical section would, thus, show fine offshore deposits grading upward into delta-front silts and sands, locally slumped, grading upward into coarser-grained delta-front sands, capped by even coarser, cut-and-fill, distributary-channel deposits. The capping distributary channel deposits may be locally interbedded with coastal, marsh, or tidal-flat deposits, depending on the type of delta involved.

The prototype sedimentary architecture of a fluvial-dominated delta in which deposits have not been reworked much by basinal processes is the so-called Gilbert

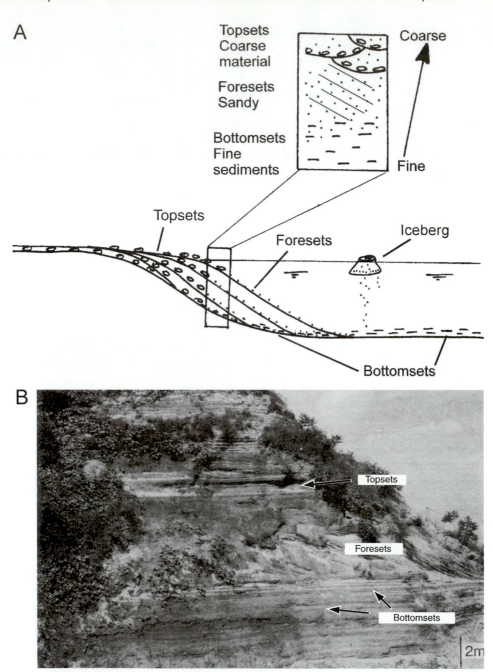

FIGURE 8.12 Representations of Gilbert-type delta. *(A)* Diagrammatic model. *(B)* Photograph of a small Gilbert-type delta in a late Pleistocene, coarsening-upward, glaciolacustrine succession (Scarborough Bluffs, north shore of Lake Ontario, Canada).

delta, first described in large Pleistocene lakes of the western United States. Gilbert deltas have subhorizontal silts and clays offshore, grading toward shore into inclined (up to about 33°), coarsening sands and gravels, which in turn grade into subhorizontal, coarse-grained, fluvial gravels and sands at the top (Fig. 8.12). The resultant vertical section shows, from the bottom up, layers changing from fine-grained bottomsets into upward coarsening, inclined foresets and finally to subhorizontal topsets with fluvial deposits. This tripartite distinction of the delta persists in almost every other delta, although the geometry may become somewhat obscured because of intense wave reworking. In any case, a deltaic body is usually recognizable because of the outward (hence vertically in cross section) changes in deposits, varying from cross-bedded sands and, locally, gravels of the distributary channels (topsets), to sands and silts of the delta front (foresets), to basinal silts and clays (bottomsets). The foresets may be affected by slumps or deformation even on gentle subaqueous slopes of only a few degrees.

Deltas in cold climates have these same features but with some peculiarities:

1. Deltas do not occur and beach and shoreface settings are rare in Antarctica because there is little or no overland running water, the coasts are mostly bounded by ice or are rocky, and the shelf is deep (about 500 m).

2. Glacier-fed, meltwater streams carry a lot of sediments throughout the melting season and, in many cases, braided outwash discharges directly into the basin. This leads to the development of braided deltas, which have multiple, closely spaced, and irregularly active sediment point sources. These deposits retain the overall deltaic architecture, but with more irregular stratification and an abundance of sandy gravels.

3. Cold-climate rivers have high flows and carry large amounts of sediment for only the brief ice breakup period. So their total sediment input into lakes and seas is generally small relative to other river systems. However, the energy of the receiving basins is also small since they are covered by ice for most of the year. Waves, tides, and currents capable of reworking the sediments can act only during the open season, primarily during the few pre-freeze-up (autumn) storms. However, sediments may be turbated and partly reworked by grounding ice blocks moved by wind and currents. Rarely do cold-climate deltas develop birdfoot morphology; most commonly they are arcuate or funneled (tidal) shape.

4. River mouths are greatly affected by ice rafting, scouring, and sediment reworking. The shallow banks of rivers and nearby shores are generally boulder strewn, and sandy bars and shoals with disseminated ice-rafted boulders and pebbles form on the delta top. In shallow tidal seas such as James Bay, the delta top forms a wide, shallow (few meters deep), sandy platform, locally dissected by shallow river flood channels (Fig. 8.13).

5. The deltaic sediments are not much bioturbated, as they contain little bottom fauna.

6. The deltas may be bordered by grassy marshes where the fauna consists primarily of numerous insect larvae, small snails, some worms, and rare pelecypods.

7. The rivers carry a lot of nutrients to the sea, where they are utilized by plankton and small benthic organisms, which attract fishes and in turn seals and white beluga whales. Beluga whales also use the estuarine areas of northern rivers during molting to scratch off their old skin on gravelly beds.

Beaches. Nearshore settings proximal to tidewater glaciers are similar to those of other cold coastlines, except in areas of active iceberg calving. In the latter case, shores may be affected by large waves due to calving ice blocks, and by grounding icebergs. Present-day cold-climate beaches develop at intermediate to high latitudes. The areas previously occupied by glaciers may contain a considerable amount of coarse debris, and the beaches are typically gravelly. Gravelly beaches do, in fact, develop preferentially in previously glaciated areas as well as tectonically and volcanically active regions, where coarse material can be transported to the coast.

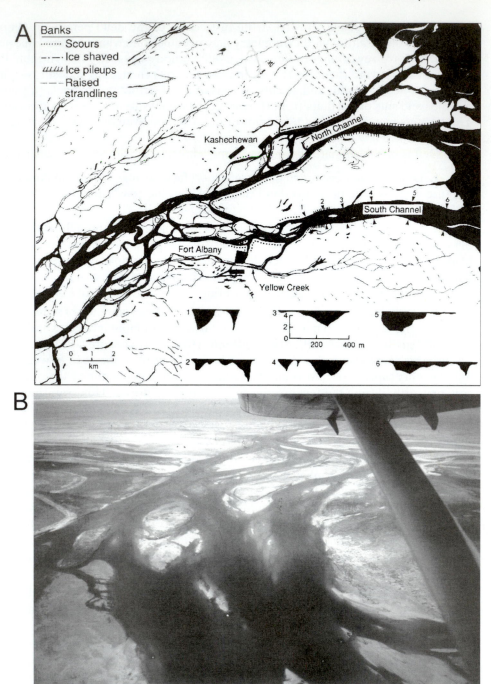

FIGURE 8.13 Representations of subarctic deltas. *(A)* Albany River, James Bay, Canada (compilation from air photographs; from Martini, Kwong, and Sadura 1993). *(B)* Photograph of an anastomosed river mouth, James Bay.

Coastal processes in cold, polar seas and cold lakes are similar to those in other areas, except for the presence of seasonal ice cover, and ice blocks and icebergs moved about by marine currents and winds. For most of the year, ice cover of the water or an accumulation of ice at the foot of the beach (ice foot) drastically reduces the wave and tidal energy expended on shaping the shores. This leads to an

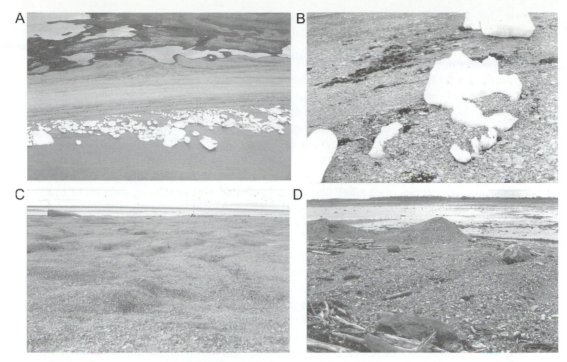

FIGURE 8.14   Photographs of arctic and subarctic beaches. *(A)* Ice floes (blocks) pushed against shore by wind (Foxe Basin, Canada). *(B)* Blocks of ice half buried in gravel (Foxe Basin). *(C)* Hummocky, irregular topography of a gravel beach, due mainly to ice push and ice-block melting. Some local disturbance may be due to Inuit who bury walrus meat to mature (Foxe Basin). *(D)* Conical mounds of sandy gravel formed by material washed from the surface of ice floes in between ice blocks or holes within the ice. Note ice-push boulder (James Bay, Canada).

apparent contradiction, because cold-climate beaches are low-energy systems (when the annual cycle is considered), yet they are generally coarse grained, mainly gravelly. Nothing happens on beaches when their sediments and surface are completely frozen (formation of an ice foot). Beaches can be reworked only during the short ice-free period when waves can impinge onshore. Furthermore, beach deposits are scoured and pushed by grounded ice rafts, and locally modified by localized release of the ice foot.

Going from arctic to subarctic to cold-temperate beaches, there is an overall irregular, but detectable, decrease in particle size, better particle rounding, and, reflecting an increase in sand, better and more frequent development of coastal dunes. Beaches in very cold, high-latitude areas (Fig. 8.14) (1) may locally develop the beachface zonation of cold-temperate gravelly and sandy gravelly settings, but only in embryonic fashion; (2) have generally angular clasts; (3) show scours, pitting, and conical deposits, all related to the grounding of ice blocks or to material deposited by waves on shorefast ice during breakup that is released as the ice melts; (4) may be modified by cryoturbation (a frozen-ground process) and develop cracks, reorientation of flat clasts into circles, solifluction lobes, and other features; and (5) have few organic remains of a few mollusk and mammal species.

Tidal flats develop along the shores of tidal seas where wave action is subdued because of shallow, wide shores, or where protected by barriers. The recurring (diurnal or semidiurnal) onshore pileup of seawater distributes sediments according to the power of the tidal currents and wave reworking. The power of the tidal water-wedge fluctuates from a minimum during low and high tides (slackwater periods), to a maximum during intermediate tidal stages. Note for example that the power of the ebb (decreasing tide) flow starts at zero when the tide is high and increases as the water accelerates down the gently sloping tidal flat as sea level drops. As water depth on the flats decreases, it may be accelerated by strong coastal winds, and supercritical flow regimes may be achieved. These flows may be able to plane off bedforms, such as sand ripples, formed under deeper-water, lower flow regimes. Note also that if the slope of the flats is steep enough, or if the flats are partially barred, channels develop in the lower part of the flats.

Everywhere, the lower part of the tidal flat has the highest kinetic energy, because it is subjected to the fastest ebb tidal flow as well as prolonged wave activity. The kinetic energy of the flat decreases landward and reaches a minimum at the coastal marshes. Those upper parts of the flats are inundated only during the highest tides, and incoming and outgoing tidal flows, normally slow, are further retarded by vegetation growing there. In fact, this upper part remains exposed for a longer time during each tidal cycle, and vegetation can colonize the upper tidal flat and flourish in the lower marshes. Low kinetic energy means that, under normal conditions, only the finer-grained particles in suspension are transported there. The fine particles settle during slack water at high tide. The deposited mud is cohesive and not readily removed by the slow-moving ebb flow. Moreover, the fine sediments are rapidly colonized by plants that further trap the mud. A consequence of all this is a net accumulation of mud in the upper parts of tidal flats and marshes and deposition of coarser sand and even gravels in the middle and lower parts. A fining landward sedimentary succession thus develops.

Tides and waves imprint the deposits. For example, tides characteristically lead to generation of recurring mud laminae formed during slack high-water periods, and tidal currents and waves characteristically develop, respectively, asymmetric and symmetric sand ripples.

Paralleling the lithofacies zones is a biofacies (flora and fauna) variation. The biofacies zones grade landward from a very productive, mollusk-rich, shallow subtidal zone; to an intertidal zone rich in burrowing mollusks, worms, and crustaceans; to zones in the upper tidal flats characterized by numerous snails that consume the growing algae; and to increased colonization by plants and insect larvae in marshes. The type of intertidal flora and fauna depends on climate and the salinity of the particular area. Large wetlands and peatlands may develop inland from these tidal flats, such as to the southwest of James Bay and Hudson Bay in Canada. It is the succession of the sedimentary facies (various sediments, structures, and faunal assemblages) that characterizes recent and ancient tidal flat deposits, rather than any single feature.

The tidal flats of cold seas have the same overall characteristics as those elsewhere, except for (1) their limited distribution, and (2) some characters specific to cold settings and the presence of ice (Figs. 8.15, 8.16).

1. Tidal flats develop in cold, low-arctic to subarctic seas with open coasts, such those of James Bay and Hudson Bay, parts of the Labrador coast in Canada, and in embayments such as in southeastern and southwestern Baffin Island. Farther north, tidal flats are poorly developed because there is little fine-grained sediment, and tidal action is strongly reduced by ice cover for most of the year.

2. The intertidal zones of cold tidal flats are generally scoured by the keels of ice rafts and ice-pushed boulders. Ice rafting during breakup forms numerous erosional and depositional features, and local unsorted (diamicton) deposits may

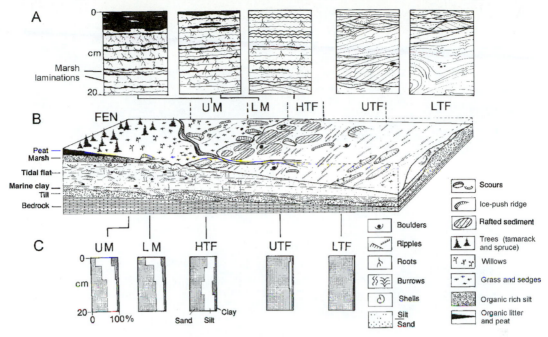

FIGURE 8.15   Subarctic tidal flat (west coast of James Bay, Canada). *(A)* Structures seen in shallow box cores. *(B)* Sedimentation model. *(C)* Vertical particle-size variation in shallow box cores. (Horizontal scale = order of kilometers; vertical scale = order of meters; UM = upper marsh, LM = lower marsh; HTF = high tidal flat, UTF = upper tidal flat, LTF = lower tidal flat.)

develop. Where there is a mesotidal range (2–5 m between high and low water), or macrotidal range (in excess of 5 m), large boulders can be rafted to form boulder barricades near the low water mark. Where fine-grained material is present, the lower bouldery and sandy parts of the intertidal zone grade shoreward into wide (order of kilometers), thick (order of meters), muddy upper tidal flats. These become colonized by plants, such as the grassy *Puccinellia phriganodes, Carex subspathacea,* and *Hyppuris tetraphilla,* and develop permafrost. In the arctic, permafrost may also be present in intertidal areas and shallow (less than 2 m) offshore waters. Its presence limits the amount of sediment that can be readily reworked by waves and ice erosion.

In an idealized transect from the arctic to the subarctic and to cold-climate regions of the Northern Hemisphere, there is an overall increase in sandy and muddy tidal flats. There is also a shift in the colonization patterns of burrowing organisms (worms and mollusks, commonly *Macoma arctica*). They are mostly concentrated in the subtidal zone in the arctic, where they are represented by only a few species. To the south they expand to colonize the whole intertidal zone, and there is an increase in number of species. This reflects the harsher arctic conditions where the tidal flats are covered for most of the year by ice and are scoured during breakup. In addition, waters can change drastically to become highly saline during freeze-up and brackish during breakup and ice melting.

Ice-rafted material, diamicton lenses in otherwise sorted ripple cross-laminated sands, ice scouring, cold-loving organisms, cryoturbation features, and stratigraphic position over glacial deposits are diagnostic features of cold, tidal-flat deposits.

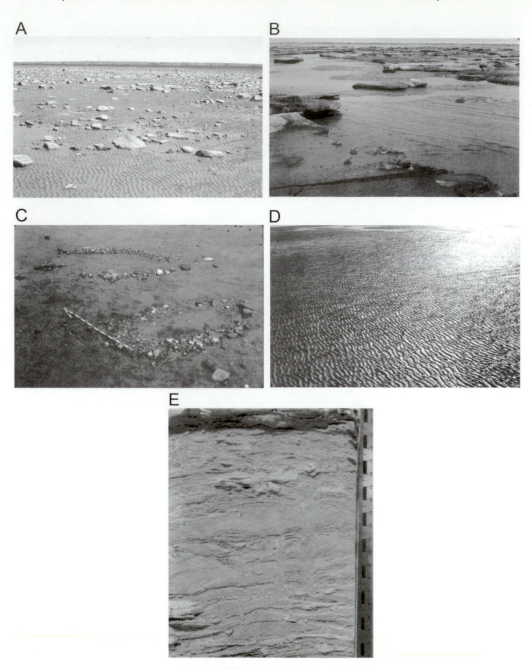

FIGURE 8.16 Photographs of subarctic tidal flats (James Bay, Canada). *(A)* Sandy tidal flats with disseminated ice-rafted boulders. *(B)* Ice blocks peeling off and scouring tidal flats during spring breakup. *(C)* Ring of stones deposited from an ice floe (block). *(D)* Sandy tidal flats with ripple marks and few boulders. *(E)* Vertical view of box core of a sandy tidal flat showing ripple cross-laminations, mud drapes, and burrows by the mollusk *Macoma balthica*.

Shoreface environments extend from the base of the beach or tidal flat to a depth where storm waves may affect the bottom (Fig. 8.17A). The effect of waves and wave-generated longshore currents decreases in intensity from the upper to the lower part of the shoreface. This is shown by an overall particle fining from the

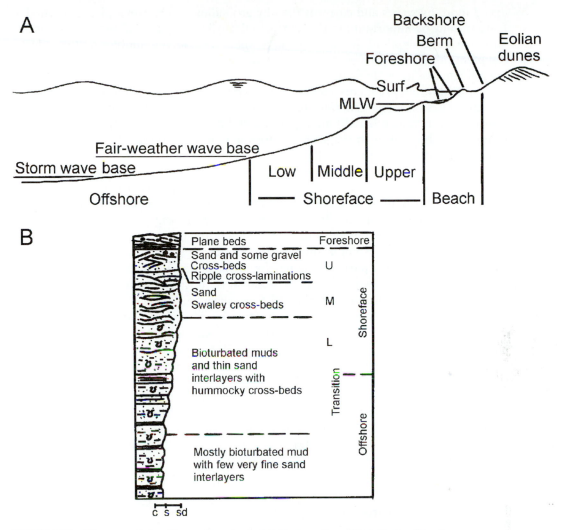

FIGURE 8.17   Diagrams of shoreface environments. *(A)* Cross section; *(B)* vertical sedimento-logical model. (MLW = mean low water level, U = upper shoreface, M = middle shoreface, L = lower shoreface; c = clay-size particles, s = silt-size particles, sd = sand-size particles.)

beach outward, and in the development of subaqueous sand dunes and ripples in the nearshore area, and fine sand swaley or gently inclined laminae (called **hummocky cross-lamination** [HCS]) in deeper parts. Parallel to this is a well-defined variation in organisms living on and burrowing into various parts of the shoreface sediments. In areas where the coast is prograding, this lateral relation is translated into deposits showing a coarsening upward sequence (Fig. 8.17B).

Cold-climate seas develop coastal deposits similar to those of other seas, except that they may be intensely disturbed by ice piling against the coast (shorefast ice). The presence of ice also reduces the influence of waves. A model has been derived from the Beaufort Sea in North America that divides the shoreface and adjacent shallow shelf into four zones: (1) A **nearshore zone** (upper shoreface) to 4 m water depth is dominated by waves during the summer and ice push and freezing during winter. (2) The **inner shelf** (middle and lower shoreface) seaward to the boundary between the shorefast ice and the polar pack ice, at about 20 m water depth, is influenced more by waves and currents than by grounding ice. However, ice-generated features may be important in this zone as well. (3) The **outer shelf,** between 20 and 50–70 m water depth at the shelf break, is where ice gouging jumbles up the older sediments and homogenizes the gravelly mud. (4) Farther offshore in deeper water the influence of ice gouging decreases.

Offshore environments include the deeper parts of the shelf and the basin troughs. These regions generally receive fine terrigenous sediments blown in, or transported in suspended sediment plumes, and biogenic material from organisms living at the bottom or in suspension. Organic-rich, fine-grained mud is the most typical deposit of these settings. In places, though, sandy material is transported offshore by sediment gravity flows such as hyperpycnal flows and turbidity currents. The turbid underflows can generate thick, laminated deposits (rhythmites, turbidites), particularly in large lakes and small seas where much sediment is derived from glaciers or major rivers. In some cases the glacial origin of the offshore deposits can be recognized by the surface textures of the particles, or by the particle composition itself. For example, offshore rhythmites in glacial lakes commonly have a coarser-grained lower part containing glacial-silt clasts. Additionally, in glacial or other cold-climate settings, icebergs or seasonal ice floes can transport a considerable amount of ice-rafted material of all sizes offshore. This ice-rafted material is deposited in clusters, while top-heavy, floating ice blocks overturn as they melt. Falling coarse particles may penetrate the water-saturated bottom mud, deform laminae, and remain as isolated outsized stones (lonestones, dropstones).

In areas receiving little terrigenous material, biogenic activities and biogenic deposits prevail. In tropical seas, large carbonate buildups develop in reefs, carbonate banks, and carbonate tidal flats. The same occurs, to a much smaller extent, in cold seas. In shelf areas, bioclastic carbonate deposits are formed, as well as locally flourishing bryozoan concentrations and worm tube reefs. In the deeper offshore areas of cold seas, the delicate carbonate skeletons of floating microorganisms are dissolved, and there is preferential accumulation of silica skeletons generating silica ooze (very fine, powdery deposit).

Sedimentary Architecture

Sedimentary architecture refers to the lateral and vertical relations among associations of lithofacies that have formed and substituted for each other in an area over time as the glacier advanced and retreated and/or the basin water level rose (transgression) or fell (regression). It may involve relations between glacial and nonglacial depositional systems or just among lithofacies of one depositional system. Three examples are given here: one deals with large systems (fjords and large lakes) where both glacier- and land-related systems can be important, and two others deal with settings where the glacier-related system dominates, the first in marine shelves bounded by ice and the second in small glacial lakes.

Fjords and Large Lakes   A fundamental difference between fjords and large glacial lakes is that the water level of the first is controlled by global sea level, and the level of the second is dictated by more local phenomena. These include damming and opening of outlets by the glacier itself; isostatic depression and rebound of one part of the lake relative to another, and thus possible opening and closing of outlets; and migration of the crustal forebulge[1] that can develop in front of the glacier and migrate in unison with its melting and retreat. As a first approximation, when a glacier initially encloses a body of water and fluctuates within it, the features of land-

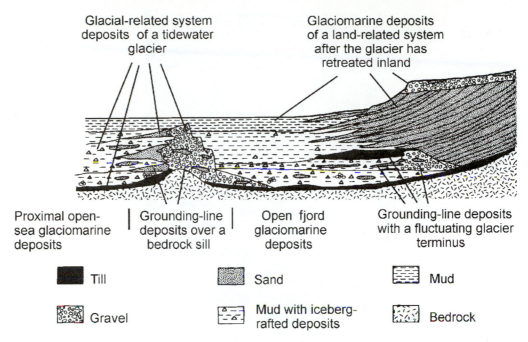

Glacial-related system deposits of a tidewater glacier

Glaciomarine deposits of a land-related system after the glacier has retreated inland

Proximal open-sea glaciomarine deposits

Grounding-line deposits over a bedrock sill

Open fjord glaciomarine deposits

Grounding-line deposits with a fluctuating glacier terminus

Till

Sand

Mud

Gravel

Mud with iceberg-rafted deposits

Bedrock

FIGURE 8.18   Diagram of glacier-related system (sedimentary deposits) in a fjord. (After Powell 1981.)

related and glacier-related depositional systems of a fjord and a large lake can be considered similar, and similar sedimentary successions may develop.

When the terminus of the glacier is stable, a static model predicts a glacier-related depositional system characterized by basal till under the glacier, grading outward into a coarser-grained, grounding-line sediment assemblage, and into increasingly finer-grained, better-sorted deposits farther away from the source (Fig. 8.18). Laterally or on the opposite shore of the basin, there could be a land-related depositional system dominated by delta and minor beach sediments, becoming increasingly finer grained offshore. Hyperpycnal flows from river floods and turbidite flows generated by slumping of loose material down slightly inclined slopes can be derived from both the glacier and the land, and spread into offshore areas where they interact. As the glacier advances and retreats within the basin, a complex facies architecture develops, particularly in the glacier-related depositional system. Deposited material may be deformed or outright removed by an advancing, grounding glacier. Alternately, rapid changes occur when the glacier retreats. Where there is little meltwater coming from the glacier, basal till is directly overlain by deepwater sediments (laminites and rhythmites) (Fig. 8.19A). This represents forced regression, that is, a quasi-instantaneous water deepening from a continental[2] to a glacier-free offshore basin setting. Where a lot of meltwater is generated, the basal till is overlain by the grounding-line assemblage (Fig. 8.19B), which may be in turn overlain by offshore rhythmites. The rhythmites become increasingly finer and thinner as the glacier retreats from the area. In this case there is a more orderly transition from terrestrial to grounding line to offshore environments, and a more regular transgressive sedimentary succession develops. Note, however, that in both cases the transgressive successions of the glacier-related depositional system are related to the location of the sediment input point that is here governed by the position of the glacier terminus, not by water depth, which may not change at all.

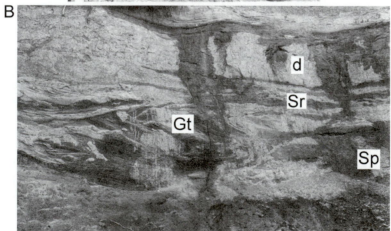

FIGURE 8.19 Photographs of glacial deposits (north shore of Lake Ontario, Canada). *(A)* Rhythmites (Cv) containing reworked glacial material and overlying till (d). *(B)* Gravelly grounding-line fan composed of slightly deformed cross-bedded sandy gravel (Gt), grading laterally into pebbly sand (Sp), capped by cross-laminated sand (Sr), capped by diamicton (d) that indicates local glacial readvance.

The overall water depth of the basins dictates, instead, the position of the sediment injection point and the development of the land-related depositional system. When water level is high, a regular facies transition may develop from nearshore or deltaic areas to the offshore. As long as the water level does not change, for example, a delta progrades offshore. Changes in water level may shift the location of the sediment input point and of the associated environments of deposition and related sedimentary facies associations. These will be shifted landward during a water level rise (transgression) or basinward during a water level drop (regression). In both cases some of the previously deposited material may be eroded.

Unlike fjords, large glacial lakes can undergo very rapid water level drops (forced regression). This happens when outlets that were dammed by glaciers are opened, at times suddenly, during a glacier retreat. Coastal terraces abandoned and left hanging in the landscape often record a forced regression. The floods related to sudden lake drainage may be recorded in extensive erosional and depositional features downstream from the outlet.

In the central part of the basin, land- and glacier-derived material interact, in part interfingering (becoming interbedded). For example, rhythmites containing glacial-silt pebbles derived from the glaciers may be interbedded with finer rhythmites formed by sediments derived from the ice-free land.

Open Sea Shelves   On ice-bounded marine shelves, the land-related depositional system does not contribute much glacial sediment, and is replaced by an oceanic system that provides biogenic material. This is the case around Antarctica and in parts of the Northern Hemisphere where ice tongues and fringing glaciers occur. This setting was much more widespread during the Pleistocene and older glacial maxima, when many ice sheets terminated into the sea. Advancing and retreating glaciers on shelves can generate a complex, terrigenous facies architecture (Fig. 8.20). During advance, the glacier may erode all or part of the pre-existent deposits. During the glacial maximum, the glacier may reach the continental shelf and deposit much material onto the continental slope. Such deposits may be locally reworked by sediment gravity flows, but for the most part they are preserved. As the glacier retreats, while the sea level is still low, sediment at the edge of the continental shelf and on the shelf platform itself may be greatly reworked by waves and marine currents. At later stages when the water level has risen somewhat, new postglacial deposits are formed. In nearshore areas, much sediment may be eroded by waves and currents as the land rises by postglacial isostatic rebound and may be replaced by newly developed coastal systems.

Small Glacial Lakes   Small lakes (few to tens of square kilometers) can form on, in front of, and in valleys dammed by glaciers, in kettles left by melting buried ice blocks, and so on. They are usually small enough to be readily overwhelmed by terrigenous sedimentation when in contact with glaciers. They fill rapidly with sediments when glacier meltwater discharges into them (rates of sedimentation up to 8–9 m per month have been measured), but they may also be catastrophically drained whenever an icy dam is ruptured or weakened and the water can flow under the bounding glacier. In many lakes this may be an annual event: the lake persists until it is deep enough to lift the damming glacier and let the water flow out. The sediments of these small glacial lakes are characterized by till at the base, overlain by one or more small deltaic sequences, occasionally showing Gilbert-delta characteristics. The deltas consist of locally laminated, but more frequently rippled, sandy bottomsets, sandy gravelly foresets, and sandy gravelly topsets, usually dissected by numerous cut-and-fills. Some deeper small lakes can develop laminites and varves. Deformation sedimentary features caused by readvancing glaciers occur frequently,

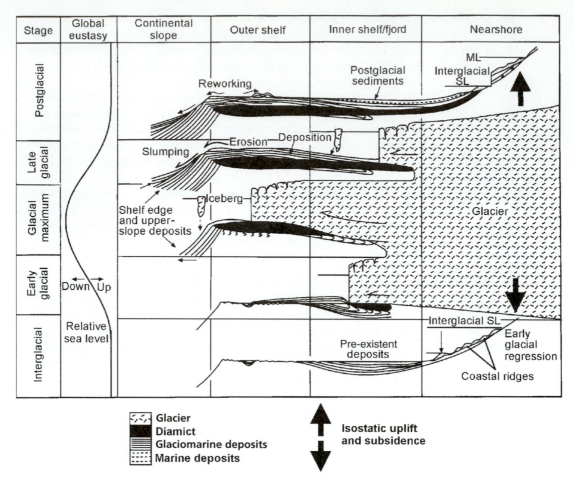

FIGURE 8.20   Diagrammatic model of sedimentation and erosion of shelf deposits during glacial advance and retreat and related marine regression and transgression (SL = mean sea level, ML = maximum sea level, half and bent arrows indicate relative sediment movements). (After Boulton 1990.)

and faults associated with the melting of buried ice blocks are common. Those parts of the lake deposits that persist while the rest is incised by streams may develop permafrost and ice-wedge features.

ENDNOTES

1. Crustal forebulge is a rise of Earth's crust generated by mantle material displaced by the weight of ice sheets outward.

2. Except for ice shelves and large ice tongues, a glacier by definition represents a continental environment.

Periglacial Environments

The term periglacial *has been used in different ways. Some authors consider as periglacial the geographic zone around (= peri) the glaciers where frost action prevails (also called* proglacial*), arboreal vegetation is generally stunted, and strong, cold winds blow. Others consider as periglacial the zone where cold processes, in particular perennial (= peri) or significant frost action, prevail on unglaciated areas or on previously glaciated* (paraglacial*) areas, independently of any direct glacial influence. Here, we accept the broader definition of "periglacial" and discuss important processes that are characteristically active in cold, but not exclusively glacial, areas, such as wind action, cold-climate weathering, and frozen ground.*

C H A P T E R   9

Eolian Sediments and Landforms

Winds are generated because of barometric pressure gradients that develop in response to temperature changes, such as those that may exist at the margin of glaciers, seas, or large lakes, or globally from the tropics to polar regions. Near glaciers, temperature and pressure gradients are steep, and strong, cold winds blow from the glaciers onto proglacial areas and outwash plains. These cold winds are called **katabatic winds.**

Air is a fluid; so the processes affecting movements of air masses, the ways in which air modifies the landscape, and how it transports and deposits sediments are the same as those governing water movements. The main differences are that air has a lower density than water but a much larger depth (height) of flow.

The lower density of air implies that winds must have much greater speed than water to be able to erode and transport a given particle size (Fig. 9.1). In strong, disruptive windstorms like hurricanes and typhoons, large gravel-size particles can be lifted by the wind, but usually the particles transported in suspension are less than 0.01 mm in diameter. The great density difference between air and rock particles allows the clasts to move more readily within the transporting fluid (air); hence they can be easily separated by size and weight, and the deposits are generally well sorted. Furthermore, the density difference means that some modes of transport are more important in air than in water. For example, transport of sand by saltation is much more important by wind, because the particles are freer to move in air and fall faster, thus impact more strongly on other particles generating greater dispersive pressure. It is common in deserts to observe sand moving in a saltation layer several meters thick. This process causes sand grains to become rounder and to acquire diagnostic surficial textures primarily consisting of numerous microscopic, V-shaped, impact marks (Fig. 9.2). Furthermore, windblown sand ripples are bombarded by saltating grains and can retain only the coarser, more difficult to move particles at the more exposed crests. These differ substantially from water-laid ripples in which the coarser particles released at the crest move down the frontal slope by gravity flows. In cross section, waterlaid ripples have an upward-fining grain-size trend; in windblown ripples the opposite is true.

In windblown sand dunes, the movement of particles down avalanche surfaces is similar to that of water-laid dunes, except that impacts between grains are more common, and well-defined, large, grain-flow tongues develop. **Grain flow** is a type of mass flow where particles are kept loose by impact among themselves (dispersive pressure).

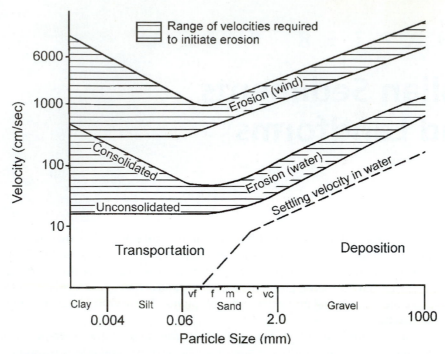

FIGURE 9.1   Diagram showing sediment erosion, transport, and deposition in relation to fluid (water and air) flow and particle size (Hjulstrom diagram; vf = very fine, f = fine, m = medium, c = coarse, vc = very coarse). (After Sundborg 1956.)

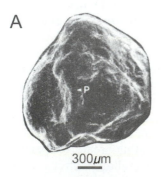

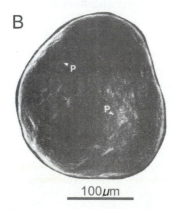

FIGURE 9.2   Scanning electron microscope photographs of characteristic windblown quartz sand grains. (A) Subrounded with impact features (P). (B) Well-rounded desert grain with impact features (P). (After Wang and Bhan 1985.)

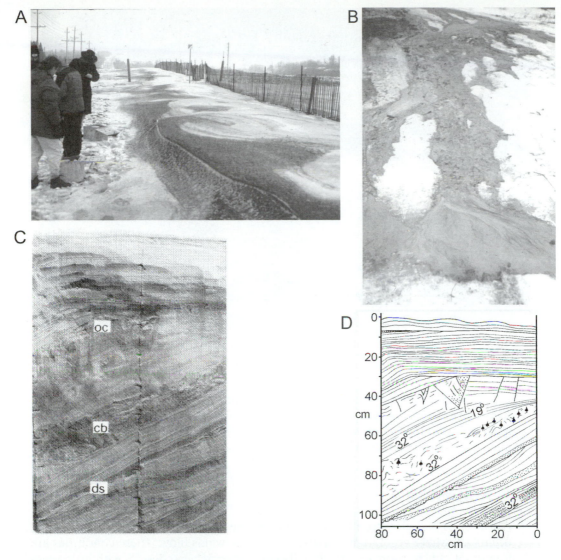

FIGURE 9.3 Photographs and diagram of terrigenous sediment and snow. *(A)* Windblown detritus on snow bank (Great Lakes region of North America). *(B)* Soliflucted sand on snow along the precipitation slope of a dune (Great Lakes region of North America). *(C)* Dune cross-beds with disturbed beds probably due to melting of snow, western Greenland (oc = open cracks, cb = disturbed bed, ds = irregular bed surface) (from Dijkmans 1990). *(D)* Explanatory diagram of *(C)* (19° = dip of layers) (from Dijkmans 1990).

One process peculiar to cold sand dunes is that sediment can be blown directly on top of snow (which is itself an ephemeral windblown sediment), or is placed on snow by granular debris flow[1] (in this case a mixture of sand grains and water) along dune slopes (Fig. 9.3A,B). Upon ice melting, the resulting interstratified snow, ice, and sediment layers form thin, massive deposits with disrupted fabric, in marked contrast to the well-sorted, better-structured windblown and grain-flow layers of snow-free sand dunes (Fig. 9.3C,D).

FIGURE 9.4　Photograph of large-scale cross-beds of a stabilized parabolic coastal dune (North American Great Lakes region).

The great thickness of wind flows (order of kilometers) permits the development of bedforms, such as sand dunes, and associated sedimentary structures, such as cross-beds, an order of magnitude larger than those of generally shallow (order of meters or tens of meter) rivers (Fig. 9.4).

WIND ABRASION

Abrasion of particles moved by wind or blown against larger exposed clasts or rocks occurs anywhere vegetation is sparse and strong winds blow. These conditions are found not only in warm deserts, but also in cold periglacial deserts with little vegetation and strong katabatic winds blowing from the glaciers. Winds blowing over outwash plains, or other periglacial sources of sand grains, may form sand dunes with typical eolian particle-size distribution (fairly well sorted) and shape (subrounded to rounded). They may also sandblast bedrock or clasts, forming **ventifacts** (Fig. 9.5A,B). In areas underlain by sediments, deflation hollows (blowouts) may develop, particularly where vegetation is sparse (Fig. 9.6). Everywhere, wind may deflate surfaces and leave coarse-grained, surficial, lag deposits.

SAND DUNES

Sand dunes typically occur in clusters. The principal types are barchan, transverse, longitudinal, star, and parabolic dunes (Fig. 9.7). Transverse and parabolic dunes, and other smaller types such as shadow dunes (accumulation of windblown sand behind an obstruction), are present on many glaciated Pleistocene terrains.

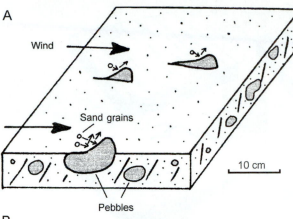

A

Wind

Sand grains

Pebbles

10 cm

FIGURE 9.5 Windblown erosion features. *(A)* Diagrammatic model of ventifact formation by sandblasting of half-buried clasts. *(B)* Photograph of ventifact on a small pebble (Cape Cod, USA).

B

Wind

3 cm

Wind

FIGURE 9.6 Photograph of blowouts at the crest of stabilized parabolic dunes (North American Great Lakes region).

Barchan dunes have semilunate shapes, concave downwind. They form in deserts with little or no vegetation and an undersupply of sand (Fig. 9.7A). Their shape results from faster movement of the thinner wings relative to the thicker, central, main body of the dune.

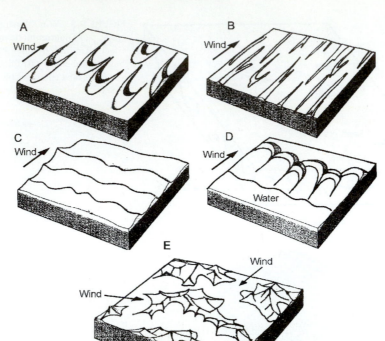

FIGURE 9.7 Diagrams illustrating types of large dunes: *(A)* barchan; *(B)* longitudinal; *(C)* transverse; *(D)* parabolic (U-shaped dunes); *(E)* star.

Transverse dunes are sand ridges that form perpendicular to the effective wind direction and have straight or undulating crests (Fig. 9.7C). They form both in deserts where there is abundance of sand and in Pleistocene glaciated terrains, particularly along the shores of lakes and seas.

Longitudinal dunes are ridges of sand parallel to the effective wind direction. They form in areas with scarce vegetation and occur frequently, but not always, on bedrock (Fig. 9.7B). Their formation indicates high wind velocity, and they are large-scale wind equivalents to the small, parting lineations that develop under supercritical flow conditions in open channel water flows. They are rare in Pleistocene glaciated areas, but common in warm deserts.

Parabolic dunes are U-shaped sand ridges, concave upwind. They are typical of Pleistocene glaciated terrains, and can form in areas where sand movement is hampered by vegetation (Fig. 9.7D). Vegetation growth is greater near the water table; hence the toe and thinner wings of the dune may be more heavily and continuously vegetated than the top of the central ridge. Because of this, the protected wings move downwind more slowly or not at all, whereas the higher, central part of the dune is continuously affected by strong winds, and migrates downwind more rapidly. The dune continues to grow and move as long as a sand supply exists in the upwind direction. Once the supply is cut off for any reason, such as vegetation growth or formation of another dune in the same path, migration ceases, and the dune is slowly colonized and stabilized by vegetation (Fig. 9.4). Any disturbance of the vegetation by fire or by human activities may restart movement of the sand.

Dune formation in Pleistocene glaciated terrains depends on conditions that make sand available for wind erosion and transport. Although dune formation is favored by drier climates, such as those of the hypsithermal (warmest postglacial period) of about 6000 years ago, dunes can also develop at other times when abundant sand is exposed to strong winds. So, katabatic winds can erode bare, unvegetated, and unstabilized ground in front of the active glaciers and pile sand downwind into large dune fields; for example, large dune fields occur both in ice-free valleys of Antarctica and around outwash plains and valley trains in Alaska. Conditions for dune formations are also favorable just after glaciation when isostatic rebound uplifts outwash plains and valley trains. These are entrenched by streams, the water table is lowered, and any existing vegetation may be disrupted. The resulting bare, sandy ground is easily eroded to provide sand for dunes. Furthermore, strong water-level fluctuations may occur in large, proglacial lakes or marine shores. Wide, sandy beaches or sandy, deltaic plains were periodically exposed during the late Pleistocene, and fed sand to numerous parabolic dunes.

During the late Pleistocene and early to mid-Holocene, extensive dune fields developed in every continent, including Europe and North America (Fig. 9.8). In North America many developed on coastal and deltaic plains when they became exposed as the glacier retreated or as land emerged isostatically from large glacial lakes or seas (Fig. 9.9). Most of these large, cold-climate dune fields are now stabilized by vegetation (Fig. 9.9B); but some have been reactivated by deforestation and trampling during historical times, particularly along modern coasts (Fig. 9.10). Some dune fields have also been leveled for agriculture. Smaller types of dunes are still active in coastal areas and along major river courses, and, in smaller measure, in the arctic, cold, desert areas.

LOESS

Loess (German word for silt) is pale brownish-yellow, unstratified, massive silt or silty-sand material, with numerous early epigenetic (formed soon after sedimentation) structures (Fig. 9.11A,B). It has a characteristic particle-size distribution, with modes (97–99.5% of population) in the silt and fine sand grades. These particles are transported in suspension in the atmosphere and can be distributed over wide, at times continent-wide, distances. Generally its composition is 60–70% quartz, 10–30% carbonates, and 10–20% clay minerals.

Loess is traditionally considered to have primarily formed in two major environments:

1. Desert or continental loess is formed of silt and sand derived from desert areas; hence it is sometimes called warm loess. The fine material is generated through abrasion due to saltation and sand-blasting processes in the desert, or has been transported, for example in Asia, into the areas by Pleistocene glacial streams. The silt and silty sand were subsequently blown over wide areas such as in central-east China (Fig. 9.11C), or out of the Sahara Desert into the Atlantic Ocean, where it forms fine-grained, reddish, oceanic deposits (ooze). Loess, being a continental deposit, develops thick soils during periods of low sedimentation. In China, this corresponds to colder periods, when more vegetation stabilized the northern Mongolian deserts and less silt was released and transported southward to the Chinese plains.

2. Loess of cold-climate environments (cold loess) is formed by silt deflated from outwash deposits, freshly exposed till, and from the rocky surfaces of tundra and arctic barrens. This material was generated by subglacial particle abrasion and

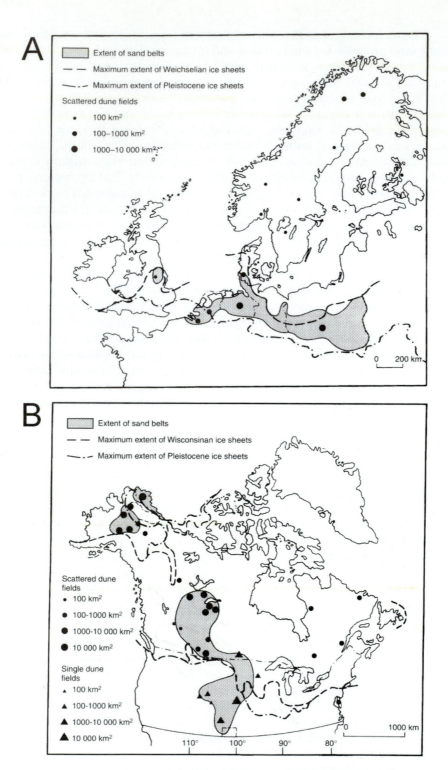

FIGURE 9.8   Maps showing the distribution of major cold-climate sand-covered areas and dune fields: *(A)* Europe; *(B)* North America. (From Koster 1988.)

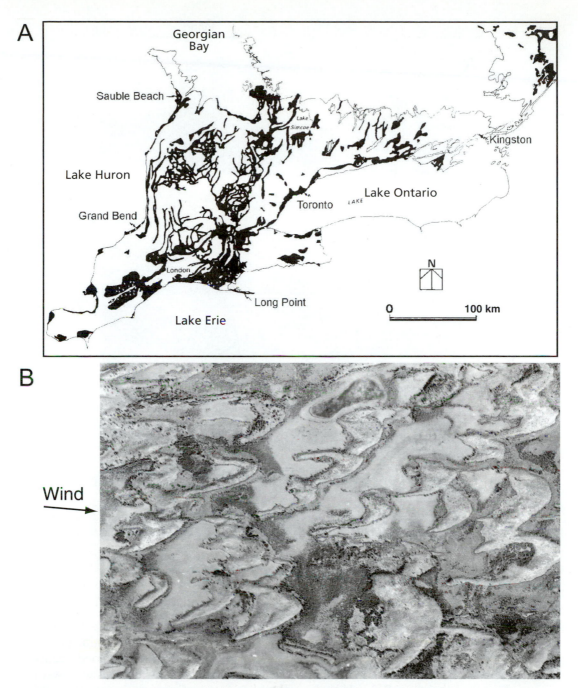

FIGURE 9.9 Dune fields of North America. *(A)* Map of southern Ontario, Canada, showing dunes (small, lunate, white symbols) formed on former outwash, deltas, and coastal sand deposits (black pattern). *(B)* Air photograph of a dune field in Saskatchewan, central-south Canada.

Wind

FIGURE 9.10   Photographs of dunes of the north shore of Lake Superior, Canada. *(A)* Parabolic dune advancing onto a boreal forest. *(B)* Oversteepened brink of a parabolic dune and precipitation slope marked by sand-tongue lines (grain flows).

either entrapped within till or transported onto the floodplain in suspension (Fig. 9.12). Extensive loess deposits rim the southern edge of former continental ice sheets in North America, Europe, and Russia, where they form very good, loamy agricultural soils (Fig. 9.13). In North America, a tongue of loess extends southward along the Mississippi River. This river was a braided stream during glacial times, and carried much coarse and fine material along its course. Deflation of this river sediment generated the southern loess.

Loess is usually slightly cemented, generally by calcium carbonate at grain contacts (meniscus cement), as hard-water droplets between grains evaporate. This cementation gives dry loess sufficient strength to support highways and buildings and to form and maintain bluffs in dry regions (Fig. 9.14A). However, loess can lose

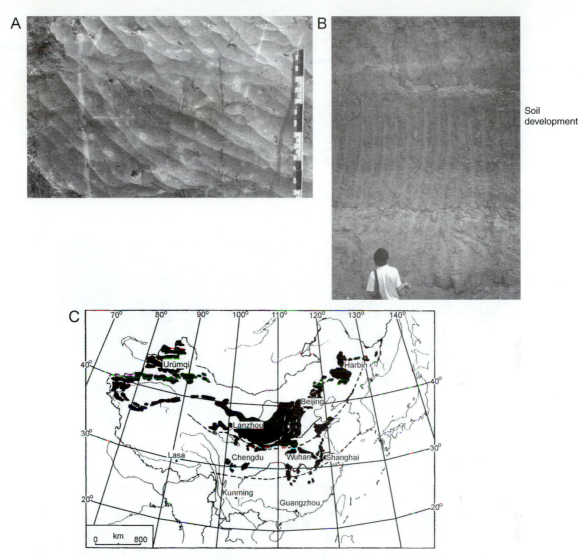

FIGURE 9.11   Loess. *(A)* Close-up photograph of an outcrop showing reduction spots along roots and near terrestrial invertebrate shells (Vicksburg, Mississippi, USA). *(B)* Photograph of Chinese loess showing a buried soil profile (Xian area). *(C)* Map showing the distribution of loess in China (after Liu 1985).

its strength and fail when water (rain, floods) enters the deposit and weakens the slight cementation, causing damage (Fig. 9.14B).

Loess mostly forms during active glacial times rather than interglacial ones, when outwash sediment, unprotected by vegetation, is continuously available for wind reworking. Thus, till and loess may be found interbedded in proglacial areas. Reworking and redeposition of silt (loess) may also occur during interglacial periods where vegetation is sparse or disrupted. Frequently the interglacial periods are characterized by variously well-developed paleosols on loess deposits.

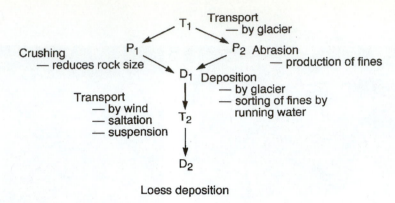

Loess deposition

FIGURE 9.12 Idealized sequential steps leading to the formation of cold loess ($T_{1,2}$ = sequential times, $P_{1,2}$ = sequential comminuting events; $D_{1,2}$ = sequential depositional events. (After Smalley 1966.)

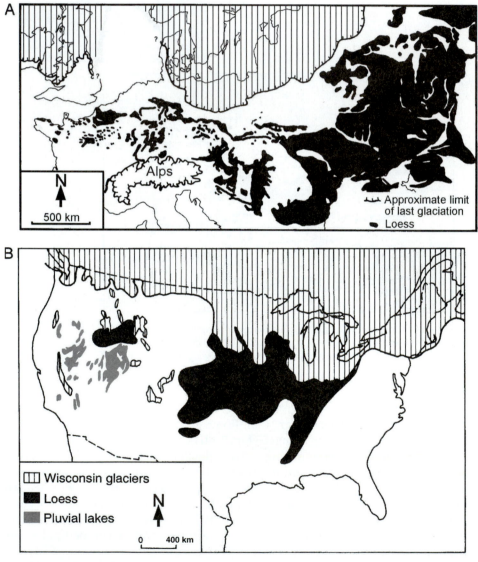

FIGURE 9.13 Maps of distribution of cold loess. *(A)* Eurasia (after Derbyshire and Owen 1996). *(B)* North America (after Washburn 1973).

FIGURE 9.14   Photographs of loess landscape (Vicksburg area, Mississippi, USA). *(A)* Vertical walls of dry loess. *(B)* Disrupted highway due to collapse of wetted loess.

ENDNOTE

1. The term "granular debris flow" has the same meaning as "grain flow," that is, a mass flow directly driven by the force of gravity and where the particles are kept separated, and thus moving, primarily by dispersive pressure (impacts between them).

CHAPTER 10

Cold-Climate and Frozen-Ground Processes and Features

COLD-CLIMATE WEATHERING

Weathering means breaking down, or comminuting (changing something into smaller pieces) through chemical or physical means. Nonglacial cold-climate rock weathering is characterized by mechanical freeze-thaw action, thermal stress, wind erosion, and, along shores, abrasion by ice blocks. Chemical and biochemical weathering processes, such as hydration of clay minerals; solution, particularly of carbonate rocks; and salt weathering are also important, although most chemical reaction rates are lower in cold regions.

Freeze-Thaw Action

Seasonal and shorter-term periods of melting and freezing affect parts of perennially frozen ground in cold areas, material below glaciers, and seasonally frozen ground of cold-temperate regions. The freeze-thaw frequency and the availability (or lack) of water and ice influence this process more than temperature extremes. Ice-crystallization forces can break rocks down; so rock clasts break down more readily in wet than in dry environments with similar climatic cycles.

Freezing-and-thawing progresses from the surface downward, and several processes help to break up the ground. (1) As interstitial water freezes during falling temperatures, the latent heat of transformation is released, warming the surroundings. A mixture of ice and water forms, and groundwater flows continuously toward the freezing front. The system remains at the pressure melting point in a dynamic equilibrium (series of micro freeze-thaw cycles) until the layer is completely frozen, at which point the system cools down (Fig. 10.1). The resultant numerous freeze-thaw cycles weaken bonds within the rocks, weathering them. (2) Melting (or thawing) occurs when the ground temperature reaches the pressure melting point of ice. If the temperature gradient between ground and atmosphere remains relatively steep, near-surface thawing is relatively rapid and continuous (Fig. 10.1). The heat for melting comes from the atmosphere; however, as melting continues, percolating water transfers heat downward, progressively melting the substrate. During melting, the pressure melting point is slightly lower than 0°C at 1 atm, because groundwater invariably contains some dissolved salt. Furthermore, water hydrolyzes and hydrates minerals, making them more amenable to disruption (weathering).

The main effect of repeated freeze-thaw cycles is the fracturing of rock into unsorted angular clasts (Fig. 10.2A). Some rocks are more susceptible to frost shattering than others. Susceptible rocks are those with internal discontinuities, such as bedding and fractures, where fluid water or water vapor can penetrate and refreeze.

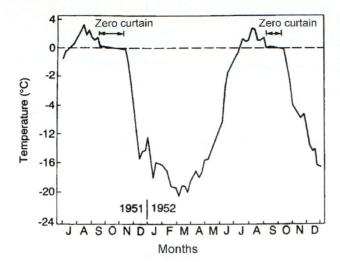

FIGURE 10.1 Diagram showing the development of the temperature curtains during the annual ground freeze-thaw cycle (Barrow, Alaska). (After Brewer 1958.)

FIGURE 10.2 Photographs of cold-climate weathering products and sediment accumulations. *(A)* Frost-shattered clast (Hudson Bay coast, central-south Canada). *(B)* Solid boulder of igneous rocks and shattered finer clasts of frost-susceptible rocks (Hudson Bay coast, central-south Canada). *(C)* Scree slope with angular clasts (Apennines Mountains, Italy). *(D)* Human-made scree deposit downslope from marble quarries (Apennines Mountains, Italy).

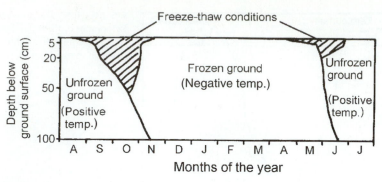

FIGURE 10.3   Diagram showing the periods and depths of freeze-thaw occurrence over a yearly cycle. (After Czeppe 1960; shown in Embleton and King 1975.)

Though frost shattering of rocks is mostly due to the expansion of ice in enclosed spaces, other processes may help. These include differential thermal expansion and contraction of rock-forming minerals, oxidation of minerals (particularly clays), hydration, and adhesion of water molecules to mineral surfaces (acting as a sort of crystallization force). Good examples occur on cold coasts or rivers where polymictic (multiple composition) glacial deposits are exposed at the surface. In such areas, large boulders are almost invariably composed of solid igneous rocks, whereas layered sedimentary rocks may be comminuted into smaller pebbles by frost action (Fig. 10.2B).

The frost-shattering process is only effective at the surface and in the upper few centimeters of the ground, where temperature and moisture variations may be extreme, and where freeze-thaw cycles occur more frequently and for a longer part of the year (Fig. 10.3). Clasts that are covered, even thinly by other mineral deposits or vegetation, are more affected by other weathering processes. Angular clasts are found within cold-climate soil profiles, but most have been moved there from the surface by cryoturbation or mass movement.

Frost shattering operates in polar to cold-temperate zones, and in cold, high-altitude areas. Remnant features indicating periglacial climates near Pleistocene glaciers and in mountainous areas are found nowadays in areas with a temperate climate. Extensive relict boulder-strewn scree slopes of frost-shattered, angular clasts help to determine the distribution of past cold-climate conditions (Fig. 10.2C). It should be noted, however, that other processes also produce deposits of angular clasts, including thermal expansion (expansion and contraction of rocks due to day-night temperature variations) in semiarid areas, solution of rock salts where they are interlaminated with carbonate deposits, and extensive quarrying by humans (Fig. 10.2D).

Mineral Solution

Water dissolves minerals and rocks at various rates depending on their structure, ambient temperature and pressure, and amount of water. Carbonate minerals and rocks are readily soluble under certain conditions. Their rate of solution is greatly affected by acids, the amount of CO_2 in solution, and temperature. In tropical and temperate areas, organic acids and CO_2 derived from decomposition of soil organic matter dictate the amount and rate of calcite solution. In cold regions, the low temperature enhances calcite solution, and the maximum rate of solution occurs at the ground surface. Solution occurs on exposed parts of the clasts, and reprecipitation just below the same clast or neighboring ones. The main products of cold-climate

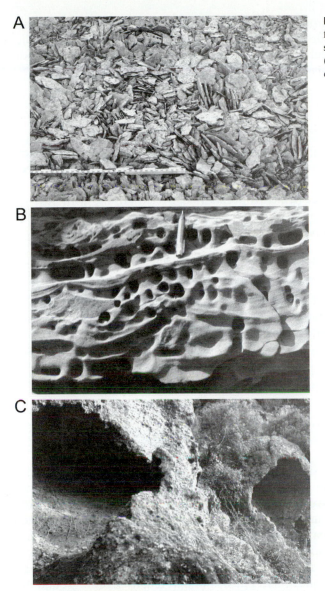

FIGURE 10.4   Photographs of weathering features. *(A)* Angular carbonate clasts showing sharp corners caused by solution (Foxe Basin, Canada). *(B)* Alveolar weathering (Italy). *(C)* Tafoni weathering (Italy).

carbonate solution are clasts with very sharp edges (Fig. 10.4A): there, the surface free energy of the material and, thus, solution is at its maximum. By contrast, dulling of carbonate clasts is produced by solution etching caused by vegetation.

Salt Weathering

Crystals, such as salts, growing in submicroscopic spaces between the rock-forming minerals or in small fissures may exert forces sufficient to break down rocks, leading to so-called **salt weathering.** Salt weathering is accompanied by other weathering processes such as hydration, hydrolysis, and oxidation, and leaves efflorescence on the rock surface. Salt weathering occurs anywhere evaporation causes salts to crystallize, both in hot and cold arid regions. Two striking and distinctive forms produced by salt weathering are alveolar cavities (few to 10 cm wide) and tafoni (meter-wide, hooded cavities) (Fig. 10.4B,C). These develop best where abundant salt is available either from a salty substrate, or from aerosols near the sea or salt lakes. Common salts in cold regions are calcareous, ferruginous, and silicic compounds and in particular mirabilite ($Na_2SO_4 \cdot 10H_2O$), a hydrated sodium sulphate.

PERMAFROST

Permafrost refers to bedrock or surficial material where temperature remains below freezing over a period of at least two years. A surficial layer, called the **active layer,** which undergoes seasonal freeze-thaw, generally covers it. Perennially frozen ground with abundant ice is **wet (ice-rich) permafrost;** frozen ground with little ice is **dry permafrost.**

Ice-rich permafrost is important for the following reasons:

1. It breaks the bedrock, preparing it for plucking by glaciers.
2. Freezing impedes groundwater movement.
3. It preserves organic matter indefinitely.
4. It reduces or prevents mass movement and landslides of perennially frozen material. It may, however, enhance mass movement of sloping active layers.
5. Growth of ice wedges and other frost features may modify or obliterate original characteristics of parent sediments through **cryoturbation.**
6. It creates typical geomorphological features and sediments.
7. It may form impermeable caps for petroleum reservoirs.
8. It may host metals such as placer gold.
9. It limits the growth of life forms, although some bacteria can survive in it.

Permafrost is widely distributed on Earth (it covers about 25.4 million km^2), and can be subdivided into continuous, discontinuous, and sporadic zones (Fig. 10.5). Typically, **continuous permafrost** has a ground temperature range from about −5 to −12°C, and the material below the active layer is perennially frozen everywhere throughout the year. Areas of **discontinuous permafrost** are characterized by patches of unfrozen ground during the summer, surrounded by frozen areas with average ground temperatures of −1 to −4°C. To the south, there is a poorly defined **sporadic permafrost** zone where patches of perennially frozen ground are surrounded by unfrozen terrain during the summer. This zone grades into the zone of **seasonally frozen ground** farther to the south still. Permafrost can vary in thickness up to about 600 m in arctic North America and 1600 m in central Siberia, to a few meters in subarctic areas (Fig. 10.6). Farther to the south in the boreal and cold-temperate areas, permafrost is replaced by seasonally frozen ground.

Type and Distribution of Permafrost

In northern portions of continental Asia, permafrost zones are roughly latitudinally distributed, and only locally affected by mountains. In western Europe, the warm oceanic Gulf Stream has distorted the latitudinal distribution of permafrost, which is present only in northern Scandinavia and high-altitude mountainous areas. In North America, permafrost distribution is broadly latitudinal in accordance with climatic zones, but it has also been influenced by the distribution of glaciers and wide glacial lakes in the past. Since deglaciation, the distribution of permafrost has been strongly influenced by the Canadian inland seas of Foxe Basin, Hudson Bay,

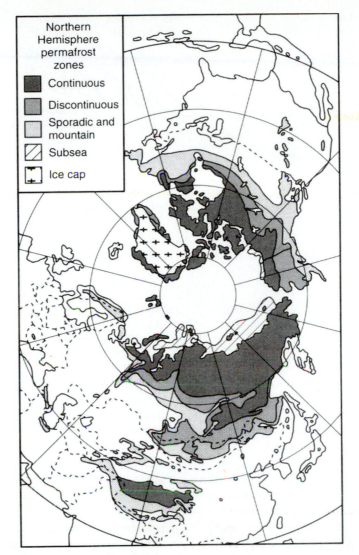

FIGURE 10.5   Map showing distribution of permafrost in the Northern Hemisphere. (From Péwé 1991.)

and James Bay, and by mountain chains. Some of the oldest American permafrost lies in those areas of interior Alaska and Yukon that escaped glaciation during the late Pleistocene. Some of the youngest permafrost occurs at relatively low latitudes around the Canadian inland seas. These seas receive cold arctic marine currents that refrigerate the surrounding, still-emerging lands and deflect the distribution of permafrost zones southward. Mountain permafrost extends far to the south, locally to Central America along the high coastal ranges. A similar south-to-north pattern of mountain permafrost occurs in South America.

Permafrost is not widespread under seas and deep lakes where water insulates the bottom. However, relict underwater permafrost exists in some areas off the northern Siberian coast, and to a more limited extent in the shallow nearshore parts of the Arctic Ocean off North America (Fig. 10.5).

Permafrost develops in cold, but not necessarily glacial, climates. Although glaciers may have frozen ground at their base and around their margins, vast per-

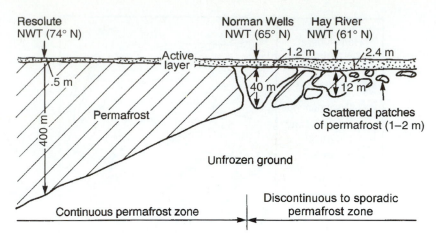

FIGURE 10.6   Diagrammatic permafrost cross section of western Northwest Territories (Canada) (2.4 m and 12 m = thickness of, respectively, active layer and underlying permafrost). (After Brown 1970.)

mafrost areas in North America and Siberia developed during the Holocene (the last 10,000 years) in nonglacial settings. For example, the Hudson Bay Lowland, now mostly covered by unconfined peat, has emerged from the early postglacial Tyrrell Sea in the last 7000 years due to postglacial isostatic rebound. Permafrost began developing as land emerged from the sea, and continues to develop as some coastlines are still emerging at about one meter per century. The Mackenzie River delta area of northwestern Canada shows another example of postglacial formation and disruption of permafrost. The area contains 8000-year-old frozen peat, in areas where peat is no longer forming. This indicates that surficial parts of the permafrost must have thawed deep enough during a warm Holocene postglacial period (about 8000 BP) to allow peat to form. At that time, large peatlands could not form farther south in Canada because it was too warm and the rate of evapotranspiration was too high to maintain the moist conditions necessary for significant peat growth. After the 6000 BP postglacial temperature maximum, however, the climate deteriorated again. As a consequence, the northern peats of the Mackenzie River delta froze, and the zone of peat formation migrated south, where extensive peatlands developed in the subarctic and boreal zones, particularly in the Hudson Bay Lowland and along the northern shores of the Great Lakes. On the other hand, permafrost responds to changing climatic conditions relatively slowly. Relict permafrost, formed during colder Pleistocene times, persists in several warming-up areas in Siberia and northwestern Canada, because there has not yet been enough time for its complete degradation.

Structure and Development of Permafrost

Frozen-ground profiles in permafrost areas may be subdivided into several horizons (Fig. 10.7).

   1. The **active layer** is a relatively thin, upper horizon subject to seasonal freezing-and-thawing. Locally, it may retain patches or lenses of perennially frozen ground (*pereletok*), or conversely, if it becomes perennially frozen, it may retain

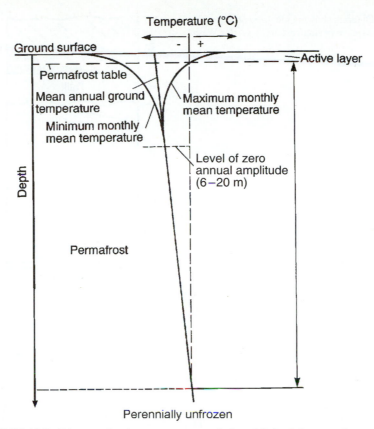

Temperature (°C)

FIGURE 10.7   Diagram showing temperature variation with depth in permafrost areas. (After Brown 1970.)

patches of unfrozen ground (*talik;* Fig. 10.8), forming what is referred to as the *suprapermafrost* zone. The seasonal base of the active layer is delimited by a *frost table* (the limit of the freezing front as it descends downward in winter). This is usually, but not always, the *permafrost table,* the top surface of the underlying perennially frozen ground. The thickness of the active layer depends first on the atmospheric temperature, and second on ground heat conduction. Thus it is thin (centimeters to decimeters) in polar or otherwise very cold regions, and thicker (up to several meters) in subpolar areas. Everything else being equal, the active layer is thinnest under insulating snow or organic cover such as peat, and in impermeable silt and clay substrates. It is thickest in well-drained sandy and gravelly and bedrock areas where there is no insulation by organic matter and thermal conduction values are high (Fig. 10.9).

2. Three generalized zones occur in the actual permafrost layer. Although these subdivisions are ill defined and not generally used, they serve well to understand some processes related to permafrost.

a. A relatively thin upper zone of perennially frozen ground (hence true permafrost) contracts and expands depending on the atmospheric temperature. The depth of zero annual thermal-fluctuation amplitude is usually obtained at depths of 10–20 m, depending on climate.

b. Below the depth of zero annual amplitude, the central main permafrost zone is unaffected by temporary changes in surficial conditions. That is, the warm summer temperature waves cannot reach it, and the amount of water

FIGURE 10.8 Diagram showing idealized stratigraphic profile of active layer in permafrost areas. (After Washburn 1973.)

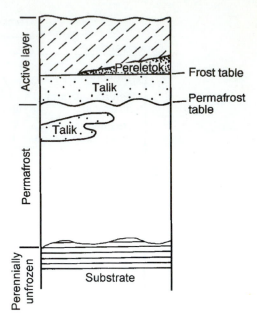

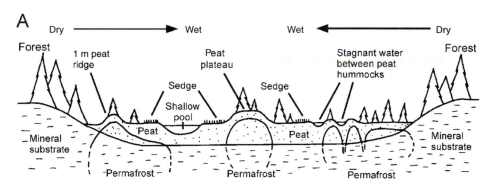

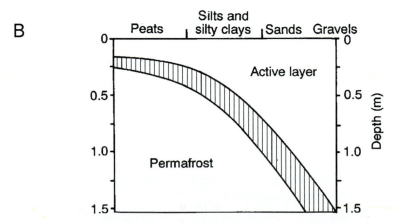

FIGURE 10.9 Diagrams showing permafrost variations associated with vegetation and substrate. *(A)* Idealized cross section of wetlands in subarctic Canada showing patchy permafrost under peat plateaus (after Brown 1968). *(B)* Graph showing variations in depth of the permafrost table for different types of sediments (after Mackay 1970).

generated at the surface during the summer is insufficient to penetrate deep underground and thus seasonally warm it up by refreezing.

c. In a basal transitional zone, the pressure-melting point is eventually reached due to the weight of the overlying rocks and the geothermal heat flux from below. In this zone the permafrost gradually melts and disappears downward.

GROUND-ICE PROCESSES AND FEATURES

Freeze-thaw processes behave the same whether they act in permafrost areas (in the active layer) or in regions with seasonal frost. Any difference results from the duration and relative effectiveness of the processes, and the dimension and persistence of features that are generated. Many such features are best developed in permafrost areas, but only a few, such as tundra ice wedges (thermal-contraction crack polygons), pingos, and thaw lakes actually require the presence of a perennially frozen subsurface layer.

Thermal Cracking

Thermal cracking may derive from contraction of the interstitial ice and the host ground, or from expansion due to ice crystallization forces. Several features are formed; a few of the most typical ones are presented here to analyze the major processes involved.

Permafrost (tundra) ice wedges are large, subvertical features that form predominantly in areas of continuous permafrost, although some may occasionally occur in discontinuous permafrost zones. As temperature drops in winter, the ground contracts and cracks, both in the active layer and in the top zone of the underlying permafrost, which is affected by temperature fluctuations (Fig. 10.10). As temperature rises in summer and early fall, the active layer thaws before the underlying top permafrost layer can expand and close the winter cracks. Water can run from the active layer into these cracks and refreeze. During the following winter the active layer refreezes, and as the cold wave penetrates downward, the permafrost cracks again, but this time within the ice fillings of pre-existing cracks.[1] The following summer, more water percolates into the permafrost cracks, and as the process repeats itself, the vertical ice lenses grow laterally into ice wedges. They may obtain dimensions of several meters wide and deep (Fig. 10.11). The annual growth of these ice wedges is recorded in ice foliations (laminae) that develop parallel to the sides of the structure. A characteristic of many ice wedges is that the host sediments are bent upward along the wedge margins (Figs. 10.10, 10.12).

Ice wedges can form either on a pre-existent, stable land surface (epigenetic ice wedges), or in areas affected by contemporaneous fluvial (floodplain) or eolian (particularly loess) sedimentation (syngenetic ice wedges). Syngenetic ice wedges may reactivate parts of older buried ones. It is not unusual to observe vertical stacking of wedges, at times penetrating into one another (Fig. 10.12).

Sand wedges form in very cold dry areas with little summer melting. Frost cracks develop as previously explained, but become filled with windblown sand and silt rather than water.

Surface ice veinlets derive from small-scale, subvertical, surficial ice cracks, which form in fine wet sediments, and are similar to sun cracking, except that the drying and contracting is caused by freezing (Fig. 10.13). In mud-cracked material, drying reduces the volume of the sediment-water mixture, causing tensional forces

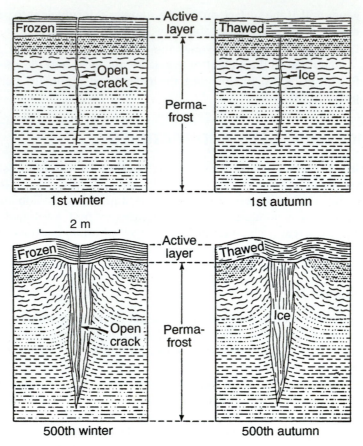

FIGURE 10.10 Diagrams showing progressive development of frost wedges over several years. (After Lachenbruch 1962.)

that generate polygonal cracks (mud cracks). In surface veinlet-bearing material, freezing does not at first reduce the volume, but as temperature drops, the interstitial ice and ground contract and cracks develop. These eventually fill with ice, forming subvertical, downward-tapering ice lenses (veins). Once formed, such veins increase in size, tending to lower their surface free energy, and molecules derived from fluid water or from smaller interstitial ice crystals migrate toward the growing ice vein. This causes further drying and cracking of the surrounding material, and more ice veins develop. Unlike mud cracks, the surface ice veinlets seldom form regular polygons; frequently they taper off laterally into needle shapes.

Ice Lenses

Ice can accumulate in lenses subparallel to the land surface and may cause several features to develop both in areas of permafrost and seasonally frozen ground (Fig. 10.14A). This is because groundwater freezes progressively as the cold wave penetrates underground. In highly permeable, granular material with large pores, such as sandy or gravelly deposits, and in shales that have extremely fine pores and very low permeability, the freezing front advances regularly and the sedimentary body may simply increase in volume (by about 9%) due to the H_2O phase change (ice crystallization). However, in materials with intermediate porosity and permeability, such as silt and peat, the freezing front advances in a punctuated fashion.

FIGURE 10.11   Photograph of large ice wedge (Fairbanks, Alaska). (From Péwé 1982.)

Under these circumstances, ice lenses develop parallel to the land surface, and have various thicknesses depending on how long the freezing front remains stationary at a certain level. The development of such lenses, their ultimate dimensions, and their disruption are related to the following conditions.

1. For an ice lens to form, the ground should freeze slowly under a gradual temperature gradient from the cold surface to the warmer subsurface.
2. The freezing front should move downward to the level where the forces of ice crystallization balance the resistance to crystal penetration through the small pores between the mineral particles (Fig. 10.14B).
3. Freezing causes drying, which generates suction (cryosuction) at the freezing front such that water moves toward the front from below, and subhorizontal ice lenses form and thicken there.
4. If the temperature gradient increases due to a lowering of surface temperature, the freezing front moves deeper and new lenses may develop at a new equilibrium level. If this process is repeated, the result is stacked levels of ice lenses.

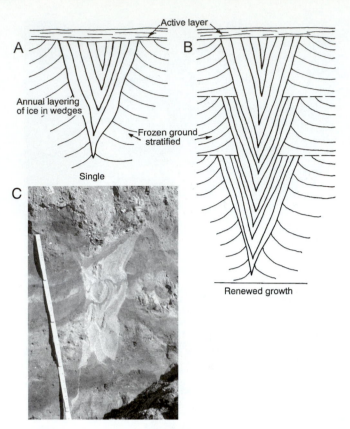

FIGURE 10.12 Diagrams showing different types of ice wedges. *(A)* Single, epige-netic. *(B)* Multiple, syngenetic wedges penetrating each other. (After Shoezov 1959; shown in West 1968.)

FIGURE 10.13 Photograph of surface frost veinlets, southern Ontario, Canada.

5. Conversely, if the temperature gradient is reduced as the ground surface warms, the freezing front moves upward and the ice lenses melt. The resulting underground voids are unlikely to be completely filled by collapsing material and persist as irregular, elongated cavities, subparallel to the land surface. A platey soil microstructure is often the result.

A
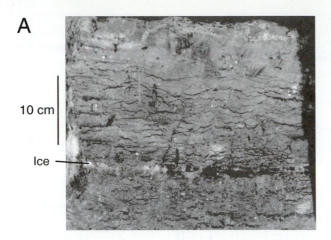

10 cm

Ice

FIGURE 10.14 Subhorizontal ice lenses. (A) Photograph of ice lenses and related planar pores (southern Ontario, Canada). (B) Diagrams showing the freezing fronts of ice lenses with a stationary front (L) and an advancing front (R).

B
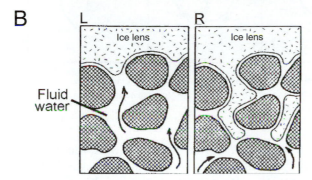

6. Variations in climate, sediment types, and surface cover, such as snow, organic matter, or manufactured features, all lead to variable development of ice lenses.

The formation and disruption of subhorizontal ice lenses lead, respectively, to differential heaving and collapse of the ground, and this forms many features typical of cold regions. Large-scale heave features include patterned ground, palsas, peat hummock, and pingos. Smaller-scale features include local frost blisters and associated potholes along roadways, formed by the formation and subsequent melting of ice lenses, and the heaving and dislocation of buildings, telephone poles, and other structures. These features are typical not only of permafrost areas, but also of seasonal frost areas.

Differential heaving of particles. In permafrost areas, many large rock clasts are displaced upward with respect to their surroundings. In cold to cold-temperate areas, new stones often appear scattered on the land surface each spring, and after a cold winter with variable low snowfall, plants may have their roots exposed and suffer winterkill, or blocks of fractured bedrock are pushed above the surroundings. These are the results of ice lens formation and associated ground heaving, and can be accomplished in two possible ways, as follows.

Frost-Pull Hypothesis   A possible sequence of events during freeze-thaw cycles in loamy (silty) soils containing disseminated pebbles is shown in Figure 10.15. As freezing prograges downward in a punctuated fashion, ice lenses develop and the ground heaves. At first a pebble is not affected much by the heaving ground

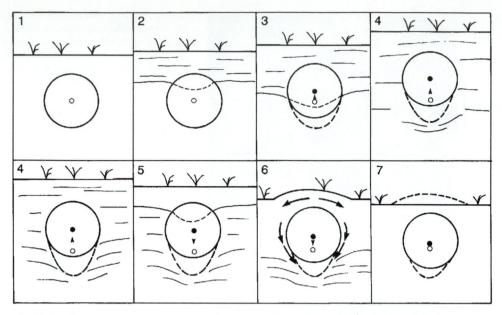

FIGURE 10.15 Diagrams showing stages of pebble heaving during a freeze-thaw cycle (arrow-heads indicate movement of pebble; full arrows indicate movement of water and some fine material). (From Beskow 1930; shown in Washburn 1973.)

(Fig. 10.15, stage 2), but, as the freezing material engulfs more of it, the pebble is lifted, however slightly, and a cavity develops below it (Fig. 10.15, stages 3 and 4). During warming stages, melting progresses from the top down and the ground subsides. At first the pebble remains unaffected, as its base is still frozen to the surrounding ground (Fig. 10.15, stages 4 and 5). However, as the material around the base of the pebble starts melting, meltwater and fine sediment infiltrates the cavity below the clast, and water refreezes, releasing some latent heat. This heat may slightly thaw the walls, causing some cave-ins and deformation of material into the cavity. Furthermore, the pebble is subjected to the weight of the overlying materials, and to differential forces associated with freezing-and-thawing patterns. If it is irregularly shaped, it may rotate to best fit the narrowing underlying space (Fig. 10.15, stage 6). When the deposit is finally completely thawed, resettling of the pebble occurs, but not completely to its original position because the cavity is partially filled by other material. In this way, there is an incremental net lift of the pebble during each freeze-thaw cycle that eventually leads to its expulsion onto the land surface. Slight variations in this process are responsible for the heaving of plant roots. In fact, it is common practice in certain agricultural operations to build earth or snow hummocks around susceptible plants to avoid formation of ice lenses.

Frost-Push Hypothesis Bedrock blocks (Fig. 10.16A) are thrust above the surface primarily by the crystallization force of ice lenses underneath them, rather than the heaving ground pulling them up. The heaved block does not resettle fully upon thawing because of partial filling of the underlying cavity by loose material and by the lateral movement of fractured bedrock due to expansion during the thaw period. As the process repeats, the block is thus progressively pushed upward.

Tundra hummocks. In low-lying tundra in polar regions, relatively small (order of tens of centimeters wide and high), vegetation-covered mounds called

A

B

FIGURE 10.16   Photographs of permafrost features (Northwest Territories, Canada). *(A)* Heaved-up large bedrock blocks. *(B)* Tundra hummocks.

tundra (or earth) hummocks generate uneven ground (Fig. 10.16B). They develop because of differential heaving and melting affected by the type and distribution of vegetation cover. In higher and dryer areas, a *Dryas* vegetation assemblage with good insulation properties allows thicker segregated ice lenses to develop and persist than in the intervening lows. The ice lenses cause more heaving than in the surrounding lows, which melt and subside every summer. The severity of the climate, the thin active layer, and the low vegetation cover under which tundra hummocks develop inhibit the growth of large hummocks.

Similar grassed-over, substrate hummocks termed *thufur* are found in subarctic regions such as Iceland and Scandinavia. Elsewhere, throughout cold regions of the world, hummocks occur on relatively steep slopes due to gravity-movement processes and differential erosion.

Palsas are found mostly in wetlands and vary from conical mounds (up to 7 m high and 50 m in diameter) to straight and winding ridges generally varying in width (10–30 m), length (15–150 m), and height (1–8 m) (Fig. 10.17). Most palsas are composed exclusively of peat, but some large ones have a core of silty material. The palsa core is characterized by the omnipresence of ice lenses, most 2–3 cm in thickness, some up to 10–15 cm. Palsas preferentially form in the subarctic, discontinuous permafrost zone, although a few are also found in the southernmost part of the continuous permafrost zone. In Iceland their southern limit approximates the 0°C annual isotherm, and in Sweden the −2 to +3°C mean annual isotherms. Developing and decaying forms are found in the same area, indicating the influence of local factors.

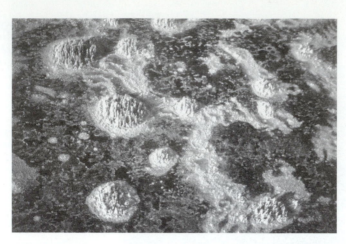

FIGURE 10.17 Photograph of palsas in wetlands (Hudson Bay Lowland, Canada).

Stages in the development of a palsa are inferred to be as follows:

1. An initial inhomogeneity on the ground (such as logs, vegetation clumps, boulder clusters, locally slight ground uparching due to hydrostatic or, rarely, gas pressures).

2. Establishment of sphagnum.

3. Growth of trees such as spruce in dryer areas.

4. Less snow on the ground beneath the trees (stopped by the crown), thus less ground insulation during cold spells, and deeper frost penetration.

5. In spring and summer, the area beneath the trees and moss (sphagnum) thaws more slowly than the surrounding wetlands.

6. Ice lenses under the dryer, longer-persisting, colder area grow as cryosuction moves available water from the surroundings to the freezing front. The mound heaves and its surface dries even further.

7. Peat and trees become better established on the mound, and an overall better insulation develops.

8. Ice lenses continue to grow beneath the mound where the thermal gradient does not change much with season, and while there is an abundant water supply from the surrounding seasonally unfrozen grounds.

9. Fully developed palsas are formed. In the discontinuous permafrost zone they can reach heights of about 1 m if confined to organic layers, and up to 8 m high if a silt-rich substrate is affected as well.

10. Any disturbance affecting the insulating vegetation cover, such as fire, allows penetration of the warm summer wave into the ground, and thus the progressive thaw of the ice lenses and palsa degradation. As melting increases, the topography of the mound becomes irregular and trees bend, generating a type of drunken forest.[2]

Peat mounds are similar to palsas, but smaller and less defined.

Pingos (Inuit word for hill; *bulgannyakh* in Russian) are large, perennial, ice-cored mounds (Fig. 10.18), generally conical in shape, occurring isolated or in clusters. They can vary in height between 3 and 70 m and in diameter between 30 and 600 m. Frequently pingos have a craterlike depression on top that may contain a small lake during the summer. Low vegetation and low disseminated shrubs cover

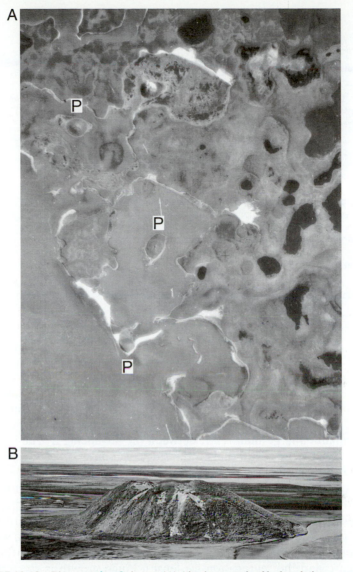

FIGURE 10.18   Photographs of pingos. *(A)* Air photograph of isolated pingos and indented coastline in the Mackenzie River delta area, Northwest Territories, Canada. *(B)* Pingo on the landscape, northern Canada.

most pingos. Internally they are composed of layered, deformed sediments, mostly silt and sand (Fig. 10.19). Many have a large, clear, massive ice body inside, generally a few meters thick, but up to 40 m thick in large pingos. Others have fine frozen material with numerous ice lenses, and still others have both. Active pingos are found primarily in the continuous permafrost zone (Fig. 10.5) in lowlands, such as the Mackenzie River delta in Canada (Fig. 10.20) and Lena River delta in Siberia, or valley bottoms as in Alaska, Greenland, and Siberia.

Some pingos consist mostly of non-frost-susceptible sand and have clear ice cores. Others have segregation ice lenses in their fine-grained interior. In both cases, the main problem is how enough water can be provided for the pingo to grow over a protracted time (order of a thousand years in some cases). Two main explanations have been proposed.

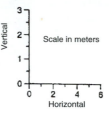

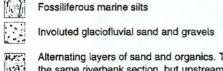 Fossiliferous marine silts

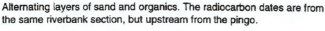 Involuted glaciofluvial sand and gravels

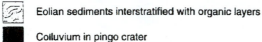 Alternating layers of sand and organics. The radiocarbon dates are from the same riverbank section, but upstream from the pingo.

Eolian sediments interstratified with organic layers

Colluvium in pingo crater

FIGURE 10.19   Diagrammatic cross sections of a pingo, northern Quebec, Canada. (From Allard 1996.)

1. Pingos can develop as closed systems in lakes that were originally deep enough to have no permafrost below the lakebed. As the lake becomes shallower because of sedimentation or drainage, permafrost grows toward the center of the lake, progressively expelling excess interstitial water from the sediments (Fig. 10.21). Eventually the lake becomes shallow enough for permafrost to form at its bottom, and a large talik is left between the advancing frozen surface layer and the permafrost table below and to the sides. The interstitial water becomes overpressured[3] and its freezing point drops. In porous and permeable sands, the water eventually flows upward in the direction of least resistance, starting the development of the massive ice core of a pingo. The pingos progressively rise as the surficial frozen sediment is heaved upward by the growing ice core. Furthermore, numerous discrete lenses could develop within the fine lacustrine sediments, which would increase vertical heaving of the mound. The talik waters feed both the central, massive ice core and the segregation ice lenses. Pingo growth is rapid in the first few years of formation (order of a few meters per year), but it slows down to a few centimeters per year later on. This is probably caused by several factors, including depletion of talik water, increased height and thus weight of the pingo itself, and thickening of the surface frozen layer.

Bulging of the frozen top of the pingo eventually generates fractures, which are avenues for water percolation and penetration of the warm wave. This damages the frozen core and leads to partial collapse of the top and, eventually, to the disruption of the whole structure.

These pingos require continuous permafrost to form. They are common in the Mackenzie River delta area, and are frequently referred to as Mackenzie-type pingos.

The localized occurrence of deformed lacustrine sediments, and a preserved morphology, often consisting of circular or irregular rims, help in recognizing paleo-pingo forms. Relict pingos have been tentatively recognized throughout Europe, often in association with other frost paleofeatures.

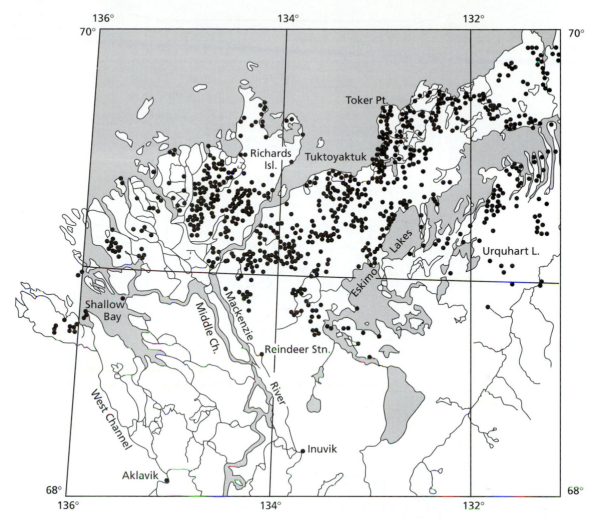

FIGURE 10.20 Map showing distribution of pingos in the Mackenzie River delta area. (After Mackay 1973.)

2. Pingos can also develop as open systems, when water is supplied to the frozen core mainly through artesian flow from subpermafrost or intrapermafrost aquifers (Fig. 10.22). Hydrostatic pressure is responsible for the initial bulging of the frozen surficial layer. Subsequent heave may be in part caused by the growth of a massive icy core and segregation ice lenses. In open-system pingos, as opposed to closed-system pingos, there is no limitation in water supply, unless the conduits freeze. Furthermore, open systems can occur at the base of slopes where artesian activities may be modulated by interruptions of aquifer flow. These pingos occur in both the southern part of the continuous permafrost and the discontinuous permafrost zones. They are common in Alaska and eastern Greenland and are frequently called Greenland type. In certain arctic lowlands, pingos form prominent relief features. They constitute landmarks and, near villages such as Tuktoyaktuk in the Mackenzie River delta, are used as communication antenna sites.

Relict open-system pingos are theoretically distinguishable from the closed type because they may not contain lacustrine sediments and may preferentially

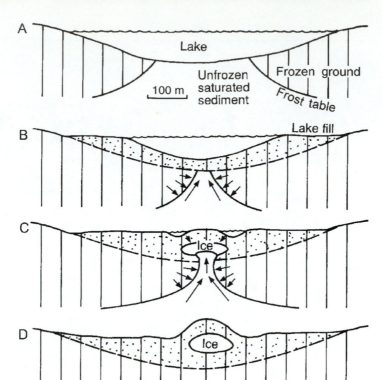

FIGURE 10.21 Diagrams showing progressive development of closed-system (Mackenzie) pingos. (After Müller 1959; Mackay 1972.)

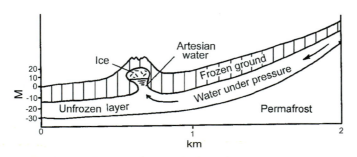

FIGURE 10.22 Diagram showing model of formation of open-system (Greenland) pingos (scale is approximate).

develop multiple, composite structures. The recognition of Pleistocene remnants of various types of pingos is difficult but important in determining the previous extent of permafrost zones.

Massive ground ice bodies tens of meters thick (some measured in excess of 40 m) and covering several square kilometers are reported from Siberia, and have been encountered in numerous drill holes in northwest Canada, particularly in the Mackenzie-Beaufort coastal plain (Fig. 10.23). The ice is massive and clear in places; in others it has foliations and mineral partings. It occurs under relatively thin Pleistocene glacial deposits. One old, now no longer favored, Russian hypothesis proposed that the massive ice bodies developed from amalgamation of gigantic ice wedges where they occur in continental Siberia. A second explanation is that they developed from ground ice segregation (ice lens formation) and amalgamation dur-

FIGURE 10.23   Photograph of large mass of foliated ground ice. (From Péwé 1982.)

ing coastal plain submersion and emersion. This may have occurred in northern Canada. A third explanation is that the ice bodies are remnants of Pleistocene glacier ice, preserved in permafrost, beneath a thin ablation till cover.

Patterned Ground

The features formed in frozen terrain (permafrost or seasonally frozen ground) usually occur in groups and impart a diagnostic pattern to the ground, referred to as **patterned ground** (Fig. 10.24). Patterned ground is usually associated with cracks and/or changes in particle orientation to form unsorted features, or particle orientation and segregation to form sorted features (Figs. 10.25, 10.26). Symmetrical forms develop in flat, homogeneous areas (circles and some nets), but they become distorted to a degree in sloping terrains (steps and stripes) due to solifluction. All these features are larger and more extensive in permafrost areas but are not exclusive to those regions.

Circles form in silts, sands, and gravels on flat areas.

Unsorted circles develop in muddy as well as gravelly deposits (Figs. 10.24B, 10.25, 10.26A,B). For example, angular carbonate pebbles from thinly bedded bedrock are frost shattered and rearranged by freezing-and-thawing into a stable position, that is, oriented vertically, with their faces parallel to the direction of lateral stresses. These unsorted circles are identifiable by the orientation of the flat clasts.

Sorted circles usually develop in stratified and poorly sorted material (Fig. 10.25). Irregular vegetation growth helps in the formation of sorted circles by differentially insulating the ground, and thus promoting differential heaving and

FIGURE 10.24 Photographs of patterned grounds in northern Canada. *(A)* Raised beaches and permafrost ice wedges (air photograph). *(B)* Circular mud boils.

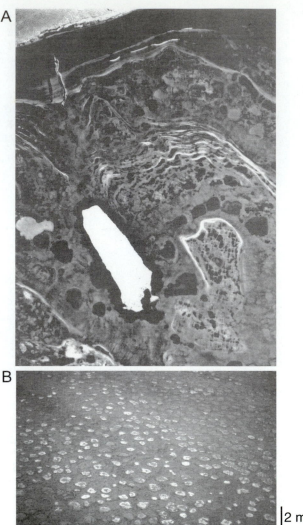

melting throughout the year. The following is another possible way sorted circles (and some mud boils) could form. If the frost table prograsdes downward irregularly, differential pressures are generated in the talik that may be present in the ground. Water is forced to flow toward the surface, dragging fine particles as well as some coarse clasts to the surface, where they accumulate, forming a slight rise. Succeeding water-saturated slurries flow outward, the large clasts slide sideways, and sorting occurs. Finer particles remain at the center of the structure and coarser clasts move outward, generating sorted circles.

Polygons develop where large ice wedges occur and bound them. The more uniform the substrate terrain is, the better hexagonal shape the polygons acquire. They occur in groups, imparting a pattern to the area (Fig. 10.24A). Where the bounding ice wedges are large and the sedimentary layers have been bent upward significantly, the polygons have distinct raised rims and are known as *low-centered polygons*. In other areas, the central parts of the polygons are domed; these are called *high-centered polygons*. During permafrost degradation, the preferential

Type	Subtype	Shape	Occurrence	Size	Sediment composition	Vegetation	Remarks
Circles (mud boils)	Unsorted		Flat areas Singular or in groups	Height: 0.1–0.2 m, slightly domed	Silt – pebbles	Yes	They are active if the center is bare.
	Sorted		Flat areas Singular or in groups	Height: 0.1–0.2 m, slightly domed	Silt – boulders (at the rim)	Irregular	They are inactive if the border stones have lichens.
Polygons	Unsorted	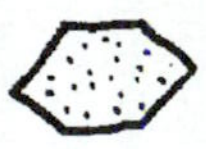	Flat areas Often have ponded centers	Height: 0.1–0.2 m	Silt–gravel	Yes	They are inactive if the centers are depressed (ponded).
	Sorted	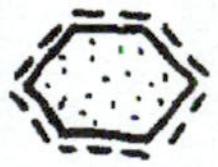	Flat areas Often have ponded centers	Height: 0.1–0.2 m	Silt – boulders (at the rim)	Optional	They are inactive if the centers are depressed (ponded).
Steps (solifluction)	Unsorted		On slight slopes (5–15°) Grouped	Height: 0.3–0.6 m Width: 1–3 m Length: 8 m	Stoney soils	Yes (downslope)	They develop from polygons, circles, or nets.
	Sorted		On slight slopes (5–15°) Grouped	Height: 0.3–0.6 m Width: 1–3 m Length: 8 m	Cobbles – boulders (at the downslope rim)	Yes (downslope)	They develop from polygons, circles, or nets.
Nets	Unsorted		Flat areas Grouped	Height: 0.5 m	Clay – gravel	Yes, thick	Intermediate shape between circles and polygons (e.g., earth hummocks).
	Sorted		Flat areas Grouped	Height: 0.5 m	Gravel – cobbles (at the rim)	Yes	
Stripes	Unsorted		On slight to moderate slopes Grouped	Width: 0.3–0.6 m	Clay – cobbles	Yes	Can be evenly spaced or 3–5 m apart
	Sorted	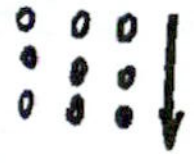	On slight to moderate slopes Grouped	Width: 0.3–0.6 m	Clay – cobbles	Yes	

FIGURE 10.25 Principal features of permafrost patterned grounds. (Information from Washburn 1973.)

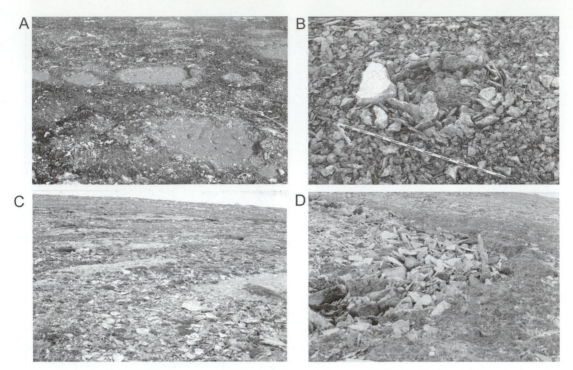

FIGURE 10.26 Photographs of permafrost features (northern Canada). *(A)* Unsorted circle in muddy area. *(B)* Unsorted circle in gravel. *(C)* Steps associated with raised shorelines with superimposed sorted mud patches. *(D)* Bouldery frontal lobe of a step.

thawing of ice wedges can change low-centered polygon terrains into high-centered polygon terrains.

Nets (Fig. 10.25) may develop where groups of irregular hummocks interface with each other, usually on subhorizontal terrains. They can be of both the sorted and unsorted type. A typical unsorted net is associated with a thick mat of ground-covering vegetation.

Stripes and steps. On slopes, repeated freeze-thaw cycles, and the fact that the surface thaws while the underlying material is still frozen, leads to a slow movement of material downslope (solifluction) (Fig. 10.25). Usually the thin, surficial material slides, and long, multiple stripes develop along the slope. This pattern is particularly evident where there is a binding cover of grassy vegetation. On steep slopes, the movement of the surficial material may occur in tongues, generally with the downslope, steep noses marked by large stones, often oriented vertically if the boulders are flat (Fig. 10.26D).

Involutions

Stratified sediments containing silty or organic layers often develop irregular folds referred to as **involutions** (Fig. 10.27). These features are thought to be the result of either (1) differential heave associated with different grain size, or (2) density-induced deformation occurring either during permafrost degradation or during periods of summer thaw in the outer layer. They are commonly preserved in ancient sediments that were previously perennially frozen.

FIGURE 10.27   Photograph of recent involutions in coastal sediments (Hudson Bay Lowland, Canada).

Thermokarst

Thermokarst forms when permafrost melts and the active layer thickens. It develops anywhere the ambient temperature rises enough for ground ice to melt, and where the insulating ground surface has been disturbed by natural (fires, slumps) or human-induced events. Naturally occurring thermokarst features preferentially form along the southern margin of the discontinuous permafrost zone, and are best developed in lowland areas where much ground ice occurs and where there has been a long-lasting warming trend. Thermokarst features are caused primarily by ground subsidence due to melting of underlying permafrost in flat areas, and/or by erosion and backwearing of frozen slopes such as coastal bluffs, and river and lacustrine banks.

Thaw lakes are a type of thermokarst feature. They develop through ground collapse and bank backwearing. They occur widely on raised coastal plains of Alaska and northern Canada underlain by ice-rich, silty deposits. They have various dimensions, from small ponds to circular or elongated lakes generally a few meters deep. They may form (1) in response to disruption of vegetation cover, which causes local ground melting; (2) from remnant tidal ponds in emerging coastal lands, such as in southwestern Baffin Island along the Foxe Basin (the Great Plain of the Koudjuak); or (3) in more established, inland permafrost zones, they may be related to low-centered ice-wedge polygons where ponded water in low-lying areas fosters increased melting of the substrate (Fig. 10.28A).

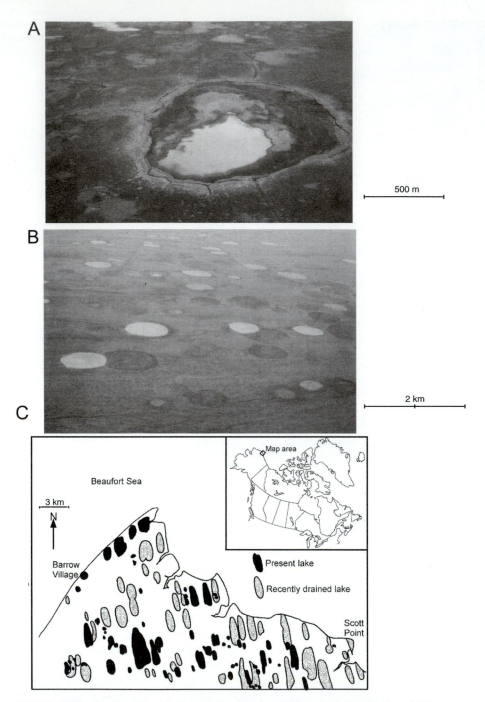

FIGURE 10.28   Photographs and diagram of thermokarst lakes. *(A)* Lakes developed in low-centered polygons (Great Plain of the Koudjuak, western Baffin Island). *(B)* Round lakes (Great Plain of the Koudjuak). *(C)* Schematic map showing oriented lakes (Alaska). (After Carson and Hussey 1962.)

Once established, thaw lakes enlarge by wave backwearing of the shores, generally becoming circular where the substrate is homogeneous (Fig. 10.28B). Well-developed elongated thaw lakes occur in coastal areas of Alaska, parallel to the prevalent wind, indicating a preferred direction of shore backwearing (Fig. 10.28C).

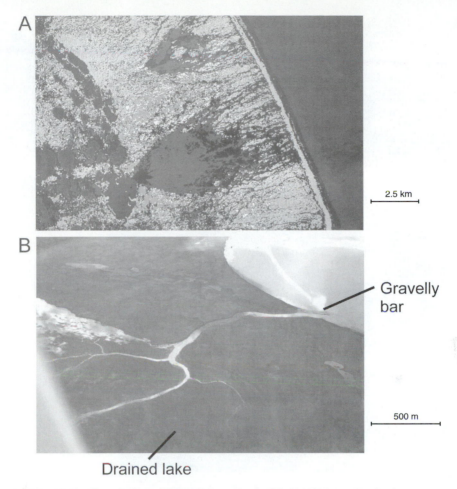

A

2.5 km

B

Gravelly
bar

500 m

Drained lake

FIGURE 10.29   Breached round lakes (Prince Charles Island, Northwest Territories, Canada). *(A)* Slow water discharge and development of large vegetation fan (fan-shaped dark patch; satellite image). *(B)* Sudden shore breach and catastrophic release of water to a second lake, causing formation of a gravelly bar.

The lakes can amalgamate with others as they enlarge, and their banks are breached by thermokarst backwearing or by drainage creeks. The breach may release water from the lakes slowly, generating downlake fans of more intense and varied vegetation growth (Fig. 10.29A); or, particularly for deeper lakes, the breach may enlarge and deepen rapidly and catastrophic draining can rework coarse-grained sediments (Fig. 10.29B). Fine-grained deposits usually cover the lake bottoms, which are partially reworked from the substrate by waves and mixed with organic matter. In the Great Plain of Koudjuak in southwestern Baffin Island, which supports one of the largest migratory geese populations of North America, small pingo-like, ice-cored islands in the lakes constitute ideal nesting sites protected from incursions by foxes.

Alas (a Russian term) is a wide depressed area (Fig. 10.30A) that may contain an active or drained lake partially filled by lacustrine deposits. These thaw depressions are common in Siberia where thick silty deposits contain a lot of segregated ice. They have not been recognized, at least to the same extent, in North America.

Thermokarst mounds form a pattern of rounded mounds a few meters high separated by narrow depressions, which gives a characteristic graveyard aspect to

A

B

FIGURE 10.30 Photographs of thermokarst features. *(A)* A Siberian alas (from Washburn 1973). *(B)* Thermokarst mounds (Fairbanks area, Alaska) (from Péwé 1982).

the land (Fig. 10.30B). The mounds contain undeformed sediments, so frost heaving did not form them; rather they are elevated remnants of the pre-thermokarst surface, formed upon melting of the bounding linear ice wedges.

Indented coasts develop along arctic shores, such as along the Mackenzie River delta in northern Canada, where pingo-bearing coastal lakes are progressively breached and the pingos partially disrupted (Fig. 10.18A).

SURFACE ICE: ICINGS

In cold regions, icings (*Aufeis, Nalyedi*) develop where water rises and freezes at the ground surface where springs exist, or on rivers and lakes (Fig. 10.31). In Siberia, icing can reach large proportions, with thickness on the order of 7 m over areas of 60 km^2. The water comes from artesian springs from thawed horizons in the permafrost, or from partially frozen rivers, where water escapes to the ice-covered surface through cracks. Icings preferentially develop during the freeze-up to midwinter period. They are dangerous for structures or transportation routes. They may also

FIGURE 10.31   Photograph of icing (Alaska). (From Péwé 1982.)

contribute to winterkill of plants due to suffocation, and, where thick, they can contribute large amounts of water to spring floods.

SLOPE MASS MOVEMENT: PROCESSES AND FEATURES

Solifluction and Soil Creep

Solifluction is the process by which the surficial part of the soil is detached and slowly moves downslope en masse. This process is not unique to, but is common in, cold regions because the surficial soil may thaw while the substrate is still frozen. The frost table constitutes an ideal detachment surface for downslope movement of the surficial materials. In addition, heaving of the terrain due to ice lenses in cold areas and swelling of clays in warm zones leads to further slow downslope movement (creep) of the surficial material. In cold-climate areas, the first has been named **gelifluction,** the second **frost creep.** Two forces drive these processes: gravity, which pulls the material downslope, and crystallization forces of ice, which act perpendicular to the direction of cold-wave penetration, generally perpendicular to the ground surface. Whereas solifluction relates to mass slipping along detachment planes, frost creep leads to downslope particle readjustment (Fig. 10.32A). The net result is a complex downslope movement of the surficial material (Fig. 10.32B).

Slopewash

Slopewash occurs where water flows on slopes. The flow may be widespread (sheetwash) or concentrated in small channels (rills) or larger channels. In cold regions, the source of water may be rainwater as everywhere else, but also important is the melting of snow and of the ice of the active layer. Because the substrate may still be frozen, water cannot penetrate deep into the soil, and much runoff occurs. Water flowing in channels may lead to locally faster ground melting (due to the temperature [thermal erosion] and kinetic energy of the water flow), hence increased en-

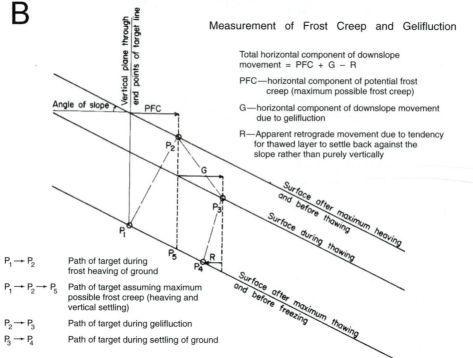

FIGURE 10.32   Gelifluction. *(A)* Photograph of gelifluction tongues along a steep slope of the Altai Mountains, western Siberia. *(B)* Diagram illustrating soil movements during gelifluction. (After Washburn 1967.)

trenchment, and, in places, breaching of lakeshores. The result can be a rapid drainage of a lake, increased erosion along the channel, and transport of much sediment onto adjacent lowlands (Fig. 10.33).

Rock Glaciers

Rock glaciers are a special type of periglacial slope mass movement, which occur in mountains. They are masses of generally coarse-grained, slow-moving (order of 3 m/yr), debris-flow tongues, a few hundred meters wide and long, and up to 40 m thick, in which the particle-supporting matrix is ice (Fig. 10.34). They can form as

FIGURE 10.33   Photograph illustrating flood erosion and fan deposition in raised beaches (Foxe Basin, Canada).

FIGURE 10.34   Photograph of a rock glacier (BC, Canada).

true permafrost features with interstitial congelation ice, or they may derive from ice-cored moraines when found adjacent to a glacier. Rock glaciers develop preferentially near to, and are indicative of, the snowline. True permafrost rock glaciers most commonly form in relatively dry, continental areas where lack of sufficient precipitation has inhibited formation of glaciers, in places like the Kazakhstan Mountains, and preferentially in the Eastern Alps (Switzerland) rather than Western Alps (France).

Cryoplanation

In cold regions, it is thought that the continuous erosion of higher ground and associated sedimentation in lower areas leads to a lowering and smoothing of the land-

FIGURE 10.35   Frost-weathered bedrock features from northwestern Spain (Galicia). *(A)* Tor. *(B)* Blockfield on the Serra de Maira. (From Pérez Alberti 1993.)

scape. There, cold-climate weathering processes such as widespread frost shattering, combined with rockfall, solifluction, soil creep, and slopewash, generate a smooth land morphology. This process is referred to as **altiplanation** (that is, formation of a plateau) or **cryoplanation** (that is, formation of a planar surface due primarily to nonglacial cold-climate processes, regardless of altitude). Depending on degree of development, **tors** or isolated upstanding remnants of the original substrate sculpted by frost and other processes may remain. When formed in layered rocks, they generally have rugged outlines (Fig. 10.35A), but where the substrate is composed of massive, crystalline rocks like granite, the remnant bedrock knobs and the boulders strewn on the land are round (Fig. 10.35B).

DISTRIBUTION OF MID-LATITUDE PLEISTOCENE PERMAFROST

Both glaciers and permafrost require a cold climate for their formation. Once developed, glaciers refrigerate their surroundings; hence permafrost develops in those subaerially exposed lands. However, striking differences exist between the distribution of permafrost features of the formerly (Pleistocene) frozen lands of Europe and Asia and those of North America. Primarily, this reflects the different latitudes involved and the quite different insolation regime.

1. During the Pleistocene, periglacial environments and permafrost developed at the margin of continental ice sheets both in high-latitude, nonglaciated areas such as northern Alaska and Yukon and parts of Siberia, and, mostly, along the southern mid-latitude margin of the ice sheets. Relative to the present-day situation, Pleistocene periglacial environments of mid-latitudes were characterized by the following:

a. More frequent daily freeze-thaw cycles due to larger insolating differences, thus daily temperature extremes

b. Higher insolation and a higher angle of incidence of solar radiation, leading to stronger differences between differently oriented slopes

c. Greater regional thermal atmospheric gradients between the cold glaciers and equatorial areas, leading to the development of stronger winds

d. Differences in global air-mass movements (jet stream path, winds due to the presence of glaciers), thus differences in type, quantity, and distribution of atmospheric precipitation

e. Stronger differences between (1) drier and colder continental areas with fully developed permafrost, and (2) more humid, coastal areas adjacent to warmer seas where frozen ground was poorly developed

All this led to more intense, albeit areally more restricted, periglacial processes during the Pleistocene in mid-latitudes than at the present time. The legacy of Pleistocene frozen-ground processes in these regions includes, among other things, angular scree slopes in mountainous areas, ice-wedge pseudomorphs, patterned ground, and pingo scars.

2. In Europe, numerous and varied remnant permafrost features have been observed over a wide area, hundreds of kilometers away from the former edge of the ice sheet (Fig. 10.36A). In Asia, vast expanses of permafrost developed, and some still persist under cold climate. In North America, permafrost developed around the ice sheets and mountain glaciers; however, the record of former permafrost features is sparse and limited to a narrow strip (about 80–200 km wide) along the edge of the maximum ice extension. In other words, in North America, permafrost and other periglacial features were not as widely distributed as in Europe and Asia.

a. One of the reasons is that, at their maximum, the North American ice sheets extended much farther south (to approximately 35° N latitude) than those of Europe (approximately to 46° N latitude) and Asia (approximately to 60° N latitude); hence much steeper atmospheric climatic gradients existed at their margins. Furthermore, during deglaciation, large lakes and seas developed at one time or another in front of the large Pleistocene ice sheets, and these were particularly widespread in North America. This impeded the formation of permafrost over wide areas. Similarly, in Europe, permafrost features are not reported from those areas covered by the glacial Baltic Lake and other large lakes that covered northern Germany and other northern countries.

b. Another reason for differences in the distribution of permafrost features is related to the distribution of windblown silt (loess) on the various continents. This cold-climate silt deposit was ideal for the development of thermal-contraction cracking and ice lenses. Loess deposits are thick and

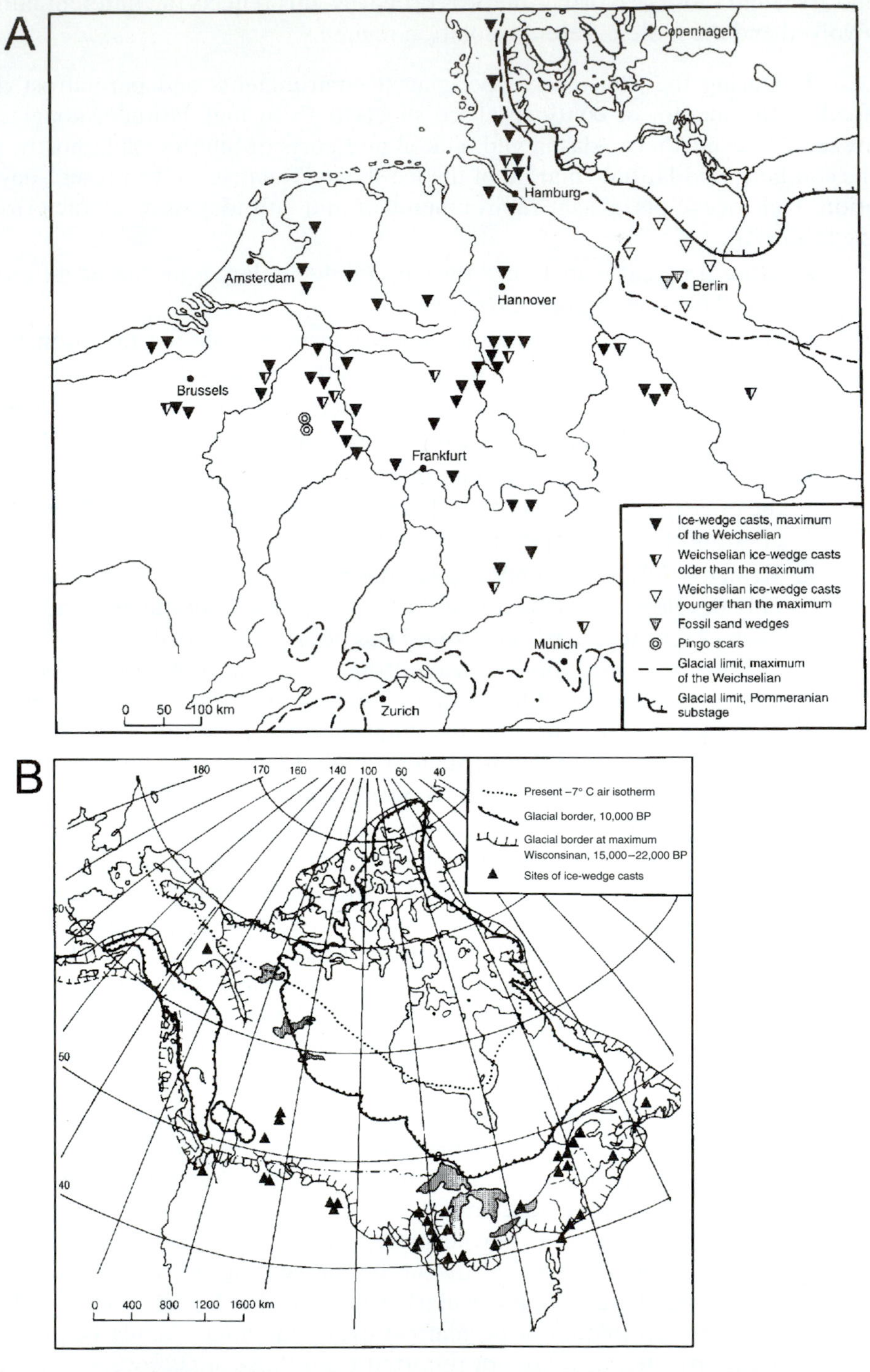

FIGURE 10.36 Maps showing distribution of fossil permafrost features. *(A)* Parts of Europe (From Karte 1987; shown in Ehlers 1996). *(B)* Glacial borders and ice-wedge casts in parts of North America (From French 1996).

widely distributed in Europe and Asia, and they keep a good record of the permafrost conditions. Loess deposits are less developed in North America (Fig. 10.36B) because much of the silt was trapped in glacial lakes and seas. Furthermore, an extensive loess area is located at low latitudes, away from the glacier as glacial debris was redistributed along the Mississippi River drainage. Thus, this loess in southern regions may have been affected by cold, but not permafrost conditions.

c. Finally, differences in permafrost development between Europe and North America can be, in part, attributed to different atmospheric circulation that occurred over the two continents during the glacial periods. The presence of the Cordilleran Ice Sheet in western North America led to a split of the jet streams (air currents in the high atmosphere). One stream carried precipitation to the southwestern and northeastern United States, and one to northwestern Alaska. Between them was an area of cold conditions (an average of 6° lower than at present) across central North America. Permafrost primarily developed along this cold middle corridor. In Europe the jet stream was diverted south across the Mediterranean area. Precipitation-bearing air masses crossed this area from west to east, whereas cold, dry masses moved from east to west across central Europe in front of the glaciers. As a result, temperatures plummeted, in some areas 15° below present values. This, plus the exposure of much of the wide, shallow marine shelves of western Europe during deglaciation, minimized any maritime influence on the continental areas, and permafrost developed widely.

ENDNOTES

1. Cracking does not necessarily occur ever year but on average about 30–40% of the time, and for reasons not completely understood, many of the earlier cracks are reactivated; they are the weakest part of the system.

2. Drunken forests are generally considered to be the result of frost heave rather than thaw degradation.

3. Water becomes *overpressured* when the pressure exerted on a unit fluid volume is greater than that due to the fluid weight itself. This occurs when the fluid is trapped in impermeable or low-permeability settings and subjected to external forces (such as weight of rocks, tectonic forces, or crystallization forces).

PART 4

Geology

CHAPTER 11

Quaternary Stratigraphy

PRINCIPLES

Stratigraphy is the study of the time-space relation of rock bodies, in particular of layered rocks. The term means "description of strata," such as those of sedimentary rocks derived from sediments most often laid down in near-horizontal fashion (strata), and other types of rocks (igneous and metamorphic) interlayered with them. Here, the stratigraphy of Quaternary glacial deposits is emphasized.

Stratigraphy uses the following three major basic principles:

1. **Uniformitarianism.** This states that the natural physical laws, which govern present-day events at Earth's surface, are the same laws that acted in the past. Another way of putting it is that *the present is the key to the past.* The concept relates to the fact that sedimentary rocks are formed from sediments, and that ancient sediments developed in much the same way as recent ones.

2. **Original horizontality** of rock layers. Most sediments deposited on basin floors are horizontal or have a very low angle of incline (only a few degrees). The resulting sedimentary rock layers are therefore horizontal or nearly so. Exceptions occur, for instance in sediments deposited against glaciers, or in steep deltas at the head of fjords. However, these sediments usually contain slump features, attesting to the particular setting in which they were deposited.

3. **Superposition.** This principle states that in an undisturbed sequence of sedimentary rocks, the bottom layer is the oldest and the uppermost one is the youngest. Although apparently trivial, this principle has far-reaching importance, because by knowing the relative age of layers, we can develop evolutionary sequences of fossil assemblages contained within those beds, without knowing the actual age in years of each fossil.

Associated with the stratigraphic principles are several biostratigraphic principles:

1. **Identification.** This relates to the fact that beds of sedimentary rock contain not only an accumulation of mineral grains, but perhaps also fossils. Once established, *fossil assemblages* can be used to identify any number of sedimentary successions, since stratigraphic units may contain the same fossil assemblage, regardless of the lithology (rock type) of the layers.

2. **Zonation.** This relates to the fact that fossils succeed one another in a definite, recognizable evolutionary sequence. Therefore, layers containing such fos-

sils can be subdivided into zones, each containing fossil assemblages signifying particular stages of evolution.

3. Correlation. This relates to the ability to correlate sequences of strata by position and age from one area to another. A stratigraphic sequence can be compared with another, a few meters, kilometers, or thousands of kilometers away, to establish the relative time of formation of strata at different localities.

To describe rock relations in space and time, stratigraphy involves five distinct but overlapping procedures:

1. Defining geological units
2. Tracing these units (space distribution and correlation)
3. Inferring time relations (time correlation and dating)
4. Determining controlling factors
5. Establishing a system (calibration and classification)

These procedures and basic stratigraphic principles can be applied to glacial sediments in much the same way as for other sediments.

DEFINING UNITS

Sedimentary units can be subdivided into suitable entities for mapping and analysis. Different criteria are used in these subdivisions, depending on the objectives of the work. In general, the sedimentary sequences are subdivided into *rock units,* and into *time-stratigraphic units* that can be laterally correlated with similar units at other localities (Table 11.1). *Rock (lithostratigraphic) units* are bodies (layers, beds) of distinct rock types; *biostratigraphic units (zones)* are bodies of strata containing particular fossil assemblages; *time stratigraphic (chronostratigraphic) units* refer to rocks formed during a specific interval of time, independent of rock type and thickness; and finally, *geological-time (geochronologic) units* refer to abstract time denominations corresponding to the time-stratigraphic unit: for instance, rocks of the Devonian system were deposited during the Devonian period. The fundamental working rock unit is the **formation,** which is defined as a group of strata formed under uniform or regularly alternating environmental conditions. The formation is usually what geologists map; therefore it should be thick enough and widespread

TABLE 11.1   Time and Rock Units[1]

Geochronologic units	Chronostratigraphic units	Lithostratigraphic units
Eon Era Period Epoch Age	Eonothem Erathem System Series Stage Zone	Supergroup Group Formation Member Bed

[1]Chronostratigraphic (time-rock) units represent the rocks that belong to a geochronologic (time) subdivision. Lithostratigraphic (rock) units have no time connotation; they are usually based on purely lithological characteristics.
After Shenk and Muller 1941.

enough to be recognizable and mappable over a relatively large area. Distinct units within formations are usually called **members.** Assemblages of formations are called **groups,** and assemblages of groups are called **systems.** Most of these units, and invariably the larger ones, are subdivided by major time breaks (unconformities) in the geological record.

The basic premise in the division of sedimentary successions is that the resulting units must be readily recognizable in and adjacent to the area where they were defined **(type section and locality).** However, away from the type section, the units may grade into or interfinger with others. Where the original units defined at one locality are difficult to recognize with distance from it, or disappear (not deposited or eroded off), a new type section is established and new units are identified that may differ in age and/or lithology.

SPACE DEFINITION AND CORRELATION

Space definition and correlation involve physically tracing the units defined in one place into other areas. This can be done in several ways, such as tracing strata and identifying their mutual interrelations as well as the time breaks between them.

Tracing Strata

Tracing can be done either directly or indirectly. Direct tracing can be done by following individual beds or distinctive units along cliff faces or in seismic profiles (Fig. 11.1). Indirect tracing can be done by comparing and recognizing characteristic individual beds or sets of beds between separate exposures.

While tracing the individual units, the geometric relations of the strata can be described with the following geometric terms, some of which derive from relations seen in seismic profiles (Fig. 11.2A).

Onlap shows successive layers extending up an inclined surface. It usually indicates a transgression.

Downlap shows successive inclined layers ending at an underlying surface. It usually indicates a buildup of sediments outward into a basin.

Toplap shows inclined layers prograding upward and asymptotically into an overlying surface. It usually indicates a subaqueous nondepositional surface of a sedimentary deposit.

Truncation shows inclined layers cut off at their tops. It usually indicates an erosion surface formed during regression.

Breaks in the Record

Breaks in sedimentation are common in all stratigraphic successions. They can be related to small, local, or major widespread fluctuations. Short time breaks caused by local variations are called **diastems.** Breaks caused by major changes that delete a considerable amount of the geological record are called **unconformities** (Fig. 11.2B). There are several types of unconformities, which describe relations of layers above and below.

1. A **nonconformity** develops when sediments are deposited over a substrate of crystalline rocks. In this case, there is a large time gap between the formation of the crystalline substrate and the overlying sedimentary rocks.

2. An **angular unconformity** occurs when a set of sedimentary layers overlay another set that are tectonically deformed, indicating a major time gap. Minor deformations, though, may also occur during deposition, such as in glaciotectonic units at the base of a glacier.

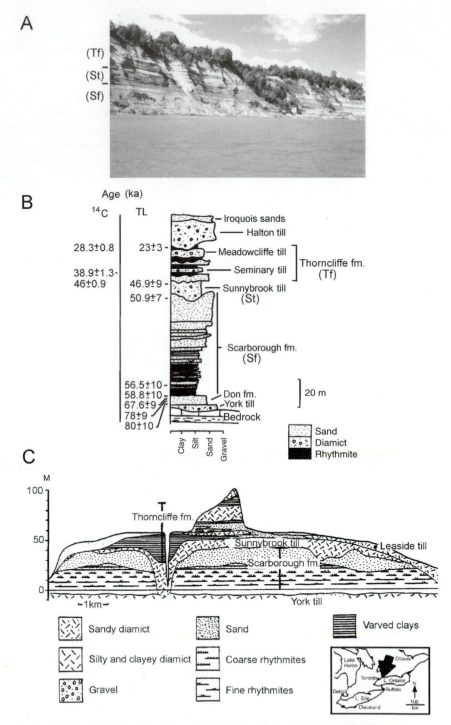

FIGURE 11.1 Scarborough Bluffs, Ontario. *(A)* Photograph of layer-cake deposits. *(B)* Generalized vertical section subdivided into lithological units (After Eyles and Westgate 1987). *(C)* Cross section showing lateral distribution of lithological units (After Karrow and Morgan 1975).

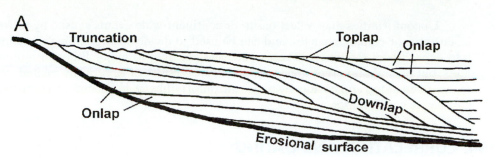

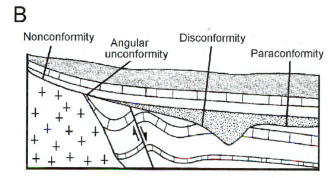

FIGURE 11.2 Stratigraphic relations between units. *(A)* Diagram showing names used to describe the geometric relation of strata (After Mitchum, Vail, and Sangree 1977). *(B)* Diagram showing major types of unconformity between stratigraphic units (After Dunbar and Rodgers 1957; Prothero and Schwab 1996.)

3. A **disconformity or erosional unconformity** exists when sequences of concordant strata (that is, strata with similar attitude) are separated by a physically identifiable erosional surface. Erosional events are more common in some areas than others. For instance, much erosion may occur in terrestrial overland settings and little of the original sediments may be preserved. However, in basin margin settings, such as alluvial fans and deltas, or in the deeper parts of depositional basins (lakes, seas), sedimentation prevails and erosion is at a minimum. Therefore, at a given time, erosion may be occurring in certain places while sedimentation continues in others, giving rise to unconformities and correlative conformities. Significant erosion of deep depositional basins occurs only during a major tectonic upheaval and uplift.

4. A **paraconformity** occurs when the succession of strata looks continuous and all beds have the same rock type, but the geological record is missing, as demonstrated by missing fossil assemblages that are known to exist in other areas. In this case, either some layers were never formed or have been eroded.

Diastems and unconformities can be used for correlation if they can be adequately distinguished either because of difference in lithology or structure of the rocks above and below, or because of features like soils. Soils form on a relatively stable surface not affected by much sedimentation or erosion. So they mark surfaces of unconformities exposed inland. Soils have been used extensively in establishing and correlating stratigraphic units, such as in the Quaternary glacial and interglacial successions of the European Alps, Europe, and North America.

Unconformities may reflect basin- or continent-wide events caused by tectonic, eustatic, or climate changes, and can be used to divide stratigraphic successions into **depositional sequences.** These sequences are akin to major lithostratigraphic units, but by definition they are composed of successions of genetically related strata, bounded at top and bottom by major unconformities or, in basinal areas, by their correlative conformities.

TIME CORRELATION AND DATING

Time correlation is finding out which units, regionally or worldwide, are of the same age. Dating is finding out how old they are. This can be either relative (that is, older/younger relations) or numerical (that is, ages in standard increments such as years). Time correlation and dating can be achieved with various methods and various degrees of precision and accuracy (Table 11.2). The method used should be the one best suited to the specific age and type of sediment/rock being examined and the resolution required. Here, a few examples are provided of the most useful or more recent methods applied to Quaternary deposits.

Time Correlation

Time correlation gives age equivalence but no actual date, relative or numerical. It depends on identifying and tracing unusual and widespread strata (key layers) formed by relatively intense but short events.

Ash falls (tephrochronology) are sudden, almost instantaneous events that deposit layers usable for accurate and precise correlation wherever they can be correctly identified in separate sections. Individual layers from particular eruptions may be characterized by a distinctive mineralogy or geochemistry. For example, the

---

TABLE 11.2   Correlation and Dating Methods

Type of Result					
Numerical Age →	Calibrated Age	Relative Age	Correlated Age →		
Sidereal	Isotopic	Radiogenic	Chemical and Biological	Geomorphic	Correlation
Historical records	^{14}C	Fission track	Amino acid racemization	Soil profile development	Tephro-chronology
Dendro-chronology	K-Ar and $^{39}Ar/^{40}Ar$		Obsidian hydration	Rock and mineral weathering	Paleo-magnetism
Varve chronology	Uranium series		Lichenometry	Progressive landscape modification	Stable isotopes
		Experimental methods			
	Uranium trend	Thermo-luminescence	Rock varnish chemistry	Geomorphic position	
	Cosmogenic isotopes	Optically stimulated luminescence			

After Colman, Pierce, and Birkeland 1987.

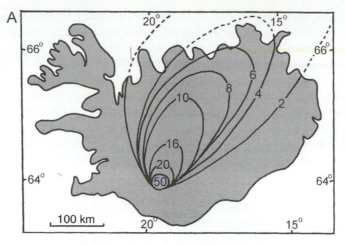

FIGURE 11.3   Volcanic terrain in Iceland. *(A)* Map showing distribution and thickness of a tephra layer from an eruption of Heckla 2800 years ago (After Einarsson 1994). *(B)* Soil profile showing three distinct ash layers in northern Iceland.

may be characterized by a distinctive mineralogy or geochemistry. For example, the stratigraphy of the last 5000 years in northern Iceland is based on the identification and correlation of variously weathered major ash layers (Fig. 11.3).

There are some limitations to the use of this method because ashes from any one volcanic eruption are not uniformly distributed onto the surrounding terrain, and at times it is difficult to distinguish the ash layers even if they are there.

Rapid sea level changes. A rapid rise in sea level can flood large areas of continental shelves, depositing thin layers of characteristic deepwater sediments on shallower-water deposits. The change from shallow to deep sedimentary facies may thus be quasi-contemporaneous across the area affected. During the Quaternary, the rapid melting of continental ice sheets caused global sea levels to rise at an estimated rate of about 100 m per 10,000 years. In some enclosed seas or large lakes, water levels rose a few meters almost instantaneously due to meltwater megafloods.

Paleomagnetic reversals have occurred throughout Earth's history. Similar patterns of reversals can thus be traced from one section to another. Like the other methods of correlation, approximate age equivalence must first be established since patterns of different age can be similar. The Quaternary polarity time scale is now dated with fairly precise $^{40}Ar/^{39}Ar$ radiometric dating and is consistent with oxygen isotope stages (see below).

Stable isotopes exist in well-defined ratios in the oceans and atmosphere, controlled by dynamic processes. These ratios change as the importance of the various processes changes through time. Patterns of changing isotope ratios can then be traced from one section to another. In the Quaternary, only oxygen isotope changes have sufficiently narrow time discrimination to be useful.

Oxygen isotopes. Oxygen comes in three stable isotopes, of which ^{16}O is the lighter and dominant (99.8%) and ^{18}O is the heavier and less common (0.2%) one. During water evaporation the heavier molecules ($H_2^{18}O$) are left behind; so rainwater and freshwater are enriched in $H_2^{16}O$. From this, two main correlation/dating procedures were developed using annual variations in ice layers and longer-term variations in seawater and the carbonate precipitated from it.

1. The first procedure is based on the premise that the isotopic composition of water vapor depends on the temperature contrast between the atmosphere and the water from which it evaporated. So the vapor and the precipitation that derive from it are isotopically heavier in winter, when temperature contrasts are less, than in summer. Ice cores from glaciers show isotopically light summer layers alternating with isotopically heavy winter layers. If the layers are undeformed, such a pattern can be used to correlate and date ice cores in the same way as annual rings can be used to date wood. Though mid-latitude mountain ice caps have only young ice, the ice sheets of Antarctica and Greenland contain much older ice. The most spectacular result of oxygen isotope analysis of ice cores from Greenland was that the last interglacial showed extreme fluctuations in ^{18}O values, suggesting cold to warm alternations within decades, in contrast to the present interglacial.

2. The second procedure considers that there is normally a dynamic balance in the oceans between loss of $H_2^{16}O$ by evaporation and gain of $H_2^{16}O$ supplied by rivers. But if water is locked up in large continental ice sheets (or lakes), then ocean water becomes relatively enriched in $H_2^{18}O$ (Fig. 11.4). So the relative concentration of the oxygen isotopes of the oceans is related to the surficial water temperature and the amount of water trapped in the inland glaciers. Another thing to keep in mind is that calcium carbonate deposited inorganically has an isotopic ratio of $^{18}O/^{16}O$, close to that of the water from which it precipitated (expressed as a deviation, $\delta^{18}O$, from a standard). Values of organic carbonate may be close to or significantly different from those of the surrounding waters, depending on the physiology of the organism secreting the carbonate. Foraminifera secrete stable calcite whose oxygen isotopic ratio is close to that of the seawater in which they live. The oceans are stratified. Near-surface seawater, down to the thermocline, roughly follows seasonal changes of temperature, so the oxygen isotopic ratios of planktonic foraminifera reflect both climate and ice-volume effects. Deeper water, instead, is colder and controlled by deep underflows of polar water, and skeletons of benthic foraminifera acquire different isotopic ratios than those of planktonic foraminifera. Temporal oxygen isotope shifts through time derived from benthic foraminifera are practically independent of temperature and reflect mostly ice-volume changes.

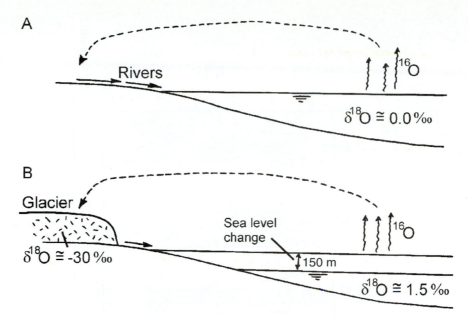

FIGURE 11.4   Diagrams showing relative concentration of ^{18}O in seawater depending on the amount of ice that traps the lighter ^{16}O inland. *(A)* Interglacial condition: light ^{16}O evaporates from the oceans and is carried landward and poleward. However, as much water evaporates as is returned by overland and ground flows, so the ^{18}O ratio remains constant in the oceans. *(B)* Glacial condition: the evaporated water enriched in ^{16}O, thus with a negative ^{18}O ratio, is trapped in inland glaciers, and the ocean waters become enriched in ^{18}O, so their ^{18}O ratio increases.

Measured detailed benthic isotopic shifts can be closely matched from different seas and allow precise correlation of marine sediments. These oxygen isotope variations are used to define **oxygen isotope stages** that reflect ice volumes and, by inference, the past warm and cold alternations of interglacial and glacial periods.

The isotope stages are numbered backward, with the interglacials being odd numbered (the present interglacial is 1) and the glacials even numbered (the last glacial is 2). A detailed pattern of oxygen isotope changes extends back to isotopic stage 103, and is calibrated with independently dated magnetic properties of rocks back to about 2.58 Ma (Fig. 11.5).

Dating

Dating can be of two types, relative dating and numerical dating.

Relative Dating   This technique gives older/younger relations but no age in years. Like time correlation, relative methods require calibration with suitable numerical methods to obtain actual ages in years. Many of the procedures used in time correlation can also be used for relative dating.

Biostratigraphy uses evolutionary changes in fossils to define relative time intervals. Large changes define the geological **periods,** and smaller changes refine these into shorter relative time units like **series, stages,** and **zones** (Table 11.1).

For pre-Quaternary times, relative dating of sedimentary rocks is routinely done by utilizing associations of fossils that evolved rapidly and were widespread and numerous. However, few organisms evolved rapidly enough during the Quater-

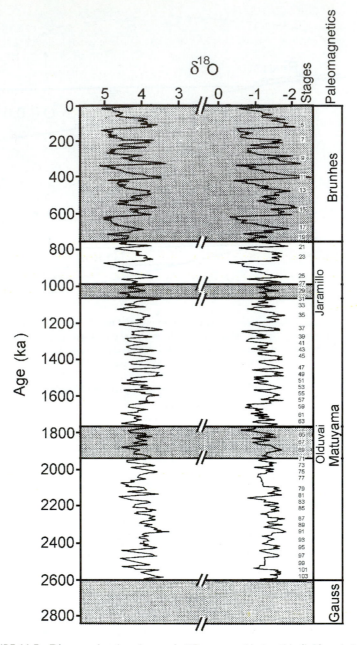

FIGURE 11.5   Diagram showing changes in ^{18}O measured in benthic (left) and planktonic (right) foraminifera, correlated with paleomagnetic stages. (After Ehlers 1996; Shackleton, Berger, and Peltier 1990.)

nary to erect useful biostratigraphic units, and so these are relatively coarse. Also, environmental gradients were steep and migrated during Quaternary glacial advances and retreats. Most changes in fossil biota in one area were caused by organisms tracking migrating environments and were not due to evolutionary changes.

On land, the most useful biostratigraphy comes from pollen and macrofossils of flora as the plants recolonized recently deglaciated or never-glaciated cold areas. Mammals are also useful because they spread quickly, although their fossils are few and their distribution irregular. The best biostratigraphy in Eurasia comes from rodents like voles, lemmings, and their relatives. Water voles show two especially striking transitions, at about 3.25 Ma and another at about 800,000 to 600,000 BP, which can be used for European correlation.

In the sea, microfossils such as foraminifera are routinely used, considering their presence or absence, number of individuals, and shape. Some forms reflect cold-loving populations, others are warm loving, and their distribution can record paleochanges in selected basins.

In every case, correlating relative time units based on different organisms in different environments remains a big problem; for example, it is difficult correlating time units based on planktonic foraminifera in the sea with those based on small mammals on land.

Amino Acid Racemization   Many organic molecules exist in two forms that are mirror images of each other and are called stereoisomers. Levorotatory (L) forms rotate light to the left, while dextrorotatory (D) forms rotate light to the right. Living organisms incorporate only L-amino acids. But, after death, the protein-bounded L-amino acids undergo a slow conversion to D-amino acids until the ratio reaches 50:50. The D/L ratio increases from 0 to 1 over time. This conversion (or **racemization**) takes place at a rate that increases with ambient temperature and with the presence of acids or alkalis in the system. The rate of change, therefore, varies from place to place and from time to time and cannot be used to numerically date samples everywhere. Nevertheless, within one system it can give relative ages for samples with the same burial history and can be used to correlate nearby units containing samples with similar D/L ratios. Individual localities can be calibrated with carbon dating of the same samples back to about 40,000 BP. The main problem is in determining the temperature history of the analyzed samples, a difficult undertaking for terrestrial Quaternary materials given the wide range of temperatures they likely experienced.

Fluorine in Bones   In general, the fluorine content of bones increases over time, as fluorine substitutes for OH^- ions in hydroxyapatite as it changes to fluorapatite. The transformation depends particularly on the rate of water infiltration, its temperature, and the fluorine content. So, the method gives only an approximation unless a local standard can be established by independent dating of several samples. Nevertheless, the fluorine method proved that the Piltdown human-skull fragments were only about 500 years old, and not the hundreds of thousands claimed.

Obsidian Hydration   Obsidian is a volcanic glass that typically contains 0.1% to 0.3% water when it forms. An exposed surface absorbs more water to form perlite, with about 3.5% water. Water absorption decreases obsidian density and increases the refractive index of the surface. If the growth rate of the surface hydration layer can be established, then the thickness of the hydration layer can be used to work out an age. The accuracy of the age depends on the constancy of the hydration rate. Although limited in application, obsidian hydration rinds have been calibrated with K-Ar dating of lavas in the Yellowstone region, and the results have been used to revise the ages of late Quaternary moraines.

Lichenometry   Lichens colonize bare rock surfaces and grow very slowly. The radius of lichen colonies can provide relative ages for, and correlation of, a series of exposed surfaces. The method is limited to the life span of individual colonies (that grow more slowly and last longer in colder conditions), around 9000 years in the frigid north and around 3000 years in temperate deglaciated areas. It has been used extensively for dating moraines, with the assumption that boulders at their surfaces are colonized soon after emerging from under the ice.

Soil Profile Development   Soil profiles take time to develop, depending on the substrate, the climate, the vegetation, and the slope gradient. Under stable climatic conditions, soil profile development progressively slows as equilibrium between gains and losses of ions is reached between the various horizons. In tropical Costa Rica, for example, podzols on volcanic ash can form within 400 radiocarbon years, and they then take about 2000 years to reach equilibrium. After this the B-horizon simply gets progressively thicker and richer in iron oxides.

Numerical Dating   This method gives quantitative estimates of age. It requires that a constant rate of change be recorded as increments or as accumulated gains or losses. The incremental units are normally calendar years. Numerical ages were once called absolute ages, but the term "absolute" is now falling into disuse because it is inappropriate.

Direct Counting Methods   The simplest methods count annual events from the "present" time (BP = Before Present) (generally taken to be 1950), or in relation to the birth of Christ (BC = Before Christ; AD = *Anno Domini:* that is, the year of God, meaning after Christ) and give ages in calendar years.

Historical records, like manuscripts, artwork, and photographs, record events and the appearance of rocks and landscapes at noted dates. For example, Pliny recorded the effects of the eruption of Vesuvius in AD 79, and diaries and court records extend as far back as AD 1000 in China. Sketches and photographs of European Alpine glaciers are routinely used to show in detail how they have receded since the early nineteenth century. But historical records are sporadic, sometimes inaccurate, and extend backward in time only a short way.

Dendrochronology is the study of annual tree rings. Tree rings form patterns related to variations in climate during growth, and the patterns of a living tree can be counted backward from the present. A minimum estimate for the age of a glacial deposit is the age of the oldest living tree growing on it. By overlapping the patterns of living trees with those of dead trees, the chronology can be extended further back (Fig. 11.6). Trees with a long life span, like the bristlecone pine *(Pinus aristata)* of the southwestern United States, have locally extended dendrochronology backward to over 7000 years BP. The most important use of dendrochronology is in calibrating radiocarbon ages to true calendar years (see below).

Varve chronology is based on the rhythmic repeated couplets of coarser- and finer-grained sediment deposited annually in glacial lakes. Like tree rings, varves and their patterns can be counted backward through successively older lakes. In Sweden, a chronology back to about 17,000 BP has been established, although its accuracy is best for the last 12,000 years. The main problem is establishing that rhythmites represent true varves and are not multiple summer inflow events. If enough organic matter is preserved in the varves, they can be used to calibrate radiocarbon ages in the same way as tree rings, but further backward in time. Alternatively, the age of an isolated varved sequence can be worked out with radiocarbon ages of the organic matter.

Isotopic Decay Methods   Isotopic decay methods are based on the constant decay of an unstable radioactive isotope and the ability to measure the amount of original isotope lost or the amount of daughter isotope produced.

The constant decay of radioactive isotopes is unaffected by the physical or chemical environment. Thus, a particular isotope decays at a constant rate, whether

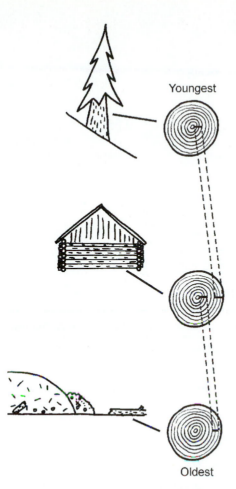

Youngest

Oldest

FIGURE 11.6   Diagram showing methodology of counting and correlating tree rings to establish age and past climates of live and dead trees. The dating (number of annual rings) and climatic information (pattern and thickness of rings, reflecting conditions required for growth) can be extrapolated backward in time by correlating similar ring patterns. The process starts by comparing a living tree with progressively older ones found either in constructions or buried in sediments. Overlapping patterns suggest that different trees lived at the same time, and this can be used to extend dating to older times. (After unknown; see website for update.)

it is at high or low temperature and pressure, in a solid, liquid, or gas, or subject to chemical changes such as weathering.

Isotopic gains or losses are normally measured as isotopic ratios in the sample because it is difficult to measure the actual isotope amounts, though comparatively easy to measure ratios accurately by using a mass spectrometer.

The half-life of the radioactive isotope determines the applicability of various isotopes. Radioactive isotopes decay exponentially, so that during each successive time increment half the preceding amount decays. At six half-lives 1.56% of the original isotope is left, and this decreases by only 0.78% in the next increment (Fig. 11.7). Due to this small change, analytical errors surpass the incremental changes after about six half-lives and the dating errors become large.

Because of this, only a few naturally radioactive isotope decay paths are useful for geological dating. These are ^{14}C, U-Pb (uranium series), Rb-Sr, K-Ar, Sm-Nd, and Lu-Hf. The first two (^{14}C and some of the uranium series) have half-lives short enough to date Quaternary materials fairly accurately.

In the **radiocarbon method,** ^{14}C is produced by cosmic bombardment of ^{14}N in the upper atmosphere. The ^{14}C then reacts with oxygen to form stable carbon in CO or CO_2, which then mixes rapidly throughout the atmosphere. A dynamic equilibrium ratio is set up between the rate of production of ^{14}C and its decay back to ^{14}N over time. During life, organisms (plants and animals that eat them) incorporate or-

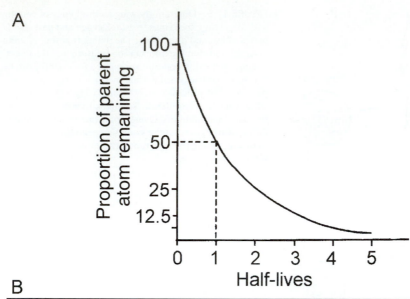

A

B

Isotope	Material Dated	Potential Range (years BP)	Half-life (years)
Carbon 14	Wood, charcoal, peat, CaCO$_3$ shells	< 60 ka	5770
Protactinium 231	Deep-sea sediment, shell, coral, travertine	< 120 ka	32.4 ka
Thorium 230	Deep-sea sediment, shell, coral, travertine	< 350 ka	75.2 ka
Uranium 234	Coral	< 250 ka	250 ka
Chlorine 36	Groundwater	< 500 ka	301 ka
Beryllium 10	Deep-sea sediment, polar ice	< 15 Ma	1.6 Ma
Potassium 40	Volcanic ash, lava	1−> 4500 Ma	1,300 Ma
Rubidium 87	Igneous and metamorphic rocks	10−> 4500 Ma	48,800 Ma

FIGURE 11.7 Radioactive dating. *(A)* Diagram showing half-life decay of radioactive isotopes. *(B)* Table reporting half-lives and the potential useful dating range of various isotopes.

ganic carbon in their tissues with the isotopic ratio of the contemporary atmosphere or water (Fig. 11.8). When the organisms die, the incorporated ^{14}C decays and the radioactivity of the dead organism declines. If the original amount of ^{14}C in the living organism is known, then the residual activity in organic remains can be used to determine its age.

The half-life of ^{14}C is 5730±40 years, so the method is reasonably accurate within the last 35,000 years or so (six half-lives). The method depends on knowing the original ^{14}C/^{12}C ratio in the living organism (since the decay product ^{14}N is a common gas).

All organic matter can be dated with this method. The best results however, are obtained from material rich in organic carbon, such as charcoal. Care should be taken to sample and date other materials because of potential contamination. Contamination can occur either through exposure of specimens to groundwater at burial or even during their initial formation in the basin through water mixing. For example, this may occur in oceanic upwelling zones where deep, cold waters mix

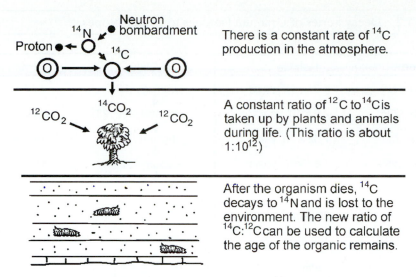

There is a constant rate of ^{14}C production in the atmosphere.

A constant ratio of ^{12}C to ^{14}C is taken up by plants and animals during life. (This ratio is about $1:10^{12}$.)

After the organism dies, ^{14}C decays to ^{14}N and is lost to the environment. The new ratio of $^{14}C:^{12}C$ can be used to calculate the age of the organic remains.

FIGURE 11.8   Diagram showing formation and cycle of ^{14}C isotopes in the atmosphere and organic matter. (After West 1968.)

with warm surface waters. These waters have different carbon-isotope ratios that are reflected in the carbonate skeletal material formed. Errors varying from a few hundred to a few thousands years are possible. Furthermore, the measured ^{14}C ages need correction (calibration) to change them into calendar ages. Originally it was assumed that the atmospheric $^{14}C/^{12}C$ ratio remained constant through time. But checking ^{14}C dates against tree rings shows that this is not true and the $^{14}C/^{12}C$ ratio has changed markedly in the last 7000 years. Uncalibrated ^{14}C ages gave systematic errors in archaeological studies before the calibration problem was solved.

Radiocarbon dates are recorded with reference to a zero year (AD 1950) and using the original 5570 ± 30 years half-life. These dates can be readily converted to the new 5730 ± 40 years half-life by multiplying by 1.03.

The **uranium series method** is based on the shorter-lived isotopes within the complex U-Pb decay series (Table 11.3). Skeletal and chemically precipitated carbonate materials such as corals, mollusk shells, bones, travertine, cave deposits, and coatings on stones can be dated with this method. This is possible because uranium decay isotopes are virtually absent from water, and therefore the carbonate material can take up uranium but none of its decay products. So, provided a sample has remained a closed system, the amount of daughter isotopes measured will give a measure of its age.

The method depends on measuring the ratios of specific parent/daughter isotopes present in a material. The $^{230}Th/^{234}U$ pair is the commonest one used and gives a range roughly back to about 500,000 years. Other isotopic pairs used are $^{234}U/^{238}U$ (back to about 6 Ma), $^{231}Pa/^{235}U$ (back to 12–10 ka), U/He (back to 2 Ma), and $^{226}Ra/^{230}Th$ (back to 10 ka).

The ^{210}Pb isotope has a short half-life and is particularly useful in dating fine-grained lacustrine and marine sediments of the last 200 years or so. Lacustrine deposits can also be dated by the ratio of ^{230}Th and ^{231}Pa (Table 11.3). These two isotopes deposit quickly from solution at a constant ratio, but since their decay constants differ, their ratio changes with depth (age) in the sediments. Use of the isotope ratio rather than the concentration of either one alone is preferred. First, because it is easier to measure ratios than accurate absolute values. Second, because

TABLE 11.3 Decay Series of Uranium Isotopes to Lead Isotopes

Nuclide	Half-life	Nuclide	Half-life
Uranium 238	4.51×10^9 years	Uranium 235	7.13×10^8 years
↓		↓	
Uranium 234	2.5×10^5 years	Protactinium 231	3.24×10^4 years
↓		↓	
Thorium 230	7.52×10^4 years	Thorium 227	18.6 years
↓		↓	
Radium 226	1.62×10^3 years	Radium 223	11.1 days
↓		↓	
Radon 222	3.83 days	Lead 207	Stable
↓			
Lead 210	22 years		
↓			
Polonium 210	138 days		
↓			
Lead 206	Stable		

After Brigham-Grette 1996.

assumptions of uniform rate of sedimentation are unnecessary to establish the age of the deposits. The only assumption required is that the parent ^{235}U and ^{238}U are thoroughly mixed in the lake or ocean so that ^{230}Th and ^{231}Pa are produced everywhere in a constant ratio.

Fission track dating requires uranium-bearing minerals (like zircon, sphene, apatite) or glasses (like obsidian). A fission track marks microscopic radiation damage in a crystal or glass caused by the breakdown of radioactive uranium and thorium isotopes. Of these, the rare isotope ^{238}U has a fission half-life short enough to produce enough tracks to measure. The tracks produced are a few micrometers (10^{-6} m) wide and less than 20 μm in length: they are longer in lower-density minerals and glasses than in dense minerals like zircons. Etching of polished sections makes the tracks visible, and thus they can be counted microscopically.

The age of a mineral or glass can be calculated from the amount of uranium it has and the number of spontaneous fission tracks it contains. Because the relative abundance of ^{238}U and ^{235}U is constant in nature, the easiest and most accurate way to determine the amount of uranium present is to create a new set of fission tracks by irradiating the sample in a nuclear reactor with a known dose of thermal neutrons. The induced track density resulting from induced fission of ^{235}U (a common isotope) is a function of the amount of uranium present in the sample and the neutron dose it got in the reactor.

Fission track dating of minerals and glasses from igneous rocks gives only cooling dates, since tracks fade and disappear (anneal) on heating. Tracks are stable in most transparent minerals below about 50°C (though this temperature varies with the individual mineral and its history and is lower for glasses). Fission track ages are thus extremely useful in working out thermal histories. In glacial studies, fission track dating can give an eruption age for volcanic ashes, since all are hotter than 50°C when they erupt (U-Pb dates from the same mineral give a crystallization age that might be very much older).

Most radiometric dating methods used in the geological record (*U-Pb, Sm-Nd, Lu-Hf, Rb-Sr*) use radioactive isotopes with long half-lives and are not precise enough for the Quaternary. The **potassium-argon (K-Ar)** and related **argon-argon ($^{40}Ar/^{39}Ar$)** methods can be used for the whole Quaternary, up to as recent as 12 ka in fresh volcanics.

The K-Ar method is widely used since potassium is a common element found in many different minerals and rocks. Potassium has three isotopes: ^{39}K (93.3%), ^{41}K (6.7%), and radioactive ^{40}K (0.01%). ^{40}K decays to ^{40}Ca and ^{40}Ar, with a half-life of 1250 million years. ^{40}Ca is the common calcium isotope, so only ^{40}Ar production is useful for dating. The half-life is short enough to allow measurable decay in rocks as young as 1 million years, but young dates have quite large errors. Argon is an inert gas and easily leaks out of a mineral or rock. This loss takes place at different temperatures for different minerals. These cooling ages can help to evaluate the history of a rock as well as the time it was formed. Rapidly cooled lava gives ages of formation, and is particularly useful beyond the range of ^{14}C dating.

In the $^{40}Ar/^{39}Ar$ method, the amount of ^{40}K can be calculated. The short-lived (269 years) ^{39}Ar isotope is produced from ^{39}K by laboratory irradiation. The ^{39}Ar produced is proportional to the original ^{39}K, and this in turn is proportional to the ^{40}K, which can thus be calculated. Since minerals can reabsorb argon gas, inherited ^{40}Ar can also be a problem, especially in young samples and those with low K content. But the effect of this inherited ^{40}Ar can be evaluated with $^{40}Ar/^{39}Ar$ stepwise heating and K-Ar isochron modifications of the methods.

Luminescence dating. Luminescence dating uses the increase of radiation damage to minerals with age. Natural minerals have internal crystal (lattice) defects. Some form during crystallization; others are produced by irradiation. The defects trap electrons that are released by heating and give a measurable luminescence that is proportional to the number of trapped electrons. The traps are emptied by heating or prolonged exposure to light. Originally, luminescence dating was used to date pottery and burnt flint. It was then extended to date the last exposure of material to light prior to burial. It can be done on material not normally considered datable and in deposits laid down within the often-critical 40,000- to 300,000-year Quaternary window. This is the time period when other dating methods do not work well. Luminescence indirectly measures the number of electrons that migrate to traps (to particular kinds of impurities and crystal defects) as a result of adsorption of alpha, beta, gamma, and cosmic radiation by the mineral. Alpha, beta, and gamma radiation comes from radioactive isotopes and their decay products, while cosmic radiation comes from the Sun. To get a date, one needs to know (a) the total radiation absorbed by the sample since its last exposure to light (the equivalent dose, or ED), and (b) the rate of radiation (annual radiation dose) the sample was subject to. It is then theoretically possible to calculate an age. The trapped electrons can be released by heating (thermoluminescence [TL]), or by exposing the material to light of a particular wavelength (optically stimulated luminescence [OSL]). In both cases, light is given off and can be measured. The luminescence is reset (zeroed) at each exposure to light. However, a residual TL signal always remains, while a few minutes of exposure to light completely resets an OSL signal. Quartz and feldspar grains are usually used.

For TL, untreated samples and samples radiated in a laboratory are progressively heated and emission intensity curves for each plotted against temperature to give "glow curves." The older the sample is in the TL dating range, the higher the intensity of the glow curve. The higher-temperature parts of the curves are used because trapped charges are most stably stored there. Some estimate of residual TL also has to be made.

For OSL, decay curves for the samples can be measured after any thermally unstable components are removed.

The upper age limit for luminescence dating is saturation of storage within crystals. In Australia, ages of quartz grains from raised beaches (formed during interglacial periods) fitted the appropriate oxygen isotope stages back to 800,000 years. However, the relation of peak light intensity to age is not linear and depends on a number of poorly understood factors. Also, the annual radiation dose can be difficult to work out. For example, pore waters reduce the external radiation dose and the "wetness" of the sediment since deposition has to be estimated. Thus, the luminescence method needs independent calibration. This can be done by luminescence dating of materials of known archaeological age and dated by calibrated radiocarbon. Nevertheless, the most useful luminescence dates are those obtained from otherwise undatable materials. These usually have very large errors; for example, almost all TL dates over 300,000 years have errors between 15 and 25%. Young TL dates may be inaccurate because a residual dose needs to be estimated. OSL dates have no residual luminescence, and hence very young materials can be reasonably accurately and precisely dated.

For glacial material, TL inheritance is a problem. The time between release of silt from beneath a glacier and deposition of the silt may be too short to reset the TL. Because of rapid "zeroing," OSL dating would be preferable. Loess and eolian sands are notoriously difficult to date by conventional methods, but loess sections from China and northern Pakistan give consistent TL results, and sand deposits from some deserts could be dated in no other way.

Electron spin resonance. Electron spin resonance (ESR) has the same basis as luminescence, except that it uses microwave energy to measure the luminescence centers (defects) directly in situ. So, the unaltered samples can be used again. The method does assume the centers remain stable over a long period of time. But ESR dates of cave deposits do not fit U-Th dates on the same materials. Like luminescence, the method is imprecise, with large errors.

Cosmogenic isotopes. Cosmogenic isotopes rely on the formation and accumulation of isotopes produced by cosmic radiation.

Isotopes like ^{10}Be, ^{36}Cl, ^{3}He, and ^{26}Al accumulate in situ when exposed rock surfaces are bombarded by cosmic radiation. The rate of production of the isotope has to be known or inferred. Suitable rock surfaces for dating are those continuously exposed since a particular process formed them: for example, the time since exposure of glacigenic material after ice retreat. Problems with the method include estimating a mean cosmic flux and determining how much of the isotopes produced is retained in the rock. Rocks on top of well-dated moraines in California gave ages that were too young, probably because of weathering of their surfaces and hence removal of isotopes.

Radioactive ^{10}Be is also produced in the atmosphere by reaction of cosmic radiation, nitrogen, and oxygen. It attaches to aerosols, falls to earth, and attaches to surfaces. The ^{10}Be content is proportional to the rate of sedimentation and can be used to separate rapid from slow depositional periods. For example, the ^{10}Be variations in sequences of loess (rapidly deposited during cold phases) with paleosols (developed during nondepositional warm phases) have been used for correlation.

Glacial Legacy: Isostasy, Eustasy, Volcanism, and Biota

When large continental ice sheets form, they have three important effects on the whole Earth: isostatic, eustatic, and climatic. First, they load the land beneath them and cause it to subside (*isostasy*). Second, they extract water from the oceans and cause sea level to drop (*eustasy*). Third, they cool both land and sea, causing climatic belts to shift toward the equator and compress. Conversely, when ice sheets melt, the underlying land rises (or rebounds), sea level rises, and climatic belts move poleward, expanding as they do.

ISOSTASY

Isostasy is the adjustment of Earth's surface so that its gravitational potential is the same everywhere. A common analogy is an iceberg floating in water (Fig. 12.1A). Water is denser than ice. For freshwater at 5°C and ice at 0°C, densities are respectively 1.0 and 0.9 g/cm^3. So wood, ice, and water have different *volumes* (*v*) for the same *mass* (*d*):

$$d_{ice} \times v_{ice} = d_{water} \times v_{water}$$

To get the same mass as 1.0 m^3 of water, 1.11 m^3 of ice is needed. This is the basis of Archimedes' principle, which states that a floating body displaces its own weight of water. It exerts a force (mass × acceleration of gravity) equal to that of the displaced water. When an iceberg breaks off a glacier, a mass of water flows under or away from the floating ice equal to the mass of the iceberg.

So, the volume of a floating block (here ice) above an ambient fluid (here water) is proportional to the relative densities of floating block and fluid.

$$\text{volume}_{solid} \text{ (above fluid)} = \text{total volume}_{solid} \times (\text{density}_{fluid} - \text{density}_{solid})$$

$$va_s = vt_s \times (d_w - d_s)$$

From the densities given above, ice at 0°C will float with about one-tenth of its volume above water. Volume, rather than height or thickness, is a better measure because few floating objects are regular cubes.

Isostasy (floating equilibrium) can be changed by changing composition, density, or volume of the floating material, or all three together. It can also be changed by changing the composition and/or density of the supporting medium. Composition changes usually involve density change. For example, iceberg composition can be changed by adding stones (which also make it denser). A dirty iceberg will be heavier and float lower, and increasing the amount of salt in water makes it denser.

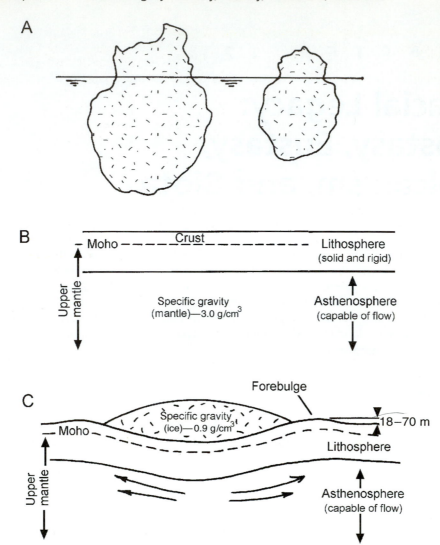

FIGURE 12.1   Diagrams illustrating the concept of isostasy. *(A)* Icebergs floating on water. *(B)* Schematic representation of lithological layers of Earth. *(C)* Diagram showing subsidence and deformation of Earth's lithosphere and migration of asthenosphere material due to the weight of ice sheets.

An iceberg floats slightly higher in seawater (density 1.03) compared to freshwater (density 1.00).

The lithosphere (the rigid outer part of Earth), analogous to an iceberg, is in floating equilibrium with the asthenosphere (the plastic upper mantle of Earth), analogous to the water (Fig. 12.1B). However, the lithosphere is not as rigid as an iceberg, nor is the asthenosphere as fluid as water. If Earth's surface is loaded or unloaded with extra weight, like sediment, an ice sheet, a lake, or more water in the sea, it takes some time to respond.

Glacial Isostasy

Unloading and loading of the lithosphere by large continental glaciers causes isostatic uplift and subsidence, respectively. The amount and rate of isostasy changes with distance from the growing or retreating glaciers. But, because the lithosphere has a certain rigidity, these changes take some time and the weight of the ice sheet is initially distributed over a wide area. In fact, the response time to glacial loading and unloading has been used to calculate the rigidity of the lithosphere and asthenosphere.

A large ice sheet loads the lithosphere and causes asthenosphere to flow away under the added weight, forming an isostatic depression, which can be complex. The ice (density 0.9–1.0) displaces asthenosphere (density 3.0–3.3). At any one point, the relative thickness (h) of ice and displaced asthenosphere is proportional to their densities (d):

$$h_{ice}/h_{asth} = d_{asth}/d_{ice}$$

With the densities given above, a 3000 m thick ice sheet (typical of the centers of North American and Eurasian ice sheets) will displace between 750 m and 1000 m of asthenosphere below it. Unlike an iceberg, both ice sheets and the asthenosphere flow under their own weight, and are incapable of supporting small-scale topography for any length of time.

These theoretical values are seldom, if ever, achieved because of the strength (resistance to deformation) of the lithosphere and the types of deformation that occur. Since the lithosphere has an initial rigidity or elasticity, the weight of an ice sheet first deforms the lithosphere elastically. The load is partly supported outside the ice margin (rather like a diving board or trampoline supports the weight of someone standing on them), and a forebulge and an ice-margin depression form (Fig. 12.1C). If the ice sheet lasts long enough (around 10,000 years), then the lithosphere starts flowing. To continue the above analogy, it is rather like standing on a diving board or trampoline made of toffee. In time, the marginal depression gets narrower and deeper.

The melting of large ice sheets causes the opposite. The lithosphere first rebounds elastically, and then isostatically more slowly as the asthenosphere flows into the rebounding area. The progressive rebound is often recorded by the shorelines of lakes and seas (Figs. 12.2, 12.3). These may be complicated by the influence of the migrating and collapsing forebulge, whose effects have rarely been considered. Isobases (lines of equal deformation of original surfaces often assumed leveled) have been used to reconstruct relative ice thicknesses and the center of ice caps, since the greater the rebound (deformation), the greater the original ice thickness.

Hydroisostasy

Changes in the size and depth of seas and lakes also cause loading and unloading (and isostatic responses), in the same way as ice sheets. For example, isostatic deformation below large reservoirs can cause earthquakes as the land adjusts to new equipotential surfaces. One example is the Pleistocene Lake Bonneville (western United States), which received a lot of water during wetter glacial times, but shrank considerably during drier interglacials. The hydroisostatic changes caused by the postglacial shrinking of Lake Bonneville are recorded by contours drawn on its former shorelines (Fig. 12.4).

Meltwater added to the oceans during deglaciation raises sea level. But it also causes isostatic subsidence of ocean floors, which increases the capacity of the oceans. If meltwater added to the oceans during deglaciation was sufficient to raise sea level by 100 m, then the absolute (eustatic) rise in sea level after isostatic adjust-

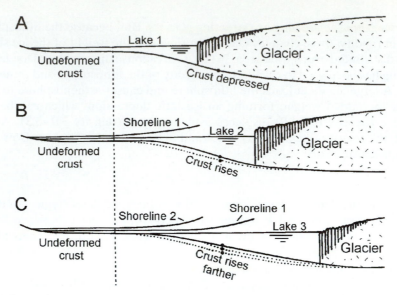

FIGURE 12.2 Diagrams showing lateral shoreline shift and isostatic uplift related to melting ice sheets. *(A)* Stage 1: proglacial lake forms a shoreline. *(B)* Stage 2: the glacier has retreated, the new lake forms a second shoreline, and the shoreline of the previous lake has been isostatically uplifted differentially with the greater uplift closer to the glacier. *(C)* The process repeats itself for the third and subsequent times until the glaciers disappear or the water is drained off. (After Flint 1957.)

ment would be less than this by about 34 m. In calculating this, we are assuming a seawater density of 1.03 g/cm^3 and asthenosphere density of 3 g/cm^3.

$$\text{Subsidence} = \text{thickness water } (h_w) \times \text{density seawater } (d_w)/\text{density asthenosphere } (d_a)$$

$$S = h_w \times d_w/d_a = 100 \times 1.03/3 = 34.3 \text{ m}$$

EUSTASY

Eustasy is the change in absolute sea level height. It is usually thought to occur in the same direction and magnitude worldwide, although this may not always be the case.

Absolute sea level is normally taken with reference to an ellipsoid of rotation (geoid), which takes the gravitational effects of Earth's spin into account. Deviations of absolute sea level from the geoid are independent of the surface features of Earth. They depend on the gravitational effects of mass variations deep within Earth. For example, absolute sea level in the northern Indian Ocean is now 180 m lower than in the North Atlantic.

Overall absolute sea level can change by near-surface processes like (1) adding or subtracting water, (2) changing the capacity of the ocean basins, or (3) changing the density of seawater. Additions and subtractions caused by changes in continental ice volume are relatively fast, on the order of thousands of years to quasi-instantaneous, such as when the very large glacial Lake Agassiz (central Canada) emptied catastrophically into the Gulf of St. Lawrence and the North Atlantic, locally raising sea level. Changing the average ocean temperature by 1°C

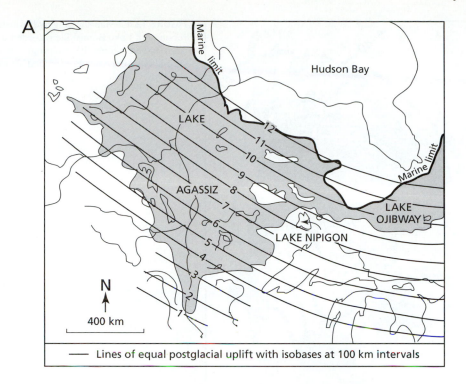

FIGURE 12.3   Glacioisostatic formation of shorelines. *(A)* Isobase map of Lake Agassiz numbered shorelines (after Teller 1995). (Although the map does not show it, there is greater isostatic rebound to the north, and thus the vertical separation between isobases increases northward). *(B)* Isostatically uplifted shorelines along the western coasts of Hudson Bay, Canada.

causes density changes that raise or lower sea level by about 1 m. However, there is no known way of measuring absolute sea levels in the past, only relative sea level.

Relative sea level is relative to the land. It is not the same as absolute sea level, and relative and absolute sea level changes may be opposite. If the land is tectonically subsiding, relative sea level can be rising while absolute sea level is dropping. Conversely, if the land is tectonically rising, relative sea level can be dropping while absolute sea level is rising.

Relative sea level changes are recorded in a variety of sedimentary deposits and landforms. In places once covered by ice, marine deposits resting on glacial tills

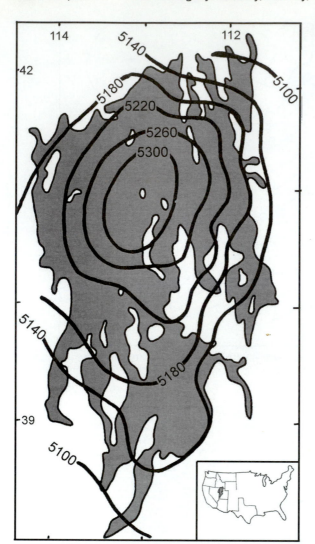

FIGURE 12.4   Map showing hydroisostatic rebound of Lake Bonneville recorded by shoreline elevations (ft) of a progressively shrinking lake. (After Benson and Thompson 1987; Forester 1987.)

record the Quaternary postglacial rise in sea level. In southern Scotland, marine clays that were deposited during the postglacial rises in sea level rest on tree stumps (and peat), which grew on glacial till as the ice retreated. These are presently covered by sand dunes and modern soils, now being eroded as rebound continues. This pattern is even recorded in a medieval Scottish rhyme: "first a wood, then a sea, now a moss and ever will be."

Relative sea level depends not only on eustasy and isostasy, but also on sedimentation. For example, the supposed site of Troy is now too far from the sea to fit the descriptions in Homer's *Iliad*. In early Quaternary times, at about 7000 years ago, the site of Troy was on an estuary, probably formed by postglacial sea level rise. When the Trojan War occurred around 3500 years ago, Troy was still coastal, and the story made sense to the contemporaries of Socrates and Plato some centuries later. Since then, sedimentation has completely filled the estuary with about 50 m of sediment, causing a relative drop in sea level (regression) (Fig. 12.5).

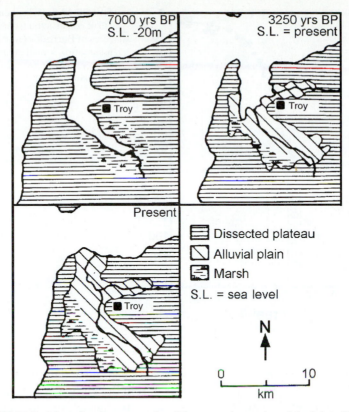

FIGURE 12.5   Maps showing progressive silting up and emersion of the inlet of Troy (S.L. = −20 m water level below present level). (After unknown; see website for update.)

INTERACTIONS BETWEEN ISOSTASY AND EUSTASY

The various isostatic and eustatic factors have been discussed separately here, but they are all related. For example, a relative rise in sea level along a coast may involve land subsidence (isostasy), absolute rise in sea level (eustasy), and a reduction in sedimentation rate related to climatic changes. Three examples are considered here, one where there is no direct influence of the glacier, but only of the rise and fall of sea level; a second where the glacier fronts a large sea or ocean; and a third in which relatively small seas and large lakes develop in the frontal moat of a retreating glacier.

1. Outside the area of postglacial rebound, but where postglacial submergence occurs, isostatic effects are much less than eustatic ones, but their interaction may be complex. For example, tropical areas have raised shorelines in part due to eustatic sea level fluctuations and, in part, due to local tectonic isostasy. It is in some of the most tectonically stable areas that glacioeustatic changes may be more accurately measured. Several sea level curves have been constructed. They generally show a rapid deepening of the sea from the start of deglaciation to about 5 to 6 ka (Fig. 12.6). A slowdown in sea level rise occurred at around 11–10 ka during the cold Younger Dryas. During the last 6 to 4 ka there has been a slow rise of sea level on the order of only a few centimeters per century. It has been estimated that sea level was approximately 120 m lower than the present level during the last glacial maximum (18 to 21 ka), and that, if the Antarctic and Greenland ice sheets were to

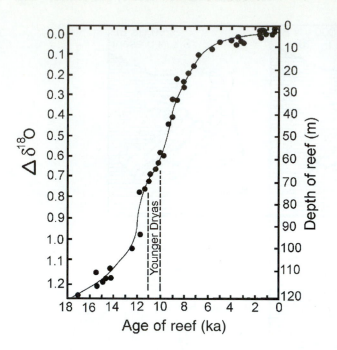

FIGURE 12.6 Glacioeustatic sea level curve constructed from information from reefs of Barbados (Δ = variation; the assumption is made that the surface of the reef is at or near sea level). (After Fairbanks 1989.)

melt completely, the total sea level rise would be about 200 m. It is unlikely, however, that the extra full 80 m rise will occur during the Holocene interglacial. However, higher sea levels can occur, as was experienced during the penultimate interglacial (Sangamon) in southern Florida (Fig. 12.7). The Florida Keys (islands) record the position of a reef tract similar to the present offshore one, but formed when the sea level was several meters higher.

2. When a glacier overrides and then retreats from a shallow marine shelf, the record reflects the combined effect of land subsidence and uplift and the change in sea level due to increased meltwater in the oceans (Fig. 12.8). When the glacier covers most of the shelf (S), sea level is low, the crust subsides (large arrows) in areas I, C, and S due to the weight of the glacier, and slight uplift occurs in the submerged bottom (area O) due to removal of water (Fig. 12.8A). As the glacier retreats, sea level rises and transgresses over the shelf (S). Subsidence continues in I and C, slight rebound of land in S as some of the weight of the glacier is replaced by water, and slight subsidence in O due to increased weight of water (Fig. 12.8B). As the glacier retreats inshore, sea level rises, and shoreline features form. Uplift occurs everywhere except in areas that continue to subside slightly due to increased water depth (weight) (Fig. 12.8C). Inshore isostatic uplift is not reduced by addition of water. After major retreat of the glaciers, absolute sea level, related to the total volume of water in the ocean, is stable or very slightly rising, but some regression and shallowing on the shelf occurs due to continuing land uplift (Fig. 12.8D). The isostatic uplift is progressively greater toward the centers of glaciation.

In deglaciated areas of Maine and Quebec, late Pleistocene deltas and other shoreline features are now found at elevations of about 80 m asl (above sea level). The complex curves showing the elevation of shoreline features over the past 14 ka indicate the complex interaction of glacial isostasy, hydroeustasy, and eustasy (Fig. 12.9). Anomalies in the curves may be interpreted in various ways. The temporal shift in lowstand from Maine (about 11 ka) and Quebec (about 7 ka) may be in-

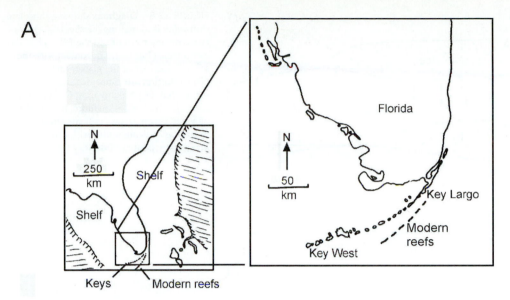

FIGURE 12.7   Southern Florida. *(A)* Map of the modern reefs and keys (islands). *(B)* Low-level air photograph of the Florida Keys.

terpreted as the effect of a northwestward migrating forebulge (about 20 m high), following deglaciation. The low rate of relative sea level change between 9.2 and 6.3 ka in Maine may have been related to variation in local rate of isostatic rebound rather than a global cooling climatic event.

3. Relative to marine settings, additional processes in glacial lakes further complicate the analysis of the postglacial interplay between water and land, and the resulting sedimentary record. For example, the amount of water in lacustrine basins is a function not only of glacial melting, but also of the elevation of the outlets. The Great Lakes of North America are a case in point.

During melting and retreat of the Laurentide Ice Sheet of North America, numerous large lakes and glacial seas developed along its southern margins. This has left a legacy of lacustrine and marine sediments along a north-south section more than 1500 km long, from the margin of the glacier to the centers of glaciation.

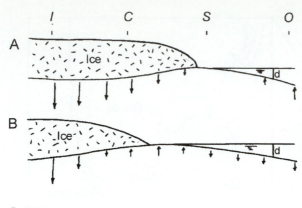

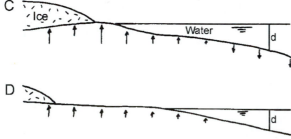

FIGURE 12.8 Diagrams showing shoreline shift related to melting ice sheets. *(A)* Glacier covering most of the shelf (S). *(B)* Glacier retreats but is still in contact with the sea. *(C)* Retreat of the glacier inshore. *(D)* Condition after most ice sheets have melted out. (*I* = inland area, *C* = coastal zone, *S* = shelf, *O* = offshore; d = depth of water; down arrows indicate subsidence, up arrows indicate uplift, with arrow length being an estimate of intensity.)

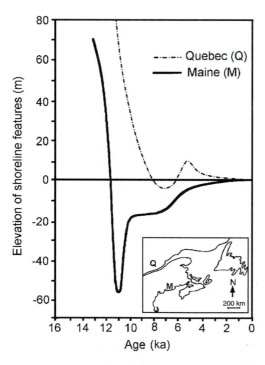

FIGURE 12.9 Late Quaternary relative sea level curves for Maine (M) and Quebec (Q), eastern North America. The curves are related to present sea level (0 elevation), and the data have been obtained from shoreline features uplifted inland or submersed in the shelf. (After Barnhardt, Gehreis, Belknap, and Kelley 1995.)

About 600 km of this section encompasses the Great Lakes area. Water level, erosion, and sedimentation patterns of these Great Lakes were a function of

a. isostatic depression due to the weight of glacier (Fig. 12.10A),

b. presence of glaciers that variously dammed and opened outlets (Fig. 12.10B), and

c. postglacial isostatic recovery: its timing, rate, and amount (Fig. 12.10C).

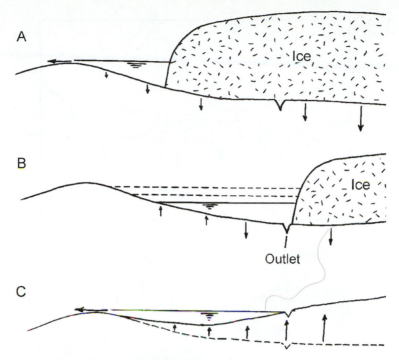

FIGURE 12.10 Diagrams showing subsidence and isostatic recovery of Earth's lithosphere related to formation and melting of glaciers. *(A)* Subsidence, glacial lake forms in front of the glacier, with potential outlets being dammed by ice. *(B)* As the ice margin retreats, depressed outlets are opened and lake levels drop, with strong regression occurring on shores distant from the glacier. *(C)* Differential isostatic recovery of the land, with faster and greater uplift in areas closer to the centers of glaciation. As the formerly depressed outlets rise, transgression occurs on distant lakeshores (arrows as in Fig. 12.8).

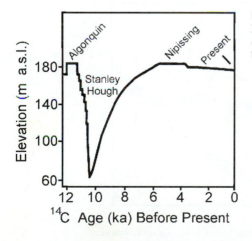

FIGURE 12.11 Relative lake levels curve for the Lake Huron basin based on elevation of shoreline features (names indicate some of the older lakes that occupied the basin). (After Lewis 1970.)

The interaction of these factors is recorded in the varying elevation of the lacustrine shorelines through time, such as those of Lake Huron (Fig. 12.11).

Four major outlets of the Great Lakes, with well-known history, form the framework for this discussion. These outlets are those at Chicago, Port Huron, Kirkfield, and North Bay (Fig. 12.12A). About 13,000 BP, large glacial lakes formed

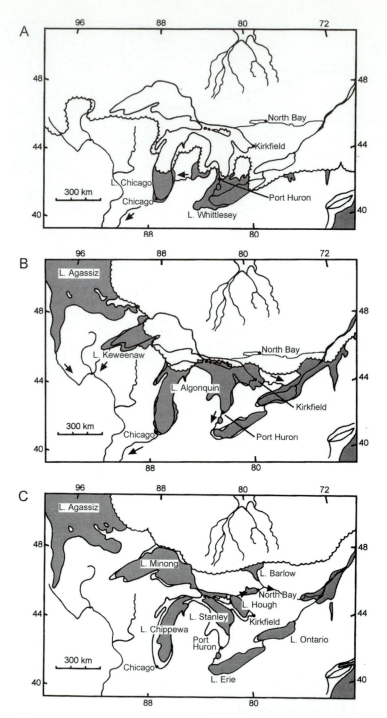

FIGURE 12.12 Map showing glacier and lake distribution in the Great Lakes area during the late Quaternary of North America. *(A)* About 13,200 BP, the thickest part of the glacier was located north of latitude 48°. Of the four potential lake outlets, Chicago, Port Huron, Kirkfield, and North Bay, only the first was active at this time. *(B)* About 12,000 BP, three outlets (Chicago, Port Huron, and Kirkfield) were active at the same time. *(C)* About 12,000 years ago, the northernmost, most depressed outlet of North Bay was freed from ice and siphoned out the waters from the lakes, and a large regression occurred on the southern shores. (Dark pattern = water; wavy line = approximate edge of ice sheet.) (After Prest 1970.)

in the isostatically depressed area in front of the glacier, all of which drained southward through the Chicago outlet. Erosion may have occurred at the outlets, beaches would have formed all around, and sedimentation, dominated either by land or glacial systems, would have occurred in the basins.

As the ice sheet retreated north, other outlets opened. A three-outlet stage (Chicago, Port Huron, Kirkfield) occurred at about 12,000 BP (Fig. 12.12B). Then the northernmost outlet (North Bay) emerged from thicker ice. It was more depressed than the southern ones, which, by that time, had already undergone considerable isostatic rebound. So the lake water was siphoned out through the northern outlet and a large regression occurred in the south (Fig. 12.12C). The glacial legacy continued after the ice retreated farther north still. The northern lands were subjected to faster and greater uplift than those to the south because, being closer to the centers of glaciation, they had experienced greater subsidence under thicker ice. In time, the differential isostatic uplift led to the rise of the northern outlets, and to transgression in the south. A three-outlet stage (Kirkfield, Port Huron, Chicago) developed again at about 6000 BP, until further tilting led to the present conditions when only the Port Huron outlet is active.

The various lake stages are marked by beach lines, which were formed, uplifted, exposed, and some drowned, depending on their location. A concept that has permeated the interpretation of the isostatic movement of these coastlines is that of the hinge line. The hinge line is considered to indicate the point (or band) south of which the coastlines were not greatly affected by differential isostatic uplift, and north of which the coastlines were asymptotically more uplifted toward the centers of glaciation (Fig. 12.12). The hinge line of subsequent lakes shifted northward as the glacier retreated in that direction (Fig. 12.13). This interpretation assumes

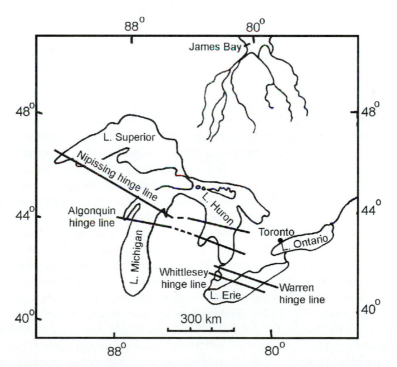

FIGURE 12.13   Map of the approximate position of the hinge lines of major lake stages in the Great Lakes region, North America. The younger hinge lines are progressively shifted to the north (dashed line = inferred position of hinge line.) (After Leverett and Taylor 1915.)

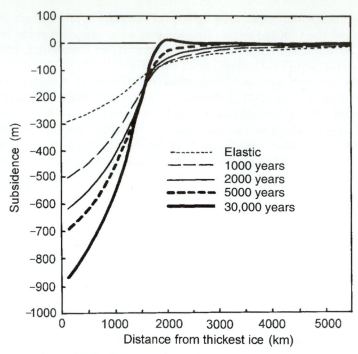

FIGURE 12.14   Diagram showing modeled formation of a forebulge under the assumed conditions of a visco-elastic lithosphere and a static, medium-thick glacier. The forebulge is identifiable after about 5000 years, and it is maximum at about 30,000 years. The basin in front of the glacier becomes shallower and narrower as the forebulge develops. (After Clark, Hendriks, Timmermans, Struck, and Hilverda 1994.)

that the lithosphere behaves elastically and that bending occurs at the hinge line separating static or quasi-static conditions in the south from a dynamic condition to the north.

An alternative interpretation is that the shorelines are deformed throughout, although less intensely tilted to the south, farther from the thick part of the glacier. In this dynamic interpretation, the hinge line would simply represent the location where the shorelines plunge below the present-day lake level, being drowned to the south and exposed to the north. The assumption made is that the lithosphere deforms under the weight of the glacier over a wide area, behaving visco-elastically. Depending on the composition of the crust and the upper mantle, complications may occur due to displacement of less viscous asthenosphere material, and to the development of a surficial bulge (topographic rise) at a significant distance from the main (thick) glacier body (Fig. 12.1C). For the bulge to develop, the thick part of the glacier must remain quasi-static for a considerable number of years. An incipient bulge is detected far from the glacier in about 5000 years. It migrates, increasing to its maximum elevation (to about 18–20 m for the Laurentide Ice Sheet) after about 30,000 years (Fig. 12.14). These times are reasonable for the North American ice sheets. As the bulge develops, the lacustrine basins and their shorelines are deformed, and a complex pattern of sedimentation and erosion occurs (Fig. 12.15). As the glacier melts, it retreats and is followed at an approximate rate of 8–10 km/century by the migrating forebulge, which modulates the postglacial isostatic uplift of the shorelines (Fig. 12.16). This is reflected in a local, apparently anomalous, land rise followed by drowning. The record of such events can be probably found in lo-

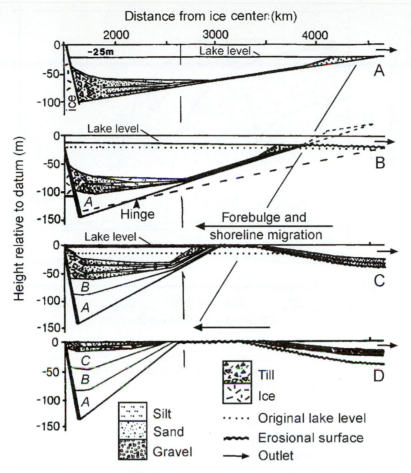

FIGURE 12.15 Diagrams showing inferred sedimentation and erosion patterns in the frontal basin as the forebulge develops. *(A)* Glacial-dominated sedimentary successions develop near the glacier, and land-dominated ones in distal parts (unit *A*). *(B,C)* In time, the sediments nearer to the glacier are buried, whereas the land-dominated sequences are continuously eroded and reformed (formation and burial of units *B* and *C*). The lake level is allowed to rise with the forebulge. The lake level is actually dictated by the elevation of the outlets. If the bulge directly affects the outlets, local deep erosion may occur. *(D)* Distal sedimentation may occur in basins formed beyond the forebulge (unit *D*).

cally deeper-than-expected erosion of lake outlets and a complex distribution of sedimentary facies. This type of analysis of the Great Lakes data has not yet been completed.

GLACIERS, TECTONICS, AND VOLCANOES

Glaciers, as a major crustal unit, affect and are in turn affected by tectonics and volcanoes. Glaciers commonly develop on mountain chains. They may cover parts of mountains, such as on the Himalayas, European Alps, and Andes, or, they may completely cover and compress mountains, such as in Greenland and Antarctica. Folded mountains are not the only ones affected. Glaciers have also formed on the

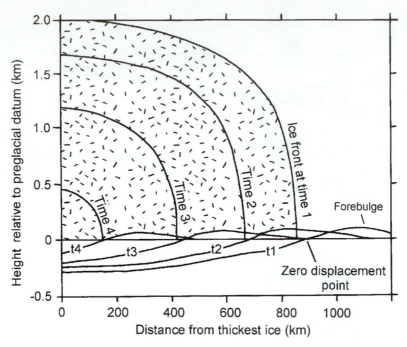

FIGURE 12.16   Diagram showing migration of the forebulge following glacier retreat. The passage of the forebulge through any location will force an uplift followed by a subsidence, thus modulating the relative lake-level curves, erosional features, and sedimentary facies distribution. (The height of the forebulge is exaggerated; t1, t2, t3, t4 = times of surface 1, 2, 3, 4.) (After Walcott 1970.)

uplifted margins of major rifts, flowing and depositing sediments in the chasms, where they are well preserved.

Glaciers also develop on elevated volcanoes, where they contribute to the formation of characteristic events and deposits, both by providing large amounts of water to the magma during eruption and thus contributing to explosive activities, and/or by constraining the volcanic deposits. Several modern examples exist.

1. Recent large eruptions of Mount Saint Helens, in northwestern United States, and Mount Pinatubo in the Philippines have shown that mixtures of volcanic ash and meltwater led to disruptive debris flows (lahars) that, diluting, change downvalley into fast-flowing hyperconcentrated flows and, eventually, into sediment-charged fluid flows.

2. In Iceland, the interrelation between glaciers and volcanoes was, during the Pleistocene, and still is, rather complex. Iceland is a volcanic island of the Mid-Atlantic Ridge. Its volcanoes mostly produce basaltic material; so it should experience quiet volcanism, generally emitting fluid lava. This is the case under normal circumstances; however, volcanoes that erupt subaqueously or under glaciers experience explosive phreatic eruptions because large quantities of water are injected into the volcanic chambers or passageways, increasing the vapor pressure in the system (Fig. 12.17). The lava is pulverized upon eruption, generating unusually large amounts of volcanoclastics. When this occurs beneath glaciers, round or elongated chasms are formed in the ice from a few hundred to more than 500 m deep, partially filled with meltwater. During the eruption, volcanic material fills the chasm, first with pillow lava — until the pressure of ice and meltwater is sufficient to restrain ex-

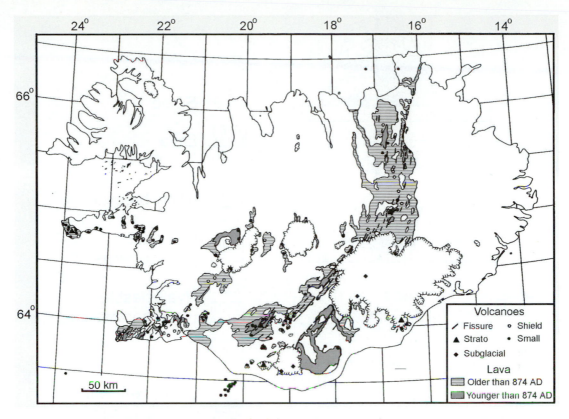

FIGURE 12.17   Map of Holocene volcanoes of Iceland. (After Einarsson 1994.)

plosive activity — and afterward with volcanoclastic materials (Fig. 12.18). When the volcanic cone becomes high enough to reach the surface of the glacier, basaltic lava flows out, at times over the glacier itself. Upon melting of the glacier, the volcanic edifices that were formed are retained (named *móberg* in Iceland, and *tuyas* in British Columbia, Canada). Partially sculpted by subsequent glacial and fluvial activities, they form most of the landscape of Iceland (Fig. 12.19).

3. Subglacial eruptions such as those of Iceland also have the effect of triggering large outwash floods when sufficient water has been melted and stored in subglacial lakes, lifting the ice. These floods (ice burst: *jökulhlaup*) occur semiregularly in Iceland, with some highly disruptive ones related to large eruptions. The path of the floods cannot be accurately predicted, nor they can be artificially canalized. Plants or animals do not colonize the unstable outwash, and until recently the large outwash of southern Iceland formed a barrier for human and animal movements. Only in 1974 was a road with permanent bridges built across the major outwash streams in southern Iceland. The bridge construction followed practices long used by ancient Romans when crossing torrents subject to seasonal floods. The costly bridge pilings are constructed strong to withstand most floods, but in order to relieve unwanted pressure on them, readily replaceable roadways are built weak so that they break when impacted by the water and debris.

Glaciers can also contribute indirectly to geological upheaval in areas well beyond their direct influence. A major case in point is the events that occurred during the late Miocene (late Messinian, about 5.4 Ma) in the Mediterranean Sea. This sea is connected with the Atlantic Ocean through the narrow, shallow Strait of Gibral-

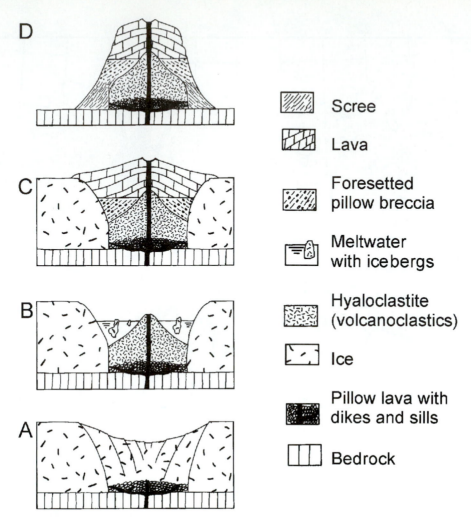

FIGURE 12.18 Diagrams showing sequence of events during a subglacial volcanic eruption and construction of a glacier-restricted volcanic buildup (móberg [hyaloclastite] mountain). The móberg construction may be interrupted at any time, thus some may have capping lava flows, others not; some may be conical in shape, others elongated depending on the character of the volcanic vent. (After Einarsson 1994.)

tar (Fig. 12.20). During the late Miocene a similar passageway existed just to the south of Gibraltar that became inactive by a combination of local tectonic uplift and a general sea level drop caused by the first large Antarctic glaciations. Once the supply of seawater was cut off, evaporation prevailed and the Mediterranean Sea dried out, depositing extensive, thick layers of evaporites, primarily gypsum. This event, known as the *Mediterranean salinity crisis,* lasted about 0.8 Ma. Upon melting of the glaciers and related sea level rise, the oceanic connection was re-established and a rapid transgression occurred throughout the Mediterranean area. In Italy, lands were inundated during the lower-middle Pliocene to a greater extent than they are now (tectonic uplift has occurred since then). This sudden influx of seawater loading on the shelf and infiltrating inland in volcanic zones led to an increase in volcanic activity in central west Italy, around Rome (Fig. 12.20). So, indirectly, glaciation in Antarctica had a significant effect on geological events and volcanism in an area in

FIGURE 12.19   Landscape of Iceland characterized by tabletop móbergs.

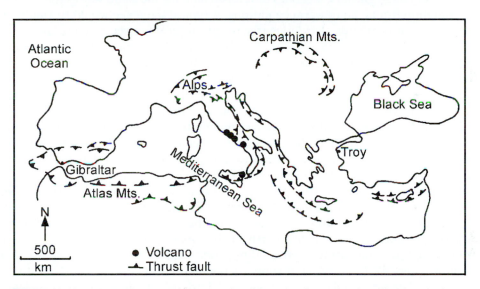

FIGURE 12.20   Schematic map of the western Mediterranean Sea and major volcanic areas in Italy.

the opposite hemisphere. This is just one example of the effects of global, climate-controlled, eustatic variations considered in sequence stratigraphy analyses.

EFFECTS OF GLACIAL CLIMATIC CHANGES ON BIOTA

The advances and retreats of the Pliocene-Pleistocene ice sheets were dictated by, and in turn affected, world climate. This, directly and indirectly, had great impact on the distribution, nature, and history of land biota of mid-latitude regions and of marine basins everywhere. The onset of cold conditions in the late Pliocene led to a great reduction of species. During the Pleistocene, extinction of organisms continued, but mostly the flora and fauna migrated and changed as the glaciers waxed and waned and as relative sea level rose and fell.

On land, the floral and faunal changes were related to climatic variations (temperature and precipitation), to opening and closing of migration corridors, and to availability of land (it was first covered by glaciers and water, and later emerged from under the ice and glacial lakes and seas). In the sea, changes occurred due to disruption and re-establishment of shallow shelf niches, and to modifications in sea-water temperature and marine currents.

Terrestrial Biozone Shift and Soils

The biota developing near a glacier terminus differs depending on its latitude. Whereas tundra develops in arctic conditions, such as in northern Canada and Europe, forests and agricultural crops can grow in temperate regions and, at times, up close to glaciers, such as in Alpine valleys. The large Pleistocene ice sheets greatly influenced the surrounding land because of very strong cold winds (katabatic winds) that blew outward from the glacier, refrigerating the surrounding lands. Outward from large glaciers, biozones developed in much the same way as today, that is, a gradation from cold to warm conditions. A typical sequence would be tundra, boreal forest or peatlands, deciduous-tree forest, tropical forests, and peatlands. Some biozones, such as the narrow zone of mixed vegetation called the Picea Parkland that developed in North America marginal to the retreating glaciers around 12,000 years ago, have since disappeared. The Picea Parkland contained tundra and boreal forest elements, the closest modern equivalent of which is found in some areas of northern Lake Huron (*Alvar* biozone). The Pleistocene biozones were compressed to various degrees from the cold glacier margin to the warm equatorial zone, more so in North America where the ice sheets reached lower latitudes (lat. 37° N) than in Europe and Asia where the ice sheets terminated farther north (lat. 45° N). Following the advance and retreat of the glacier, the biozones shifted (Fig. 12.21), as illustrated by pollen diagrams (Fig. 12.22). At present there is an overall northward shift and expansion of the warm-climate biozones (Fig. 12.23).

The generalized natural vegetation distribution of North America follows the quasi-latitudinal climatic zones north of the Great Lakes. The tundra (T) is the northernmost zone variously covered by mosses, lichens, sedges, and sparse dwarf willows (Figs. 12.23, 12.24A). It supports populations of caribou, musk ox, fox, rodents, and large populations of migratory birds (shorebirds and geese). During the Pleistocene it also supported now-extinct gigantic mastodons and mammoths. The southern limit of the tundra pretty well follows the limit of continuous permafrost. The spruce-fir (Sf) (Fig. 12.23) is the next zone to the south, and it can be further subdivided into two halves. The northern half is the taiga, characterized by growths of spruce and lichens (Fig. 12.24B). The southern half is the boreal forest (Fig. 12.24C), characterized by black and white spruce, tamarack, balsam poplar, aspen, and birch. It supports a variety of animals such as red squirrels, moose, martens, hawks, and large populations of migratory geese, shorebirds, caribou, and, to the north, polar bears. South of it are widespread hardwood growths, where they have not been destroyed for agriculture (Fig. 12.24D).

South of the Great Lakes, the natural vegetation zones (Fig. 12.23) are controlled by the irregular physiography and variable climate. Of note are the mountain regions to the west (Mt and redwood forest) and part of the east and southeast (Op). The semiarid, grassy, central plains (S, Sg, Tg) support mostly grasses, and the western coastal areas (P) and the eastern zone (Se, M) various forests. The

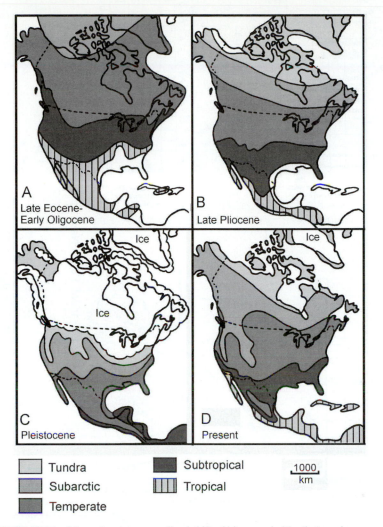

FIGURE 12.21 Maps showing generalized shift of biozones in North America, associated with advance and retreat of glaciers. (After Hare 1976.)

southeastern mangrove zone also marks the northernmost limit of the tropical coral growth offshore.

The distribution of the natural vegetation is mimicked by the distribution of the major soil types (Fig. 12.25; Table 12.1). Soil is considered here to be the weathering product of the mineral substrate on a relatively stable (no significant sedimentation or erosion) land surface. Soil development is a function of vegetation, land morphology, climate, substrate composition, ground drainage, and time. The soil profiles that develop are characterized by various horizons: O = organic-rich horizon, A = most weathered near-surface horizon, B = subsurface horizon where re-precipitation of material leached or otherwise removed from the A-horizon occurs; C = subsurface, slightly weathered horizon retaining several characteristics of the mineral substrate; D = essentially unweathered substrate. The organic layer is poorly developed in arid, warm and cold conditions (Fig. 12.26A), but it can become very thick in wetlands where peat is formed. Under wet, cold to cold-temperate conditions, organic acids produced by the decomposition of organic matter and transported downward by infiltrated water can severely leach the surficial A-horizon and

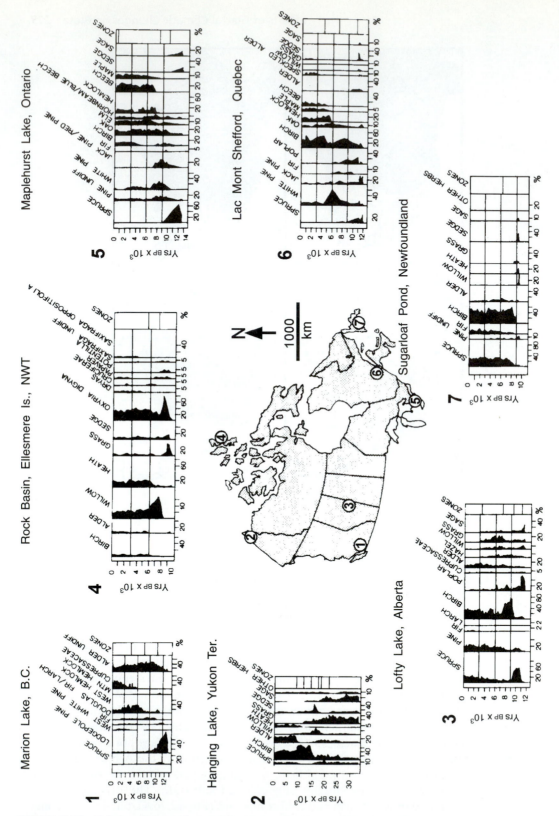

FIGURE 12.22 Pollen percentage diagrams for Canada showing variations in flora through time at characteristic locations. (After MacDonald 1990.)

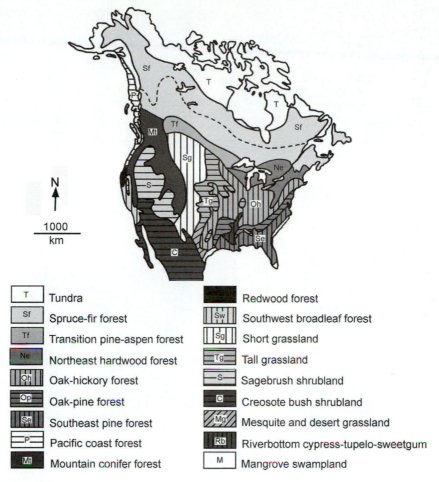

T	Tundra		Redwood forest
Sf	Spruce-fir forest	Sw	Southwest broadleaf forest
Tf	Transition pine-aspen forest	Sg	Short grassland
Ne	Northeast hardwood forest	Tg	Tall grassland
Oh	Oak-hickory forest	S	Sagebrush shrubland
Op	Oak-pine forest	C	Creosote bush shrubland
Se	Southeast pine forest	Mg	Mesquite and desert grassland
P	Pacific coast forest	Rb	Riverbottom cypress-tupelo-sweetgum
Mt	Mountain conifer forest	M	Mangrove swampland

FIGURE 12.23 Schematic map showing distribution of natural vegetation in North America. (After Hunt 1974.)

translocate newly formed minerals (clay) and ions into the underlying layer. The result may be a whitish to gray A-horizon and a reddish B-horizon with iron hardpans in places. These are spodosols (podzols) (Fig. 12.26B). In subhumid, temperate to cold-temperate zones and under grassy plains, dark-colored soils develop, with much organic material redistributed through the upper part of the soil (mollisol, chernozem, chestnut, and brown soils) (Fig. 12.26C). Under warm subhumid conditions, available water may not be sufficient to flush the soil to a great depth, and reprecipitation of calcium carbonate can occur in the B-horizon. Under arid, warm conditions, intense evapotranspiration may pump water from underground to the surface, and net movement of the ions may be upward. Precipitation of calcium carbonate may occur in nodules or layers forming calcareous soils called caliches (Fig. 12.26D).

On the whole, the soils are more poorly developed (thinner) to the north because chemical reactions are slower under cold conditions, and also because in most areas the soils are younger. Except for parts of Alaska and Yukon and a few other northern areas that escaped glaciation, those lands did not become available for soil formation until about 13 to 8 ka, or, in areas emerging from the early postglacial sea, even more recently. The climate and soil development also dictate land use,

FIGURE 12.24   Photographs showing major features of northern biozones: *(A)* Tundra; *(B)* taiga; *(C)* boreal forest; *(D)* hardwood forest.

commercial agriculture being feasible only in the lower Great Lakes area and to the south of it.

Refugia

The overall latitudinal biozone shift during the Pleistocene may have irregularities due to the presence of refugia (singular = *refugium*). Refugia are those areas that remained unglaciated during the last glaciation. They were isolated patches of land in the sea of ice, and maintained selected cold-loving species. One of the best-known refugia (Beringia refugium) developed in northwestern Canada, Alaska, and eastern Siberia. In those areas there was not enough precipitation to build up ice sheets. The summers were dry, sunny, and relatively warm. A large tundra supported hardy grasses, herbs, dwarf willow, and birch (Fig. 12.23), and was home to the last Mammoth Steppe fauna, including mammoths, mastodons, steppe bison, giant ground sloths, giant beavers, giant short-faced bears, and North American horses and camels.

It is debated whether or not Iceland had refugia as well. Physical evidence of glaciation is present everywhere on the island, and moraines have been found farther than 100 km offshore to the south. However, it is also believed that drier north-

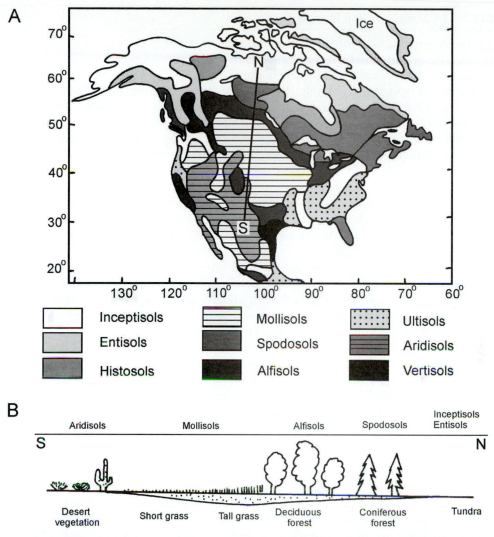

FIGURE 12.25   Soils of North America. *(A)* Generalized map. *(B)* Cross section showing vegetation and soils from desertic areas of the United States and cold areas of north Canada. (After Legget 1968; Hunt 1972.)

ern areas and some nunataks remained ice free and may have supported plant species throughout the glacial maximum. Whether or not refugia existed there, Iceland was isolated and its biota was greatly impacted by climatic changes throughout the ages. During the Tertiary, Iceland supported rich vegetation similar to the present-day broad leaf communities of the eastern United States down to Mexico. There were broadleaf trees like alder (*Alnus*), birch (*Betula*), willow (*Salix*), aspen (*Populus*), hazel (*Corylus*), beech (*Fagus*), maple (*Acer*), oak (*Quercus*), elm (*Ulmus*), and vine (*Vitis*). Conifers were represented as well, such as pine (*Pinus*), spruce (*Picea*), fir (*Abies*), larch (*Larix*), and sequoia. Close to the transition between the Miocene and Pliocene and particularly when significant climatic deterioration occurred about 3 Ma, many warm-loving plants disappeared, but temperate species such as pine, spruce, and alder flourished. Fossil shells indicate, though, that marine conditions were still warmer than at present. During late Pliocene, glaciation start-

---

TABLE 12.1   Simplified Soil Descriptions

Tundra soils (inceptisols, entisols, cryosols). Brown organic top layer underlain by gray to yellowish brown mineral layer. Permanently frozen ground below a few centimeters to 1 m from the surface. Cold-climate regions.

Acid forest soils (spodosols, podzols). Characterized by a variable, thick, humus-rich top layer, underlain by light gray to whitish leached horizon (A-horizon), underlain by a brown reddish horizon (B-horizon), which may develop an iron hardpan. These soils best develop under coniferous vegetation and a wide range of climate, from the cold boreal to subhumid temperate to hot humid regions. Where the whitish A-horizon is not well developed due to natural processes or is modified by cultivation, the soils are named Gray-Brown Podzolic soils.

Deciduous forest soils (alfisols, luvisols). Well-developed forest soils that have a subsurface horizon in which clays have accumulated. Alfisols are generally quite productive because of high natural fertility and location in temperate humid and subhumid regions of the world.

Prairie soils (mollisols, chernozem, chestnut, and brown soils). Black to gray-brown top or O-horizon that grades downward into lighter-color horizon to a zone of carbonate accumulation. The soil is friable. It develops in grassland, calcareous regions, in temperate to cool, subhumid climate. Chernozem soils form under tall grasses and chestnut and brown soils under short grasses. They acquire a reddish coloration in warm regions (red prairie, red chestnut, and red brown soils).

Desert soils (aridisols). Light brown to brown (gray desert soils), to reddish (red desert soils) coloration in warmer areas, and thin and discontinuous surface ground layers. Carbonate accumulation near surface. This calcareous accumulation may be due to upward migration of $CaCO_3$ due to evapotranspiration of soil moisture. These soils develop in cool to hot, arid climate.

After Hunt 1974.

ed, and of the higher plants, only birch, alder, and willows remained in Iceland. The warm-loving species died off, never to return even during warm interglacial periods. The alder was one of the last plants to disappear permanently at about 0.5 Ma.

The postglacial recolonization by land plants in Iceland and elsewhere can be reconstructed by pollen diagrams of peat successions (Fig. 12.27). Flourishing birch followed the initial stage of colonization by herbaceous plants and willows. When the climate ameliorated and became more humid, peatlands developed with much sphagnum. Extensive birch returned during the drier period between about 5000 and 2500 BP (Fig. 12.27). This was followed by a renewed climatic deterioration and a temporary reappearance of widespread peatlands. Europeans colonized the island around AD 850, and the remaining birch forests were disrupted for fuel, charcoal for smelters, and to create open fields for agriculture (pasture and grass growing). Whereas it is estimated that 50% of the country was covered by birch forest about 2500 ka and 25% about 1 ka, only 1% of it remains covered today. Human activities and sheep overgrazing have also disrupted the vegetation cover of many parts of the land, inducing significant wind erosion. To combat this, many species have been imported to replace the slow-growing birch. Lupines have been used to reduce soil erosion, and fast-growing coniferous trees are used for pulp and timber. Of the total number of plant species existing in Iceland (approximately 450), up to half of them may have persisted in refugia during the glacial maximum. Half of the remainder

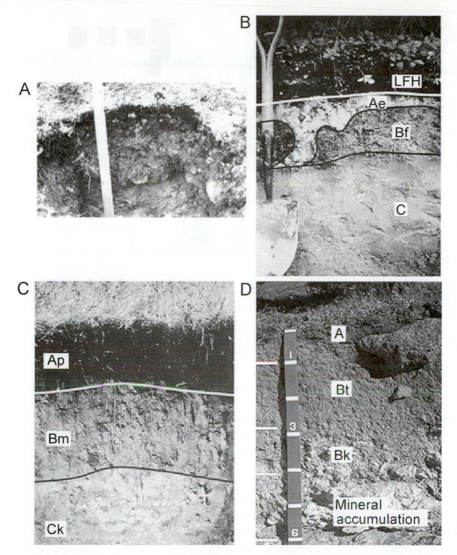

FIGURE 12.26 Examples of soils of North America. *(A)* Soil of the tundra. *(B)* Podzol of the subarctic-boreal region. *(C)* Chernozem. *(D)* Calcareous-rich soil (caliche) of semiarid areas. (From Scholle 1978.)

were reintroduced during the early Holocene by sea rafting and air transport, and the rest were imported by humans in the last millennium.

No significant land animal fossils are found in Iceland. However, marine beds indicate that most of the warm-loving shell species became extinct during late Pliocene, and the rest during the Pleistocene, and cold-loving species occupied the vacated niches. Contrary to expectations though, there have not been nor are there now many very-cold-loving species, such as *Portlandia arctica,* in the shelf around Iceland, probably because of the influence of the Gulf Stream.

Peatlands

Glacial lakes and seas inundated vast mid- to high-latitude regions depressed by the weight of the glaciers. Lands were subsequently isostatically uplifted, in places de-

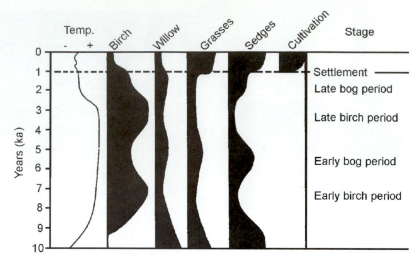

FIGURE 12.27   Generalized pollen percentage diagram for Iceland and interpreted temperature variation. (After Einarsson 1994.)

veloping vast coastal plains. Plants and animals are still progressively colonizing the emerging plains. Local biological zonations occur, from incipient colonization along the emerging coasts to fully developed forests on older, elevated, inland areas. In subarctic and boreal regions this has led to the development of unconfined peatlands. Some of these are hundreds of kilometers long and wide, such as in Siberia (the largest one), central-north Canada (the Hudson Bay Lowland [HB] southwest from James Bay and Hudson Bay), and Alaska (Fig. 12.28A).

Peatlands are wetlands that contain at least 40 cm of peat (modified plant deposit). Wetlands are areas that have the water table at or near the surface. All peatlands are wetlands, because that is a necessary condition for preservation of dead plant material and its transformation into peat. However, not all wetlands are peatlands, because in some areas there is insufficient organic matter production, or the rate of production does not keep up with the rate of oxidation and consumption by organisms, and the minimum 40 cm of organic accumulation does not develop. Microorganisms carry out most of the consumption, but migratory animals in marshes, such as geese, also play a role. Peatlands are classified based on whether their plants have access to nutrient-rich groundwater (minerotrophic) or nutrient-poor groundwater (on granitic terrains, or in areas of perched water tables) (ombrotrophic), and according to their plant communities (Figs. 12.28B, 12.29). The bog is the only ombrotrophic peatland. In most cases, its existence is related to sphagnum growths (Fig. 12.29A). Sphagnum produces a highly porous accumulation capable of maintaining a perched water layer, isolated from the mineral substratum. This water is not buffered by calcium carbonate and becomes acid. The bog plants receive their nutrients from airborne substances. Typical plants are shrubs, such as Labrador tea, and, locally, stunted black spruce. Fens are gramineous (grassy), very wet peatlands, with some growth of sphagnum, and, locally, sparse trees, generally tamarack (*Larix laricina*) and stunted black spruce (Fig. 12.29B). Swamps are peatlands with at least 25% of their area covered by trees (Fig. 12.29C). Marshes are gramineous wetlands that undergo strong variation in water table (Fig. 12.29D). In the north, coastal marshes rarely become peatlands due to the intense consumption of organic matter by migratory geese.

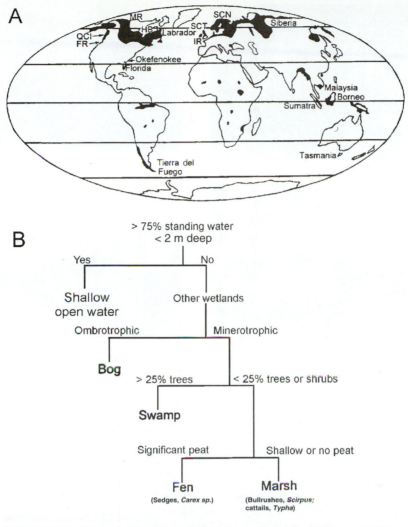

FIGURE 12.28   Wetlands of the world. *(A)* Wetlands distribution of the world (FR = Fraser River delta, QCI = Queen Charlotte islands, MR = Mackenzie River delta, HB = Hudson Bay Lowland, IR = Ireland, SCT = Scotland Highlands, SCN = Scandinavia). *(B)* Classification of cold climate peatlands (After Martini and Glooschenko 1985; based on Zoltai and Pollett 1983).

In the Hudson Bay Lowland, the distribution of peatlands and vegetation is dictated primarily by strong north-south changes in temperature, by the coastal-inland position, as well as by local topography and substratum that, in the north, are affected also by permafrost (Fig. 12.30). The "mid-boreal wetlands region" (BMh) has fens everywhere, bogs in northern parts, swamps in parts of the southern interior where some deciduous trees occur, and in areas near streams. The "high-boreal wetlands region" (BH) is characterized primarily by fens and bogs, and there is a northward increase in the influence of permafrost, with development of few peat plateaus and palsas (ice-cored mounds). The "low-subarctic wetlands region" (SL) is characterized by numerous peat plateaus associated with the increasing development of permafrost. The "high-subarctic wetlands region" (SH) has continuous permafrost except at the emerging coast. It is characterized by lichen-dominated treed

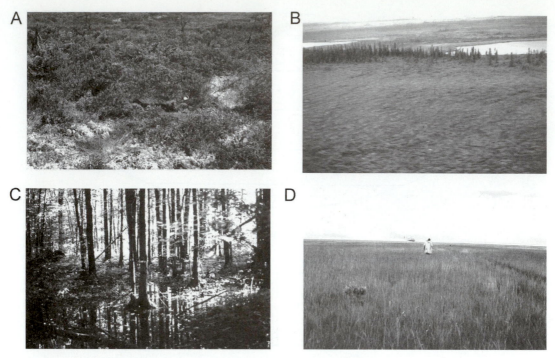

FIGURE 12.29 Photographs of peatlands: *(A)* Bog; *(B)* fen; *(C)* swamp; *(D)* marsh.

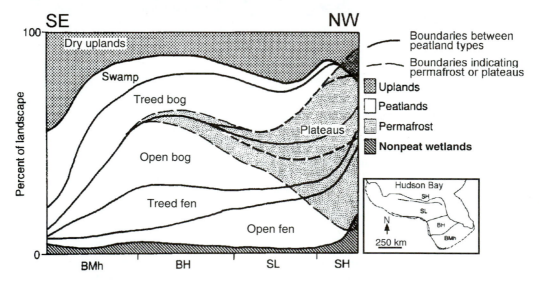

FIGURE 12.30 Diagram showing peatland development in the Hudson Bay Lowland. (After Riley 1982.)

bogs and peat plateaus. Raised beach ridges and river levees are better drained, and forests, not peatlands, develop. The main forest trees are black spruce (*Picea mariana*), white spruce (*Picea glauca*) near the coast locally with some balsam fir, trembling aspen (*Populus tremuloides*), balsam poplar (*Populus balsifera*), and white birch (*Betula papirifera*). Many of these forests have undergrowth of lichens.

Peat has developed in the Hudson Bay Lowland and in areas to the south around the Great Lakes in the last 5000 years in response to two main factors.

(1) The Hudson Bay Lowland was emerging from the postglacial sea (Tyrrell Sea), which formed in the area upon deglaciation about 7500 BP and (2) the global climate was deteriorating from the optimum postglacial temperature maximum of about 6000 BP, such that the rate of evapotranspiration had decreased and wetlands and peatlands could form. The peat in the Hudson Bay Lowland has reached a thickness of up to 3–4 m in raised bogs, about 100 km inland from the coast, and about 2–3 m in inland fens. A quasi-regular increase in peat thickness can be demonstrated from the shore inland, with the rate of net deposition in the southern part of the Lowland being about twice that in the northern part of the Lowland (Fig. 12.31). Where best developed, the peat sequence shows vertical transitions reflecting development of various peatlands, from decomposed, dark, sedge peat (fen) at the base to a sphagnum peat (bog) with occasional woody fragments.

In cold-climate peatlands, such as those of the Hudson Bay Lowland, the relatively low organic productivity is compensated for by the frozen conditions for many months of the year, so that sufficient plant material is preserved to develop peat. In warm-climate peatlands, the strong evapotranspiration and high rate of organic matter decomposition and consumption is compensated for by the large amount of precipitation throughout the year, and the high rate of organic matter productivity. Furthermore, the different environmental conditions are such that distinct plant associations and peat successions are developed in the two regions. The ages of the major peatlands, however, are similar, most having started about 5000 BP. This occurred after the rapid sea level rise due to melting glaciers, and soon after the postglacial temperature optimum when evapotranspiration rates decreased. To be sure, peats had developed in tropical regions before that time as well, but coastal ones were drowned by the rising sea level. Examples of this are found in cores taken from the shallow Yellow Sea in Asia.

Migration of Land Animals

Migrations of land organisms during the Pleistocene occurred whenever sea level dropped enough to form a land bridge, both in northern cold areas and tropical areas. These migrations had strong effects on the present distribution of land biota.

In northwestern North America (Beringia refugium), an ice-free land bridge existed between North America and Asia during Pleistocene glacial periods, when sea level was as much as 125 m below the present. During glacial periods, cold-loving plants and animals, including humans, could readily migrate between Asia and North America. But during interglacials, land migration was greatly reduced as the Bering Sea flooded. Much of the area is now 50 m below the Bering Sea.

Humans colonized North America between about 30,000 and 9000 BP, probably by following migrating herds of animals from Asia. Tool-like artifacts have been found in the Old Crow Basin of the Beringia refugium of northwest Canada, which some researchers considered made by humans; others dispute that. In either case they are reworked. The artifacts lay at the base of lacustrine sediments, whose top gave a nonfinite radiocarbon date of 32,000 BP. (Fig. 12.32).[1] The lake deposits postdate a cryoturbated ash layer of 110,000 ± 12,000 BP. If the artifact-like things are of human origin, they indicate that the area could not have been colonized before about 30,000 years ago because the technology required to survive cold conditions (specifically the eyed needle for sewing weather-tight clothing from animal skins) had not been developed. After 9000 BP, the postglacial rise in sea level cut North America off from Asia.

A corridor also may have existed between the Laurentide and Cordilleran ice sheets during part of the last glacial period, which joined the northern Beringia refugium with southern unglaciated parts of North America (Fig. 12.33A). Few

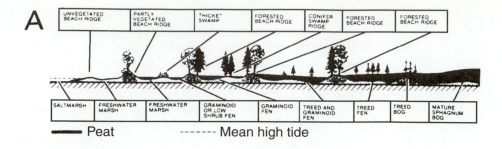

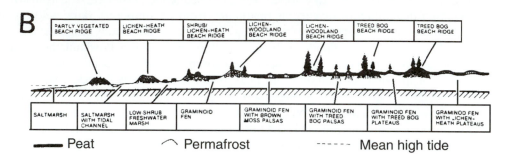

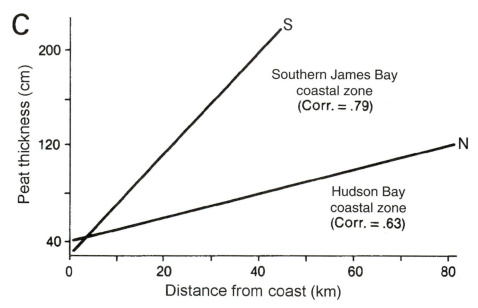

FIGURE 12.31   Progressive peat development in the emerging Hudson Bay Lowland (HBL). *(A)* Schematic cross section from the coast inland in southern HBL. *(B)* Schematic cross section from the coast inland in northern HBL. *(C)* Thickness of peat at various distances from the coast. Assuming an average 1 m/century rate of emersion of the land, the rates of peat development in HBL vary between approximately 0.1 and 0.65 mm/yr. In comparison, the rate of peat development in the tropical peats of Borneo was about 3 to 4 mm/yr. (After Cowell, Wickware, Boissoneau, Jeglum, and Sims 1983.)

human artifacts have been found along this corridor, and its usefulness as a passageway for plants and animal is still not clear, although it may be potentially important. Pollen evidence from the foothills of the Rocky Mountains in Alberta seems to confirm the existence of the corridor (Fig. 12.33B,C). The pollen diagram shows that an *Artemisia* (grass)-dominated setting with sparse tundra species exist-

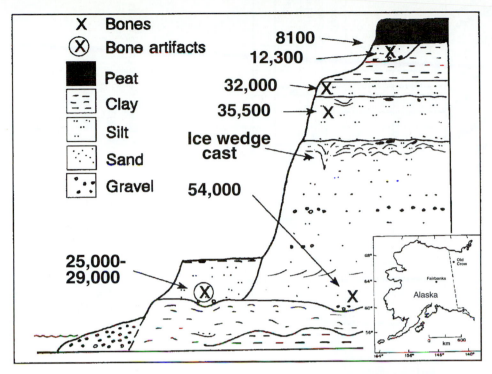

FIGURE 12.32   Generalized stratigraphic diagram of the Old Crow area in the unglaciated northwestern part of North America, showing age in years BP. (After Jopling, Irving, and Beebe 1981.)

ed from 24,000 to about 11,400 BP. An initial warming is indicated by the increase in *Salix* pollen at about 15,500 BP. Trees increased after 11,400 BP, and with the advent of *Pinus* at about 7800 BP, the present boreal forest developed.

In more recent times, migration of humans and development of their cultures have also been influenced by climatic deteriorations and associated extension of pack ice on the northern seas. A case in point is the migrations and development of the four major Inuit cultures of northern Canada during the last 5000 years or so (Fig. 12.34).

1. The first pre-Dorset archaeological finds in western Alaska date back to 5000–4000 BP (~2000 BC). At this time the Bering Strait was ice free and as wide as it is now. These people must have crossed it from Asia by boat following the migrations of sea mammals (seals, walruses, and whales). The people spread across the Canadian Arctic to Greenland from about 4000 BP to about 2800 BP.

2. A new culture (Dorset) was well established from the first immigrants, and good archaeological remains are found between 3000 BP and about 700 BP in Baffin Island. These eastern Inuits were isolated from the ancestor homeland, probably because of a climatic deterioration. Increase in sea ice cover limited and disrupted the migration of sea mammals, and the Inuit changed into hunters of land mammals, primarily caribou. They changed their tools and habits. They manufactured ivory snow knives and small soapstone lamps for light and heat, probably for turf and snow habitations (igloos). They also constructed ivory and bone sled runners. They had good artistic production of ceremonial masks and small carvings of human and animal figurines.

3. Around 1000 BP there was warming for a few centuries. This led to a decrease in sea ice, allowing a renewed northward and more widespread migration of sea mammals, followed by a new immigration of Inuit. These new Inuit (Thule

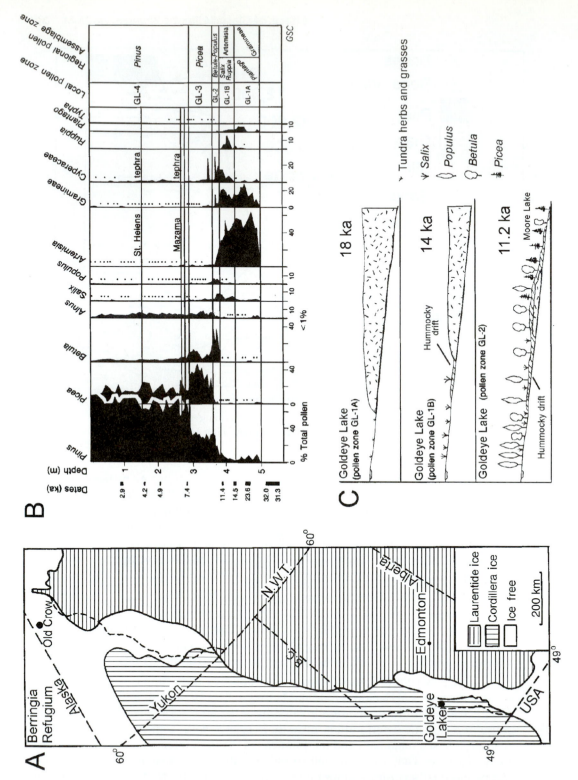

FIGURE 12.33 Plant colonization of the Goldeye Lake region of eastern Alberta, Canada. *(A)* Location map showing the joining of the Laurentide and Cordillera ice sheets and the easternmost, ice-free Beringia region in the north (the Yukon-NWT and B.C.-Alberta borders follow the continental divide; N.W.T. = Northwest Territories; B.C. = British Columbia). *(B)* Pollen diagram of the Goldeye Lake area. *(C)* Diagram showing hypothesized evolution of biozones responsible for the pollen distribution in the area during the retreat of the Laurentide Ice Sheet from the Goldeye Lake region. (From Schweger 1989.)

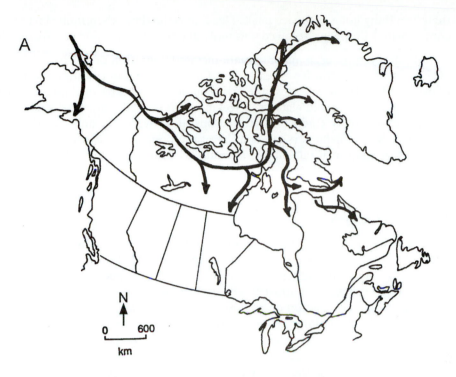

FIGURE 12.34   Inuit of North America. *(A)* Migration routes (arrows) in North America and Greenland (after Wilson 1976). *(B)* Ancient village with remnants of stone and turf houses, Foxe Basin, Canada.

culture) had a more sophisticated technology (for instance, the *umiak,* a large, open skin boat) and left good archaeological sites in the Thule area in northern Greenland. Thule artifacts are found overlapping in time with the Dorset culture: probably the two groups coexisted and may have eventually merged. The Thule villages were characterized by winter houses built of sod over a framework of stones and whalebones (Fig. 12.34B).

4. The climate deteriorated between AD 1650 and 1850 (a period called the Little Ice Age). The ice pack expanded. Northward migration of mammals was limited, and whaling was no longer widespread. The Thule people had to modify their habits and tools to adapt to the new ecology. It is probably at this time that the Inuit started once again to use the igloo routinely, and hunted seals, catching them from their breathing holes in the ice pack. This is also the time when Inuit started to have contact with Europeans, received new tools, and the modern Inuit culture developed.

In modern times, a major climate-dictated migration out of Iceland occurred at the end of the Little Ice Age. At that time, climate deterioration led to cooling of the North Atlantic waters, slight deviation of the Gulf Stream, and increasing development of sea ice and glaciers in Iceland. This caused a drastic decrease in agricultural productivity (failure of grass crops, thus livestock) and fisheries, leading to famine. Many Icelanders, up to 15% of the population, chose to emigrate to a cold part of south-central Canada (Manitoba), and reconstructed their communities there, changing their customs to adapt to the new landlocked setting and American habits.

Marine Biota Changes

In the sea, biota changes have been severe during the Pleistocene due to changes in water temperature, marine currents, and exposure of shallow shelves. In the northwestern Atlantic, the cold-temperate (boreal) shelf province has only about 60 species of shore fishes, in contrast to the 180 or so of the equivalent North Pacific province. A mass extinction of shallow marine organisms occurred in the Atlantic toward the end of the Pliocene and start of the Pleistocene, as continental glaciers started forming in the Northern Hemisphere. All purely tropical species disappeared. These western Atlantic extinctions were far more severe than those in the eastern Pacific, because the Atlantic Ocean openly connects with the Arctic Ocean whereas the Pacific does not. The Atlantic has a broad and deep connection to the Arctic Ocean, and cold, glacial water flows directly down to tropical latitudes. The Pacific, a much larger ocean, has a much narrower and shallower connection to the Arctic via the Bering Strait (which was closed during glacial periods), and the amount and effect of mixing is much less.

ENDNOTE

1. These dates should be considered as indicative, because they are contentious ones.

CHAPTER 13

The Cenozoic Ice Age

SCOPE AND LIMITATIONS

The objectives of this chapter are to briefly summarize the events that led to the widespread Pleistocene glaciation and to illustrate the major features of the Pleistocene ice sheets and their behavior during deglaciation. These can be inferred directly from glacial features and sediments, and indirectly from isotope and other analyses. The most continuous record of past glaciations exists in few continental areas at the margins of glaciations, in subsiding rifts, and in marine basins particularly beyond the shelf edge. The glacial geomorphological-geological record is best for the last Pleistocene glaciation and deglaciation. So the reconstructed features and behavior of the last ice sheets on Earth can serve as a model for understanding previous Pleistocene and older glaciations.

THE CENOZOIC ICE AGE

The Cenozoic includes two very unequal geological periods: the Tertiary from about 65 Ma to 1.6 Ma (divided into five epochs: Paleocene, Eocene, Oligocene, Miocene, and Pliocene), and the Quaternary from about 1.6 Ma to the present (divided into two epochs: Pleistocene and Holocene [also called Recent]).

Tertiary Period

The Quaternary glaciations are the culmination of over 50 million years of worldwide cooling and ice buildup. Marine sediments record decreasing deepwater temperatures and increasing glacier cover throughout the entire Tertiary period. Much of the deepwater cooling occurred in three major steps, about 36, 12, and 3 million years ago.

During the early Eocene, between 57 and 52 Ma, Earth was relatively warm, slightly warmer, in fact, than the preceding Paleocene. Mid-latitude regions, like southern Britain, northern France, and the central United States, had tropical climates and biota. The polar regions had temperate climates: forests grew near the poles, Antarctica probably was ice free, and the surrounding oceans were quite warm, around 18°C, as surmised from oxygen isotope paleotemperature analyses and fossils (Fig. 13.1). At the same time, the equatorial regions were not much warmer than at present, as a very efficient system of marine currents distributed heat toward the poles, and a gentle temperature gradient existed throughout the planet.

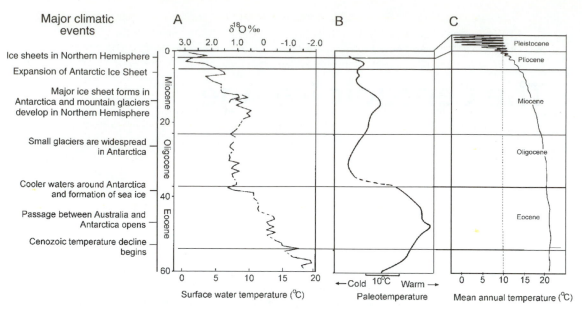

FIGURE 13.1   Diagrams showing variations in oxygen isotopes and temperature estimated (from oxygen isotopes) in the sea and terrestrial fossils inland during the Cenozoic. *(A)* Antarctica (After Goudie 1992; Kennett and Shackleton 1975). *(B)* North Sea (After Goudie 1992; Buchard 1978). *(C)* Generalized temperature change estimated from terrestrial fossils (After Woldstedt 1954; Clarke and Stearn 1960; Andersen and Borns 1994).

World climate then cooled progressively through the Eocene in and around Antarctica, and during the late Eocene in the Northern Hemisphere (Fig. 13.1). The cooling corresponds to the development of the Antarctic Circumpolar Current, which started forming around 50 Ma as Australia separated from Antarctica and drifted north. This encircling current prevented warm currents initiated at low latitudes from reaching the Antarctic coast. This not only caused increasing cold in Antarctica, but also invigorated atmospheric circulation by increasing the equator-pole temperature gradient. Bottom-water temperatures in the southern ocean dropped to about 5°C, and ice volumes increased rapidly around 36 Ma. This sudden change is attributed to the development of freezing conditions at sea level around Antarctica, the consequent formation of sea ice, and cold Antarctic currents sinking into the ocean depths. This climatic change also affected the Northern Hemisphere (Fig. 13.1).

By the late Eocene–early Oligocene, sizable glaciers existed in eastern Antarctica and may have triggered a sea level drop. Diamictites are found in Antarctica interbedded with glaciofluvial sediments containing pollen derived from a diverse temperate flora. Glaciomarine sediments of the middle-to-late Eocene age occur along the Victoria Land coast (Antarctica), and cores from the east Antarctic continental shelf show interbedded diamictite and diatom ooze of late Eocene to early Oligocene age. So, though no evidence of continental ice sheets is found until the late Oligocene, mountain and possible tidewater glaciers existed during the Eocene and Oligocene.

Lower Oligocene cooling was rapid and extensive. Many marine sites, both shallow and deep, have thick accumulations of ice-rafted debris throughout the Oligocene, indicating widespread glaciation. Oxygen isotope measurements indicate ice volumes approximately half those of the present day by the earliest Oligocene, and large ice sheets covered Antarctica by 25 Ma.

Sea surface temperatures remained relatively stable between 6 and 8°C between 36 and 12 Ma. In fact, between about 20 and 12 Ma early and middle Miocene, surface water temperatures in the southern ocean increased slightly. But this was followed by a second major cooling so intense that by the end of the middle Miocene, Antarctic ice sheets larger than existing ones were cutting deep into the continental shelf and were depositing thick diamicton all around the continental edge. Sea surface temperatures in the southern ocean dropped a further 3°C in the middle Miocene. By 7 Ma, southeastern Greenland was completely covered with glaciers, and by 5 to 6 Ma, glaciers were forming and advancing in Scandinavia and the North Atlantic. At this time, sea level had fallen sufficiently to contribute to the isolation of the Mediterranean Sea and its salinity crisis.[1]

A climatic amelioration occurred around 5 Ma at the start of the Pliocene. Trees grew in Iceland, Greenland, and Canada as far north as 82° N, where now only tundra survives. But major cooling started again around 3 Ma, and by around 2.5 Ma tundra spread over north-central Europe. Soon after, the humid environment of central China was replaced by harsh continental steppe. In sub-Saharan Africa, arid and open woodland expanded, replacing wetter forests. Some of these environmental changes may be linked to the evolution of the first humans in the late Pliocene of northeastern Africa. Like many other biological changes, these predate the start of the Quaternary period at about 1.6 Ma, during which time a worldwide glaciation occurred.

Quaternary Period

During the Quaternary (Pleistocene and Holocene) most high-latitude and high-altitude areas of the continents were glaciated, and ice shelves or sea ice covered wide oceanic areas (Fig. 13.2). Most of Antarctica and Greenland remained glaci-

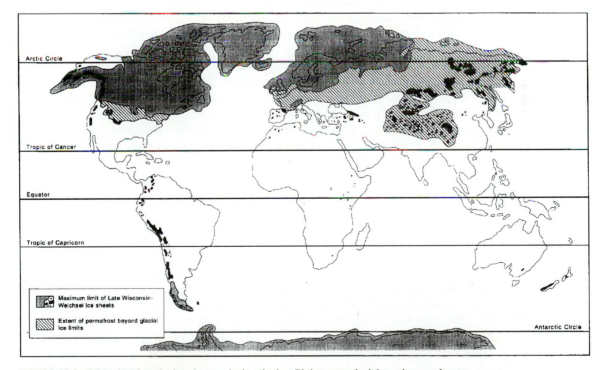

FIGURE 13.2   Map showing glaciated areas during the last Pleistocene glacial maximum, about 18 ka. (From Viles and Spencer 1995.)

ated, with only minor fluctuations, and they are still covered by ice sheets. Ice caps and valley glaciers developed and still persist in places on several mountain chains. However, the defining feature of the Quaternary is the advance and retreat of huge ice sheets over the northern parts of North America, Europe, and Asia. Glacial periods, when ice sheets were widespread, alternated with interglacial periods, when the continental ice sheets melted away. Even during the glacial periods there were fluctuations between the **stadials,** when the glaciers were at their maximum or advancing, and **interstadials,** when, locally, the glaciers retreated. Maximum glacial extent was achieved during the Pleistocene, and we are now living in the Holocene, a cool interglacial time.

Complex climatic, environmental, and biological events occurred during the highly variable Quaternary times, when worldwide temperature was at or near the threshold value for the formation and persistence of widespread glaciers. Small environmental changes may have sufficed to force a shift from fully glacial to interstadial or interglacial conditions and vice versa. These may have been related to slight temperature changes due to variations in solar energy received (insolation) at mid-latitudes, to large volcanic eruptions injecting much dust into the atmosphere, or to oceanic currents switching on and off. How many major glacial-interglacial shifts have occurred during the Quaternary, in the last 1.6 million years? Past researchers have long recognized the deposits of four major glacial terrestrial events in the European Alps, Northern Europe, and North America (Fig. 13.3A). More accurate dating of terrestrial deposits and the more continuous sedimentary record of the oceanic deposits and isotope data, particularly oxygen isotopes, indicate that many more glaciations have occurred (Fig. 13.3B). There is a good, widespread terrestrial and marine sedimentary record only of the latest two glaciations, which are still called Illinoian and Wisconsin in North America, and Saale and Weichsel in Northern Europe (Fig. 13.3A). Upper Pleistocene glaciations provide a model for understanding the possible behavior of older glaciers and perhaps for estimating what may occur in the future on Earth. The major events of the last 150,000 years are summarized in Table 13.1.

The Upper Pleistocene record provides sufficient information to understand the onset, development, and waning of a glaciation. The ice sheets or the various parts of a single ice sheet are not perfectly synchronous in their advance and retreat as their terminal parts are affected by local climatic, morphological, and geological factors. Nevertheless, major glacial advances and retreats are concentrated in certain time periods, and reflect global climatic changes. So the onset and demise of the penultimate glaciation of the Late to Middle Pleistocene are detected all over the world. A major glaciation occurred both in the Northern and Southern Hemisphere in the Late Pleistocene, between 79 and 65 thousand years ago, and the last glacial maximum occurred about 21–18 ka. Details of this ultimate glaciation and deglaciation are preserved, and thus it is possible to establish the behavior of the glacier rather accurately. From this, several concepts can be derived.

1. During the Pleistocene the average temperature of Earth was at or near the threshold value such that any minor disturbance could have pushed the system toward a glaciation or conversely toward a warmer interglacial period (Fig. 13.3B). Probably changes of a few degrees centigrade could suffice.

2. Certain events were global, hence must be related to energy input on Earth and changes in overall climate. One of the triggering effects has been ascribed to variation in insolation of mid-latitude areas, according to the **Milankovitch effect.** Once a threshold condition was reached such that centers of glaciation were estab-

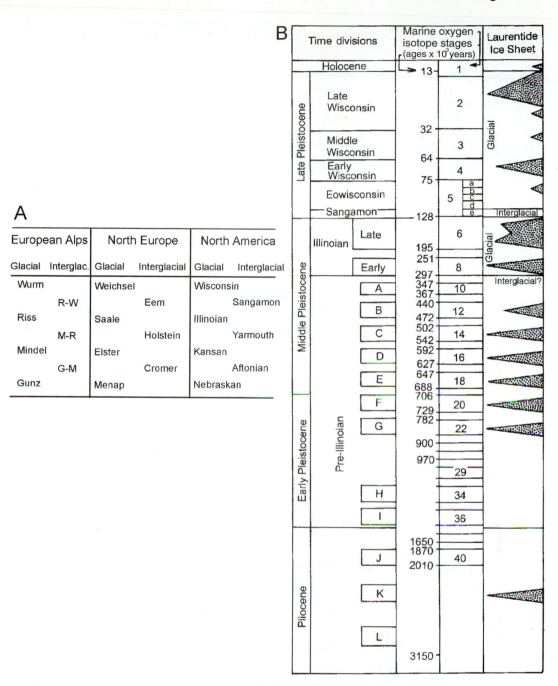

FIGURE 13.3 Diagrams showing Pleistocene glacial and interglacial periods. *(A)* Classical terminology, part of which (pre: Riss, Saale, and Illinoian) is no longer considered valid because of more accurate new dating and oxygen isotope data. *(B)* Presently used oxygen isotope stages and terminology for the Pliocene and Quaternary (stippled area = glacier advance) (After Bowen et al. 1986; Calkin 1995).

lished, the system may have fed on itself, and until the global climatic trend was reversed, it would have led to expansion of the ice. Modulation of the glaciations by the Milankovitch effect is not as regular as might be expected because glacier response time to changing climate may be long and variable. In addition, extraordi-

TABLE 13.1   Major Upper Pleistocene Glacial Events, with Particular Reference to Northern Europe and Northern America

Isotope Stage		Age (ka)	Major Event(s)	Typical Location
1		<10	Development of humans Strong variations in climate About 8–6 ka postglacial temperature optimum Disappearance of ice sheets	Global Age of temperature changes slightly from area to area 8.5 ka in northern Europe 8.0–7.5 ka in N. America
2		11–10	Local glacier readvance (Younger Dryas)	Global, but particularly well recorded in northern Europe
		17–14	Collapse of ice sheets	Northern Europe and N. America
		25–17	Development and start of fluctuating retreat of the last continental ice sheets	Ice maximum: 22–17 ka: northern Europe and eastern N. America 15–14 ka: in western N. America
3		65–25	Continuing cold conditions with persistent, but fluctuating glaciers	Global
4		75–65	Major, rapid cooling and development of ice sheets on generally ice- and snow-free lands	Ice sheets most extensive in eastern N. America; large Toba eruption about 75 ka
5	a b c d	115–75	Variable cooling trend	Global
5	e	130–115	Warmest ultimate interglacial period	Global
6		150	Penultimate glaciation	Global

nary events may have occurred. An example of the latter may have been the triggering of the strong, but relatively short-lived glaciations of about 75 ka. This event corresponds with the eruption of the Toba volcano in Indonesia. This was the largest Quaternary eruption, orders of magnitude larger than all the others. By itself, this eruption could not have led to a glaciation, because eruptions reduce the transparency of the atmosphere for only a relatively short period, perhaps decades. But during a cooling trend, the eruption may have accelerated the onset of the glaciation. A second example of environmental conditions that may have modified the Milankovitch effect is the presence or absence of snow cover at mid-latitudes. This may have been the case in North America during the middle Wisconsin (isotope stage 3), such that any increased insolation could have been compensated for by a higher albedo. After isotope stage 4, rather than achieving full climatic reversal to warmer conditions similar to those of the earliest (early Wisconsin) times (isotope stage 5), the cooling trend continued until the temperature minimum of the late Wisconsin (isotope stage 2).

3. During a climatic change, events may serve as triggers to initiate self-fulfilling catastrophic events. Some of these events are synergetic. For example, during deglaciation and the corresponding sea level rise, a glacier surging out into a marine shelf may be rapidly removed by calving. This could trigger drawdown of the glacier as it tries to replenish the removed ice, and hasten the collapse of ice sheets with sea termini.

Recent Greenland and Antarctic Ice Sheets

Most of the present glacier ice is concentrated in Greenland and Antarctica.

Greenland   Greenland contains the largest glaciers in the Northern Hemisphere, second only to the Antarctic ice sheets. It stores about 7% of the world's freshwater, and about 80% (1.7×10^6 km^2) of the land is covered by ice. The volume of the Greenland Ice Sheet has been estimated to be about 2.74×10^6 km^3; its maximum thickness is 3420 m in the center-middle part (Fig. 13.4A). Its elevation is about

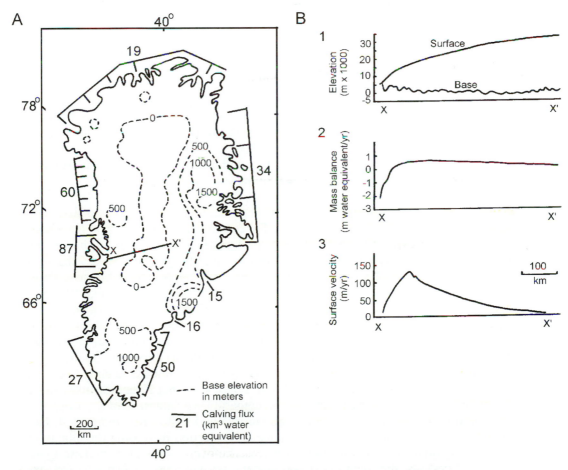

FIGURE 13.4   Greenland. *(A)* Map showing elevation of subglacial substrate, and calving fluxes. *(B)* Profile, mass balance, and observed surface velocity of the ice sheet from X to X′. (After Reeh 1989.)

3200 m asl (above sea level), and its base elevation is well below sea level in the central part. The surficial slope of the ice sheet is very low (0.005° or less in the interior and 0.05° at the margins). In the north, the termini are relatively steep, with a final height of about 30 m or so. In the south, the margins slope gently to the terminus (Fig. 13.4B). The substrate of the glacier is mountainous (mountains up to about 1500 m asl) to the east and south, and relatively flat or hilly to the center and west.

The average annual surficial temperature of the ice sheet is about −32°C. Its basal temperature reaches the pressure-melting point in many places, from −10°C toward the margins to slightly less than −30°C toward the center. The precipitation is greater to the south (250 to 60 g/(cm² × yr)) and decreases rapidly to the more arid central and northern areas (20 to 15 g/(cm² × yr)). This leads to a greater positive budget near the southern glacier margins (Fig. 13.4B). Correspondent to this, the southern part of the glacier displays a more active regime than the northern part. Numerous tidewater outlets exist and many icebergs are produced each year, particularly from the central-western part (Figs. 13.4A, 13.4B). Cold currents flowing around the southern and western part of Greenland remove many of these icebergs.

Numerous ice cores have been taken from the Greenland Ice Sheet, and they provide valuable information on the glaciology, past climatology, meteorology, atmospheric chemistry, and on global volcanic activity (Box 13.1). Much of this information is obtained by analysis of the oxygen isotopes of the ice, the chemistry of the trapped air bubbles (samples of ancient atmosphere), and by impurities within the glaciers derived either from wind transport or from meteorites. For example, the transition from glacial to postglacial Holocene condition is well defined by oxygen isotopes at 10 ka (Fig. Box 13.1.1). Volcanism is recorded in the glacier by impurities and by the fact that this material renders the ice slightly acid. This record, however, was masked somewhat during the Wisconsin because the acidity was buffered by large input of calcareous dust, probably derived from exposed shelves as sea level dropped during the glacier maximum.

The Greenland Ice Sheet in its structure and behavior constitutes a reasonable analog for Pleistocene and some older glaciers.

Antarctica   The present glacier ice in Antarctica is locked up in two thick (2–3 km) ice sheets: the East Antarctic Ice Sheet and the West Antarctic Ice Sheet, which together expand for about 11.5 × 10⁶ km² (Figs. 13.5, 13.6). Although the Antarctic ice sheets are polar with cold ice base, locally they have large subglacial lakes (Box 13.2). The East Antarctic Ice Sheet has developed from Miocene times (about 14 Ma) on the continent and has remained rather stable through the ages. The West Antarctic Ice Sheet has built up on an archipelago of islands, and its dimensions have fluctuated as considerable iceberg calving has occurred from time to time. The ice sheets are known to have been larger than at present about 11 Ma, and have fluctuated (particularly the West Antarctic Ice Sheet) during the Pleistocene. The greatest limiting factor for major variations is the extreme cold climate (mean annual temperatures of −50°C at the South Pole), such that little moisture can be transported into those regions, and global temperature fluctuations of a few degrees cannot modify the glaciers to any extent. During the Pleistocene, cold periods had the effect of enlarging the extent of sea ice cover, further increasing the albedo of the region and decreasing the temperature and moisture of the glaciers. During interglacial periods, the main effect was eustatic sea level rise, which triggered increasing calving of the West Antarctic Ice Sheet. No glacier collapse has ever occurred in Antarctica during the Pleistocene-Holocene, nor is it expected to occur

BOX 13.1   Greenland and Antarctic Ice Core Data

In the last 30 years, data from ice cores in Greenland and Antarctica has revolutionized late Quaternary studies.

First, the oxygen isotope curves for the last 160 ka of both Greenland and Antarctica have similar shapes, and agree with those from the oceans (Fig. Box 13.1.1). This suggests a mainly external forcing mechanism for glaciation, such as the astronomical one proposed by Milankovitch. Furthermore, it shows that the glacial cycles of the Northern and Southern Hemispheres were synchronous.

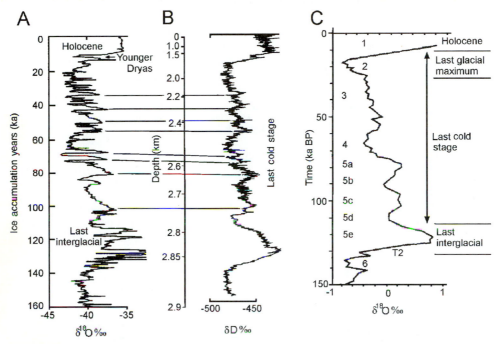

FIGURE BOX 13.1.1   Diagram showing oxygen isotope variation. (After Lowe and Walker 1984.)

Second, both areas show numerous high-frequency climatic oscillations over the last 120 ka, reflected in abrupt changes in oxygen isotope values. These record huge climate swings over both the short and long term. Some 20 interstadial events occur between 80 to 20 ka, with individual temperature fluctuations of 5 to 8°C. Each fluctuation lasted no more than 0.5 to 2 ka, and cannot simply be explained by Milankovitch radiation variations. Some feedback mechanism involving coupled ice sheet, oceanographic, and atmospheric fluctuations seem necessary. For example, the abrupt return to glacial conditions in the Northern Hemisphere during the Younger Dryas phase (involving 1.5–3°C cooling relative to today) corresponds to sharp changes in the Greenland core (Fig. Box 13.1.1).

Third, temperature variations correspond with changes in contemporary atmospheric gases (obtained from trapped bubbles). The Vostok core covers over 420,000 years of glacial history, that is, over three full glacial/interglacial cycles. In this core, both methane and carbon dioxide fluctuations mimic the surface temperature fluctuations obtained from oxygen isotopes (Fig. Box 13.1.2). The strong correlation between greenhouse gases and climate over the last 160,000 years confirms that the present logarithmic increase in these gases will have major effects on future climates.

(continues)

BOX 13.1 (*continued*)

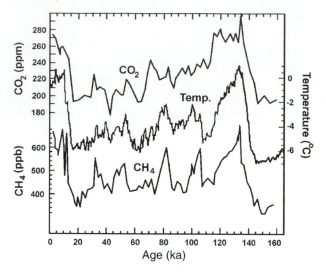

FIGURE BOX 13.1.2   Diagram showing concordant variations in temperature and greenhouse gases during the last 160 ka in the Vostok ice core (eastern Antarctica). (After Ice Core Working Group: *http://www.gisp2.sr.unh.edu/icwg/fig2.html*.)

Fourth, the anthropogenic additions of carbon dioxide to the atmosphere are over 30% in the last 200 years (Fig. Box 13.1.3). The rise in the last 50 years, due to the internal combustion engine and forest clear-cutting, is especially marked.

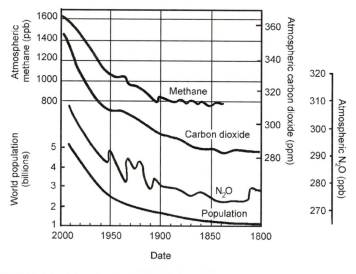

FIGURE BOX 13.1.3   Diagram showing variation and correlation between N_2O, CH_4, and CO_2, and global population variation in the last two centuries. (After Ice Core Working Group: *http://www.gisp2.sr.unh.edu/icwg/fig16.html*.)

Lastly, the cores show that volcanic eruptions have perturbed climate more frequently and severely over the last 110,000 than was previously thought. The huge Toba eruption of 71,000 years BP was probably the cause of several centuries of cold conditions. Should an eruption of this size occur now, it would have major impact on human activities.

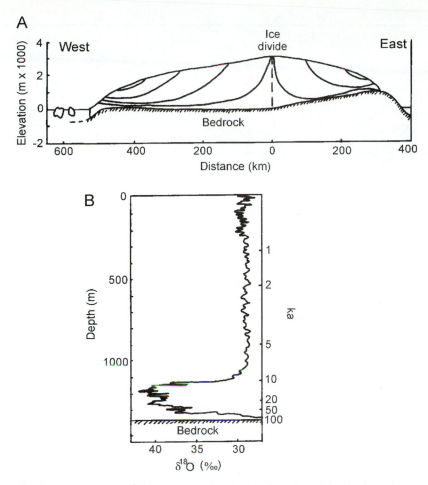

FIGURE 13.5   Greenland. *(A)* Diagrams showing idealized flow of the ice sheet along an east-west transect. Notice calving terminus to the west. *(B)* Oxygen isotope data from an ice core from the central-western part of the ice sheet, showing marked change at the start of the Holocene (10 ka). (After Reeh 1989.)

in the foreseeable future, while the continent maintains its polar location. The high-polar ice sheets of Antarctica are surrounded by deep cold ocean and locally ground down to 500+ m water depth, and are not a good analog for the Pleistocene Northern Hemisphere ice sheets. The latter ice sheets extended over shallow shelves, reached mid-latitudes, and received considerable moisture from adjacent, relatively warm oceans.

North American Pleistocene Ice Sheets

Extension and Architecture   During glacial maxima, most of North America was covered by glacial ice, from the Arctic Ocean down to Wisconsin, Illinois, and Nebraska in the northern United States, and from the Atlantic shelf edge to the Pacific shelf (Fig. 13.7). Glaciers did not cover some northern areas, such as northern Alaska and Yukon Territory, because those areas did not receive enough moisture for glaciers to form. Moisture was also cut off from the west and south by high mountains and coastal glaciers.

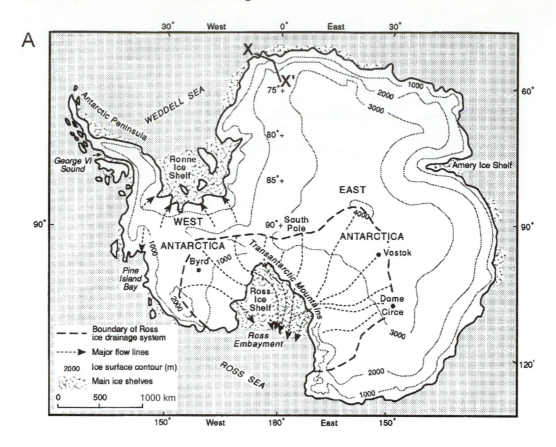

A

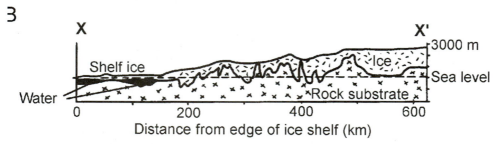

B

FIGURE 13.6   Antarctica. *(A)* Map of Antarctic ice sheets and shelves (After Clapperton 1997). *(B)* Cross section of the margin of the East Antarctic Ice Sheet from X to X'.

The North American ice was subdivided into two major ice sheets, the Cordilleran Ice Sheet to the west, and the Laurentide Ice Sheet in central and eastern North America. The Laurentide Ice Sheet was joined with the Greenland Ice Sheet to the east through the extensive ice cover of what are now Baffin Bay and Davis Strait and adjacent seas.

The Cordilleran and Laurentide ice sheets were composite entities characterized by variable thickness and activities. Several internal domes and ice divides existed, which remained relatively stable throughout the existence of the glaciers. The principal domes in the Laurentide Ice Sheet were the Labrador (LT, just east of Hudson Bay [HB]), the Keewatin (KW, just west of Hudson Bay), at the center of Foxe Basin (FB), and the McClintock Plateau (MK) west of Foxe Basin (Fig. 13.7).

BOX 13.2   Antarctic Lakes

Ice-penetrating radar has now identified more than seventy lakes below the thick Antarctic ice sheet. Most of the lakes are only three to four kilometers long, and lie beneath major ice divides under three to four kilometers of ice. Lake Vostok, however, is the size of Lake Ontario. Lake Vostok was identified in 1996 from a variety of observations. It is about 700 m deep and rests in a deep, narrow depression, possibly a rift, beneath four kilometers of ice. The extent of the lake is inferred from the flat surface of the Antarctic Ice Sheet where it goes afloat on crossing the lake (Fig. Box 13.2.1). The ice sheet thins, and then thickens, as it crosses the lake, which seems to be caused by melting and refreezing of its base.

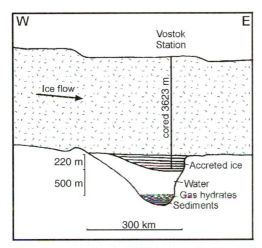

FIGURE BOX 13.2.1   Diagram showing large subglacial lake near Vostok station, eastern Antarctica. (After Bell and Karl 1999.)

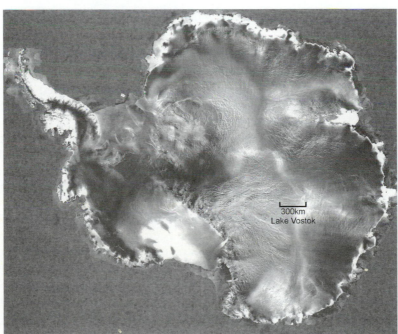

FIGURE BOX 13.2.2   Location of Lake Vostok shown on a remote sensing compilation using RADARSAT. (From: *http://svs.gsfc.nasa.gov/imagewall/antarctica/whole.html*.)

If the lake is in a rift, then geothermal vents could supply sources of energy for living organisms, like deep ocean vents do. Such a subglacial lake environment would be oligotrophic, with low nutrient levels and a low standing stock of organisms.

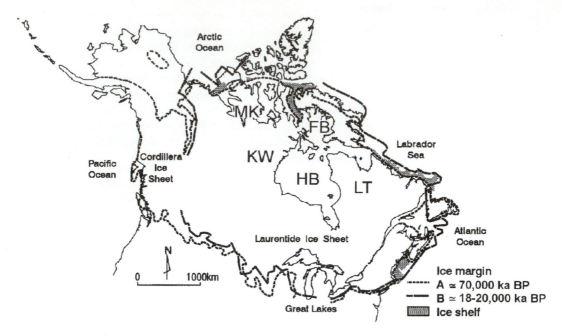

FIGURE 13.7 Map showing maximum extension of North American ice sheets (FB = Foxe Basin, HB = Hudson Bay, KW = Keewatin Plateau, LT = Labrador Trough, MK = McClintock Plateau).

Onset of the Latest Glaciation During the early part of the Late Pleistocene, during isotope stage 5 and in particular 5e, no major glaciers existed in North America, or at least no more extensive than the present-day ones at high latitudes and altitudes. Glacial advances probably began on parts of Canada during late isotope stage 5, and certainly during isotope stage 4. Glaciers extended across the northern part of the continent, and remnant glacial features are preserved, particularly in southern and eastern regions. Much of the ice probably disappeared during the subsequent isotope stage 3, but in many areas scattered patches remained and the whole region had considerable snow cover.

Relatively precise reconstructions can be made of the last Wisconsinan glaciation. Preserved glacial erosional features, erratic distribution, sediment lithology, and, indirectly, postglacial features associated with isostatic rebound point out several centers of glaciations. Some centers were in the eastern maritime regions, but the two most important ones were in the Labrador Trough (LT) and in the Keewatin Plateau (KW) regions, respectively east and west of Hudson Bay. These two areas are at mid-latitudes, so they may have been most influenced by the insolation variations associated with the Milankovitch effect. They were also located where the reconstructed Pleistocene snowline intersected the surface of continent, and received a lot of precipitation since they were quite close to the moisture sources of Hudson Bay and the North Atlantic. It is from these and other centers that the glaciers grew, joined, and spread to the north and rapidly to the south. It is not possible to accurately date the tills themselves, but the advance and retreat of the glaciers can be reconstructed by dating the interglacial sediments underlying and overlying the tills along north-south transects (Fig. 13.8).

Holocene Deglaciation The Laurentide Ice Sheet reached its maximum extent around 20–18 ka. It had extended to low latitudes, and the ensuing warming trend

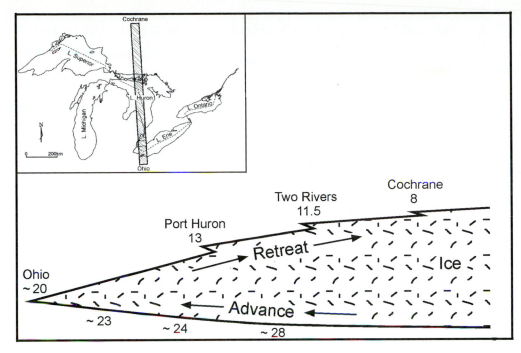

FIGURE 13.8   North-south cross section from Cochrane (north Ontario) to Ohio indicating progressive advance and retreat of the ice sheets, as recorded by approximate dates of the water-deposited sediments underlying and overlying tills (numbers indicate thousand of years). (After unknown; see website for update.)

had a strong effect on the ice that thinned and retreated rapidly, particularly starting at about 15–14 ka. The ice retreat was not regular and synchronous around the ice margin, due to local controls by latitude, altitude, and climatic regime (continental or maritime). Furthermore, the weight of the ice sheet had isostatically depressed Earth's crust into a bowl shape, and numerous large lakes formed at the margin of the glacier as it retreated. The largest of these was Lake Agassiz, which reached a size of 350,000 km² in central-south Canada. The temperate ice and the presence of basinal water promoted dynamic glaciers that experienced numerous surges, irrespective of the overall climatic trend. Surging ice tongues penetrated hundreds of kilometers into the lakes and were rapidly removed by calving and melting, creating a drawdown of ice from the main glacier body, thinning it rapidly. These rapid changes may have triggered catastrophic events, such as the collapse of ice dams and the sudden release of waters from very large lakes, such as Lake Agassiz. A similar catastrophic event possibly occurred during the deglaciation of the Hudson Strait area, and this may have led to the collapse of the ice barrier and the flood of freshwater from the Hudson Bay/James Bay area and the Ojibway glacial lakes into Baffin Bay, Davis Strait, and, ultimately, the North Atlantic. Conversely, the opening of this strait led to a rapid entrance of salt water into the Hudson Bay basin and the formation of the early postglacial Tyrrell Sea, which is the precursor of the modern James Bay, Hudson Bay, and Foxe Basin.

Synthesizing, the following is a possible sequence of deglaciation stages of the Laurentide Ice Sheet.

18,000–20,000 Years BP   The last maximum extension of the Laurentide Ice Sheet over the continent was reached at this time (Fig. 13.7). However, the ice sheet had already retreated somewhat from the eastern marine shelves, and was absent from the cold, arid areas of northern Alaska and parts of the Yukon Territory, which lay far from open oceanic waters and rested in the shadow of mountains.

14,000 Years BP Ice shelves were well defined in deep embayments in the Arctic Sea area (Fig. 13.9A). Ice covered parts of the shelf of Labrador, Nova Scotia, and the Gulf of St. Lawrence near the present mouth of the St. Lawrence River. Inland, the glacier retreated to the Great Lakes and away from the Atlantic seaboard. Terminal lobes and proglacial lakes developed in the Great Lakes region and other parts of the southern ice margin. To the west, the Laurentide Ice Sheet started separating from the Cordilleran Ice Sheet.

13,000 Years BP Marine terminations still existed on the shelves of the Arctic Ocean and in a narrow strip along the Labrador and Newfoundland coast (Fig. 13.9B). The north shore of the St. Lawrence River (Goldthwait Sea: part of what is now the St. Lawrence River valley) was still depressed, as the glacier had just left it. At this time, the main body of the Laurentide Ice Sheet started separating from the Appalachian ice cap of the Gaspé Peninsula. The eastern American seaboard (Nova Scotia, Maine) was abandoned, and the glacier retreated to the Appalachians, leaving only a few ice outliers scattered over Nova Scotia and Prince Edward Island.

Inland, the Appalachian ice cap developed lobes into the eastern valleys. In the Great Lakes region, the terminal glacial lobes were fully developed and large lakes formed, such as Lake Whittlesey, which were wider than the present-day lakes. To the west, the Laurentide Ice Sheet separated from the Cordilleran Ice Sheet and proglacial lakes started to develop.

12,000 Years BP Few ice shelves were preserved in the northeast (Fig. 13.9C). Only part of the Labrador nearshore area was covered by ice. The glacier also retreated from the north shore of the Gulf of St. Lawrence. Inland, a few remnant ice blocks persisted in Nova Scotia and Newfoundland. The Appalachian ice cap shrank greatly. The southern and southwestern parts of the Laurentide Ice Sheet were rimmed by lakes, from Lake Vermont on the east, to a large lake covering the St. Lawrence Lowland, to Lake Algonquin in the Great Lakes region, and to Lake Agassiz and the precursor of the present Great Slave Lake, respectively, in the west and northwest.

11,000 Years BP Only one large ice shelf persisted in the Arctic Ocean in the present Gulf of Boothia area. In the south, the main event was the penetration of seawater into the upper St. Lawrence River valley, replacing the existing proglacial lake. The large glacial Lake McConnell formed in the northwest (Fig. 13.9D for location).

Various changes occurred in the southern Great Lakes and Lake Agassiz. Fluctuations in glacier terminus triggered large, perhaps even catastrophic discharges from Lake Agassiz eastward into Lake Algonquin (this interpretation is not accepted by everybody). To the east, the Newfoundland ice cap started to split into smaller blocks. Along part of the eastern seaboard, local glacier readvances have been associated with the so-called Younger Dryas: a cold period noted in various parts of the world, particularly in Europe and South America. Along the eastern North American seaboard, the local, triggering event has been tentatively associated with the sudden cooling and freshening of Atlantic surface waters caused by megafloods from Lake Agassiz. Other events may also have diverted the northward-

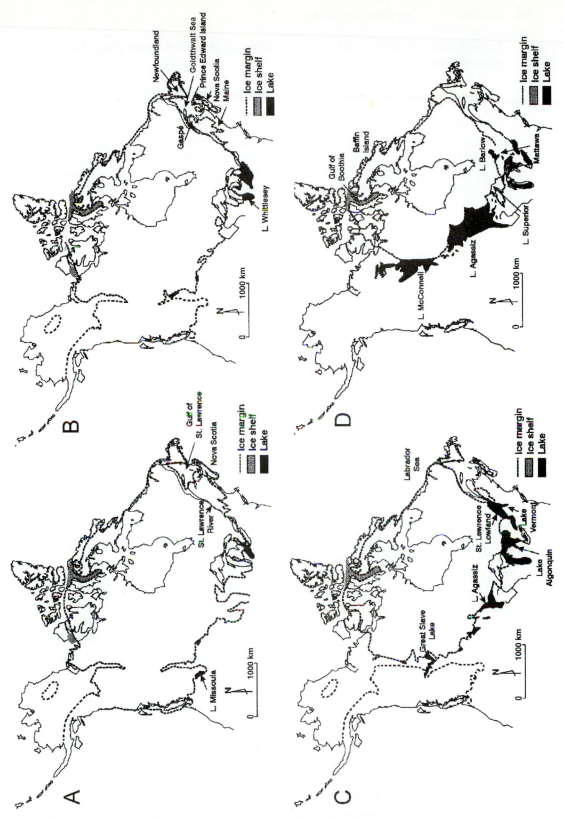

FIGURE 13.9 Maps showing stages of deglaciation of the Laurentide and Cordilleran Ice Sheets: *(A)* 14 ka; *(B)* 13 ka; *(C)* 12 ka; *(D)* 10 ka. (After Prest 1970; Prothero 1994; Dawson 1992.)

flowing, surficial, warm Gulf Stream and the southward-flowing, deep, cold North Atlantic Deep Water marine currents.

10,000 Years BP The ice shelf of the Gulf of Boothia persisted in the north (Fig. 13.9D). Some lobed termini extended into the fjords of eastern Baffin Island. Ice disappeared from the central Appalachians and eastern areas, except for a few remnant ice blocks in Newfoundland. The precursors of the Great Lakes shrank as waters flowed out along depressed northern outlets near the glacier terminus. A large delta formed at the eastern end of one of these outlets in northeastern Ontario (Mattawa). Proglacial Lake Barlow developed in northern Ontario, and Lake Agassiz and Lake McConnell enlarged in the west. A glacier advance in the Lake Superior area cut off Lake Agassiz from the eastern lakes.

9000 Years BP The Gulf of Boothia ice shelf disappeared in the north, and a large marine embayment opened in what would later become the Hudson Strait (Fig. 13.10A for location). A large lake appeared along the ice margin in Labrador. The Great Lakes basins started to refill with water when the land was tilted back to the south due to differential isostatic uplift as the glaciers retreated farther northward. Because the ice thickness was greater to the north, that land was more depressed, and, when the glacier melted, it rebounded more rapidly and to a greater extent than the southern lands. Lake Ojibway-Barlow enlarged in the north and received discharge from Lake Agassiz, which, at this time, was slightly reduced in extent. The glacier retreated from the northwest, and Lake McConnell was split into smaller, scattered lakes.

8400 Years BP The glacier retreated inland in the north, but Hudson Bay was still under the ice (Fig. 13.10A). The Hudson Strait embayment was enlarging toward Hudson Bay. Lake Agassiz enlarged to its maximum extent and discharged into Lake Ojibway, forming a continuous 3100 km-long water body along the southern margin of the Laurentide Ice Sheet. Relatively large lakes formed in the northwest, east of the remnants of Lake McConnell. The waters of the southern Great Lakes were rising slowly, but did not quite reach present-day levels. Large lakes were still present in Labrador, in northeastern Canada.

At this time the regular retreat of the glacier was interrupted by a series of glacial surges into Lake Ojibway, into the Hudson Bay Lowland (southwest of Hudson Bay), and throughout the Canadian Arctic.

8000 Years BP The Laurentide Ice Sheet split into two parts: the Labrador ice cap to the east and the Keewatin-Foxe Basin ice cap to the west and north. Possibly an ice island persisted in Hudson Bay. The sea re-entered Hudson Bay, forming the Tyrrell Sea whose shores were several hundred kilometers inland from the present ones because the land was still depressed just after glaciation (Fig. 13.10B). To the south, the Mississippi River, the main collector of meltwater during the early stages of deglaciation, no longer received waters from glacial Lake Agassiz. The river changed from its early braided form when it was carrying meltwater, to a meandering form as it drained large parts of the North American continent. Lake Agassiz was greatly reduced in size and split, becoming in part the precursor of the present Lake Winnipeg. The Great Lakes continued their slow adjustment toward their present levels. Former locations of large lakes are shown by vast areas of lacustrine deposits (Fig. 13.10B). The Laurentide Ice Sheet continued to wane until it ceased to exist as an ice sheet about 8.0–7.5 ka.

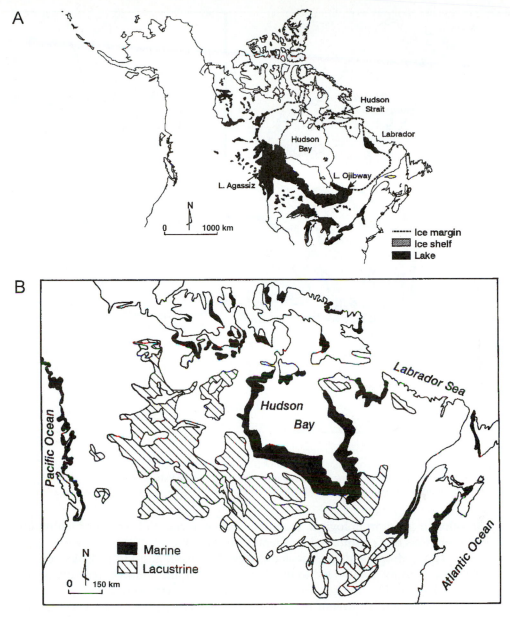

FIGURE 13.10 *(A)* Map showing late deglaciation stages and associated basinal deposits of the Laurentide Ice Sheet at 8.4 ka (After Prest 1970; Dawson 1992). *(B)* Distribution map of glaciomarine and glaciolacustrine deposits of North America (After Quigley 1980).

4000 Years BP   The last vestiges of the Laurentide Ice Sheet had waned from North America. A few remnant glaciers persist in the arctic islands, such as the Barnes Ice Cap on Baffin Island and valley glaciers in the western mountains.

Deglaciation of Southwestern Ontario: North American Great Lakes Region

Ice sheets can be affected by substrate morphology to a varying degree, and, in turn, they mold the landscape. Southern Ontario is a cratonic area where a well-

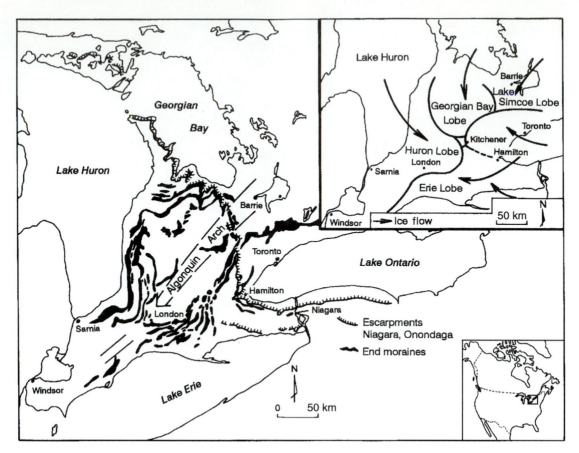

FIGURE 13.11   Maps of glacial features and lobes of southern Ontario, Canada. *(A)* Distribution of end moraines in southwestern Ontario, Canada (After Chapman and Putnam 1966). The Algonquin Arch is a bedrock-topographic high that has persisted from early Paleozoic to the present. *(B)* Glacial lobes during deglaciation (After Cowan, Sharpe, Feenstra, and Gwyn 1978).

developed geomorphological, sedimentological, terrestrial record of highly lobate terminations of a continental ice sheet is preserved.

The bedrock geology, preglacial fluvial erosion, and the modification of the fluvial valleys by glaciers led to the development of the southwestern Ontario peninsula delimited by the southern Great Lakes of North America (Lakes Ontario, Erie, Huron, and Georgian Bay) (Fig. 13.11A). The peninsula is characterized by a morphotectonic high (Algonquin Arch) that has formed the southwest-to-northeast-oriented backbone of this area from Paleozoic times. The last ice sheet started retreating from the area about 14,000 years ago. Large lakes formed in front of the ice in the basins (modified ancient fluvial valleys) depressed by the weight of the ice. The highest central part of the peninsula became ice- and water-free and has been dubbed "Ontario Island" because it was surrounded by glacial lakes to the south and by glacier ice to the north (Fig. 13.12). In the basins, the glacier remained relatively thick and experienced active retreats and advances, indicated by a series of lobate end moraines on the adjacent lands (Fig. 13.11B). The various lobes retreated from the highland toward the center of the basins (the location of the present-day lakes) in a pulsating fashion (retreats alternating with local advances), and various geomorphological features (such as drumlins, eskers, end moraines, outwash and lacustrine plains) developed in quasi-predictable successions.

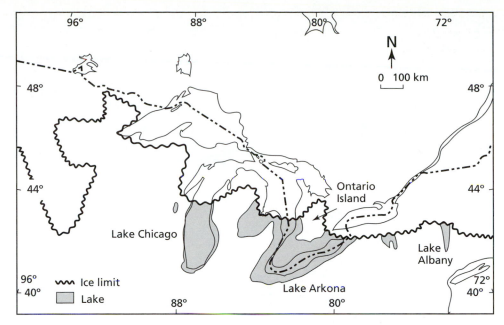

FIGURE 13.12   Map showing early stages of development of the Great Lakes of North America during deglaciation (about 13.6 ka). "Ontario Island" is ice-free land bounded by glacial lakes to the south and the Laurentide Ice Sheet to the north. (After Prest 1970.)

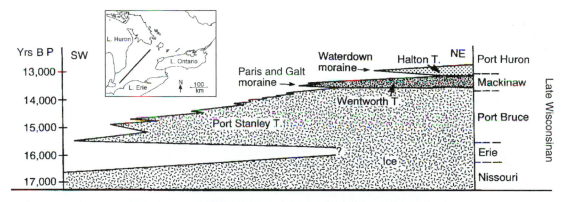

FIGURE 13.13   Generalized cross section of the more recent tills deposited by the Laurentide Ice Sheet in southwestern Ontario. The Wentworth and Halton tills and associated moraines are interpreted to be related to local glacier readvance after major retreats from Lake Ontario. Note that the Paris and Galt moraines have been interpreted as local readvance during an interstadial (Makinaw). (After Barnett 1985.)

The following examples may suffice to illustrate some of the difficulties in making detailed interpretations of the relict glacial features in a complex area such as the Great Lakes region, such as those related to the significance of recessional moraines and drumlins. First, some moraines were originally defined as recessional, implying that they formed during staging periods of the glacier retreat. However, more complex genetic systems have also been devised illustrating the possibility that glaciers may have at times retreated up to 100 to 300 km into the lake basins and even farther north of them, and then readvanced, depositing new tills and forming new end moraines, such as the Paris, Galt, and Waterdown moraines (Figs. 13.11A, 13.13). Second, the conventional interpretation of drumlins is that they

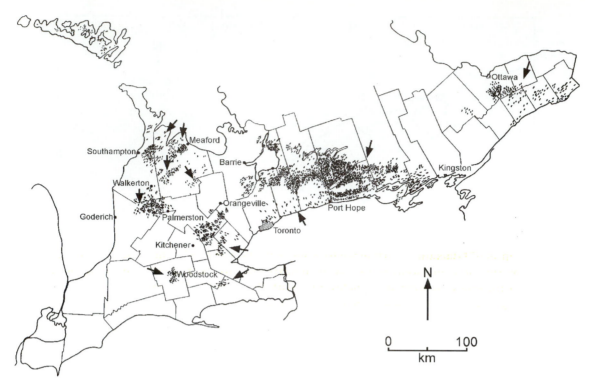

FIGURE 13.14   Map of drumlin fields of southwestern Ontario. (After Chapman and Putnam 1966.)

often form behind moraines under active glaciers where ice is thick enough to move some of the basal sediments around obstructions. According to this hypothesis, the drumlin fields of southwestern Ontario (Fig. 13.14) were formed at different times as the terminus of the glacier retreated from one place to another. But an alternative view is that meltwater megafloods have occurred in this area, some under the glacier itself. These megafloods would have been responsible for the quasi-contemporaneous formation of all drumlins. Partial justifications for this last hypothesis are the streamlined form of the drumlins, which resemble similar, but smaller water-formed features, and the presence of stratified drift inside some of the hills. This hypothesis needs further testing.

Deglaciation of the Western Cordillera of North America

The Cordilleran Ice Sheet extended from southwestern Alaska down to Washington State in the United States and from the Pacific shelf to the eastern foothills of the Rocky Mountains. Its growth and deglaciation were strongly influenced by the large latitudinal range, but mostly by the orographic effect created by mountain chains. Whereas the northern and interior part of the glacier behaved in a similar fashion to the adjacent western side of the Laurentide Ice Sheet, the Pacific side experienced considerable advances and retreats. On the whole, the ice sheet was still advancing on the Pacific coastal areas about 15 ka. The Pacific islands were probably free of ice sheets by 13.7 ka, and the glacier was retreating about 13 ka when large lakes, such as Lake Missoula, formed, and associated catastrophic floods developed, sculpting the Washington Scablands. By 11 ka the mainland coastal plains were ice free. Local

Cordilleran ice readvances occurred between 11 and 10 ka, probably representing a local manifestation of the Younger Dryas.

European Ice Sheets

Ice sheets periodically covered much of northern Europe, from the northern arctic shores down to the Netherlands and Germany (to about 52° N), and from the North Sea to the west to northern Asia to the east (Fig. 13.15). Separate glaciations occurred on the Pyrenees and European Alps (Fig. 13.16A).

Like North American ice sheets, the European ones had various centers of glaciations and domes that acted differently during deglaciation due to mountain chains and distance from the Atlantic, and, thus, from the influence of oceanic currents such as the Gulf Stream. The European ice sheets had domes in Scandinavia, Denmark, and Great Britain.

Onset of Glaciation   Allowing for some variation due to latitude and distance from marine moisture sources, the onset and expansion of the Late Pleistocene ice sheets in Europe were virtually synchronous with those of North America. Major glacier expansion occurred during isotope stages 4 and 2, with the maximum expansion of the last ice sheets reached at about 21 ka (Fig. 13.16A). Most centers of glaciations were on mountains, and the ice sheets spread out from there. However, for the last glaciation, centers have also been inferred in lowland and shelf areas in Scandinavia and northern Russia (Fig. 13.15).

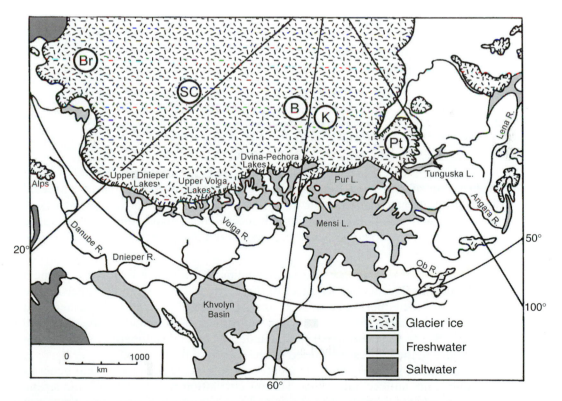

FIGURE 13.15   Map showing maximum extent of ice sheets in Europe and Asia according to Grosswald (1980) (Br = British ice dome; SC = Scandinavian ice dome; B = Barents ice dome; K = Kara Sea ice dome; Pt = Putorana ice dome). (After Rutter 1995; Benn and Evans 1998.)

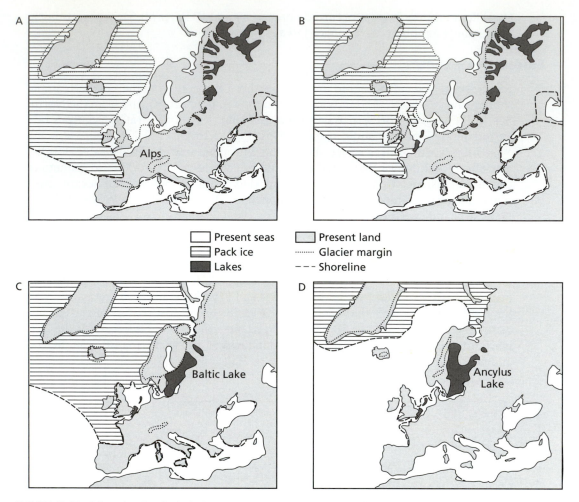

FIGURE 13.16 Maps showing deglaciation stages of European ice sheets: (A) 21–18 ka: maximum extension of the northern European ice sheets; *(B)* 15 ka; *(C)* 11–10 ka; *(D)* 9.5 ka. (After Andersen and Borns 1994.)

Deglaciation of Northern Europe The northern European ice sheet retreated slowly from about 17 ka to about 15 ka. Afterward, the rate of retreat increased, alternating with temporary glacial readvances. The collapse and disappearance of the northern European ice sheet occurred about 500 years earlier than those of North America.

21,000–17,000 Years BP At its maximum, the northern European ice sheet joined with the Barents Sea ice sheet to the north and the Asian ice sheets to the east, and, for a brief period between 21 and 20 ka, with the Devensian ice sheet that covered Great Britain (Fig. 13.16A). At its maximum, a grounded ice sheet covered the North Sea to the continental shelf. Numerous lakes developed around the glacier margins in the depression caused by the weight of the ice itself.

The glacier supported a permanent, atmospheric, high pressure that generated strong, cold, anticyclonic winds. This air circulation contrasted with the strong cyclonic winds associated with the atmospheric low-pressure area over the north Atlantic Ocean. These latter, moisture-charged Atlantic winds tracked south of the Alps into the Mediterranean Sea area. The relatively high latitude of the ice sheet and the cold glacial winds were responsible for the development of the vast periglacial regions of central Europe and for the rim distribution of cover sands and loess. Winter temperatures in central Europe have been calculated to be about 20°C colder than at present. To the south, mountain glaciers over the Pyrenees, Alps, and Carpathians were large enough to form a barrier to the north-south migration of flora and fauna.

17,000 Years BP   A major, albeit fluctuating glacier retreat occurred. This may have been influenced by changes in insolation at mid-latitudes and by an increase in iceberg calving for glaciers terminating into the sea during a rise in sea level.

15,000 Years BP   The ice sheet distribution was similar but much reduced from that of the ice maximum (Fig. 13.16B). Numerous lakes existed along the eastern rim of the glacier, and the North Sea area and central Europe were covered by tundra.

11,800–10,000 Years BP   A warmer period (11.8–11 ka) was followed by the colder Younger Dryas (11–10 ka) when glaciers experienced local readvances. The ice sheet was restricted to Scandinavia and to the Barents Sea area (Fig. 13.16C). The large, glacial Baltic Lake formed to the east. The ice cover in the Atlantic Ocean retreated northward.

9500 Years BP   The western boundary of the ice sheet became landlocked and in fjords. Much of the north Atlantic Ocean was ice free, and major glacial retreat occurred in Iceland. To the east, a large (Yoldia) sea developed, which was in communication with the northern part of the North Sea.

9000–8600 Years BP   Remnants of the ice sheet existed in the Scandinavian Mountains (Fig. 13.16D). A large lake (Ancylus Lake) occupied the area of the present Baltic Sea as isostatic rebound closed the communication of the previous Yoldia Sea with the North Sea. Smaller lakes persisted in the southern part of the North Sea that was still above sea level at this time. The English Channel was a large continental valley.

8500 Years BP   Essentially any trace of the ice sheets disappeared, and slowly the outline of the land assumed its present form as the sea rose to its final level. At about 8 ka, saltwater re-entered the area of the Baltic Sea and the Littorina Sea was formed.

European Alps   An ice sheet covered the European Alps during the Pleistocene (Fig. 13.16A). Several mountaintops (nunataks) towered above the glaciers and some valleys were ice free. The main morphological characteristics of the European Alps are dictated by the lithology and structure of the bedrock. The European Alps are a young mountain chain still tectonically active, in places rising at about 1 mm/yr. Adjacent areas are subsiding, like the Black Forest and Rhine graben in Germany. The main valleys follow tectonic lineaments. These valleys have been deepened by fluvial processes first, changed into U-shaped forms by glacial processes later, and modified again by slope and fluvial processes after deglaciation. The importance of the Alpine glaciers is related to the fact that they constituted a barrier to air circulation and free movement of flora and fauna during the Pleistocene. From a historical, scientific point of view, their importance is related to the fact that the glacial theory and the stratigraphic-climatic subdivisions of the Pleistocene in its quadripartite form were first described here (Fig. 13.3A). The classification of cold-climate periods alternating with warm interglacials was refined here during the nineteenth and early twentieth centuries, utilizing principles of morphostratigraphy and soil stratigraphy.

Asian Ice Sheets

A debate still exists on whether the northern Asian sheets covered all of the northern part of the continent, or nonglaciated areas persisted because of the low moisture content (Fig. 13.15). In either case the ice sheets had various centers of glaciation. One was located on the northern shallow Kara Sea shelf, where the ice sheet started forming as the sea level dropped and thick sea ice grounded. The glacier later expanded onto the adjacent continental areas to the south. A second major center of glaciation occurred in eastern Siberia. According to the restricted ice sheet hypothesis, the two glaciers never joined, and drainage to the north persisted. According to the widespread ice sheet hypothesis, these sheets joined and dammed any water flow to the Arctic Ocean, and extensive glacial lakes developed to the south of the ice. These lakes drained southward to the Caspian Sea, eventually spilling over into the Black Sea and perhaps, at times, also into the Mediterranean Sea. Soon after the glacial maximum, from about 18 ka to 13 ka, deglaciation was caused primarily by aridification. Global warming caused later deglaciation from about 13 ka to 10.8 ka. The slight cooling and glacier readvance of the Younger Dryas (10.8–10 ka) is recorded in Asia as well. During postglacial times, considerable peat started forming at about 9 ka, and the postglacial temperature maximum (hypsithermal period) occurred at about 6–5 ka, during which time the annual temperature was about 3–4°C warmer than the present.

The Tibetan Plateau A debate exists on whether the Tibetan Plateau was ever covered by a major ice sheet or only by scattered mountain glaciers and local ice caps. The Tibetan Plateau and the surrounding mountains (such as Himalayas, Karakorum) have been tectonically very active in the last 5 million years, and have been uplifted by several thousand meters. This uplift had the effect of changing airflows and climatic conditions in the region. Although the land lies above the snowline, many researchers believe that large ice sheets did not form because of lack of moisture. Others, instead, think that ice sheets developed, during the early Pleistocene at least, when climate was still a wet monsoonal one on the rising plateaus, as yet unsheltered by the uplifted mountain rims (Fig. 13.17). In later Pleistocene times only mountain glaciers or relatively small ice caps developed. There is little evidence for an extensive glaciation on the Tibetan Plateau itself.

Large Glaciers of the Southern Hemisphere

Widespread glaciers occur now in the Southern Hemisphere and were more widespread during the Pleistocene in Antarctica and the high-latitude islands. Glaciers also occur in New Zealand, a small one exists on the equatorial Mount Kilimanjaro in Africa at high altitude, and there are numerous small glaciers on the Andes above 5000 m asl (Fig. 13.18). A true ice cap of about 70 km^2 developed in southeastern Peru, and large ice fields occur in Patagonia (wider than 14,000 km^2) and Tierra del Fuego. The absence of more widespread large glaciers in the high Andes is due to aridity.

130,000–21,000 Years BP A readjustment of global climate occurred from the last warm interglacial (isotope stage 5e), to the first major glaciation (isotope

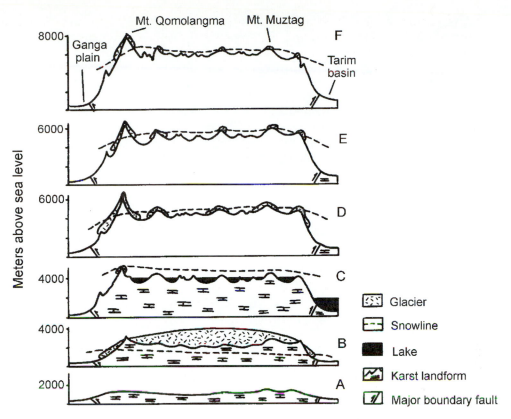

FIGURE 13.17 Cross sections showing glacier distribution and snowline elevation in the Tibetan Plateau area and surrounding mountain chains. Notice that after the formation of an ice sheet covering the whole area, as the plateau rose in elevation and was progressively cut off from moisture sources, the snowline rose in altitude and only local glaciers developed (After Han 1997). *(A)* Peneplanation at the end of Pliocene. *(B)* Middle-early Pleistocene ice sheet stage. *(C)* Early to Middle Pleistocene interglacial stage. *(D)* Middle Pleistocene glaciation. *(E)* Upper Pleistocene glaciation. *(F)* Late Pleistocene-Holocene deglaciation.

stage 4), and to the last major glaciations (isotope stage 2). Low-temperature periods have been recorded at about 65, 21–18, 14, and 11 ka, with other smaller temperature fluctuations of shorter (few thousand years) cyclicity present as well. Mountain glaciers are expected to have responded more rapidly to these climatic changes than large ice sheets, although evidence does not always bear this out. It is also of interest that glaciers of the Southern Hemisphere, which are significantly affected by moisture variations, reach their widest expanse well before the temperature lows associated with the overall global ice maximum. This may be related to the fact that cold air can carry lower moisture; thus the mountain glaciers are starved during particularly cold times. Changes in air movements have also influenced the areas receiving greater precipitation along the north-south–oriented Andes, and thus the location of shrinking and advancing glaciers. Conversely, most glaciers with marine terminations expanded during the global climatic lows because of sea level drop when it was possible to expand onto exposed shelves.

21,000–14,000 Years BP During the global glaciation maximum (21–18 ka), landlocked glaciers in the south were larger than the present ones, but not as large as those preceding them during slightly warmer but moister climatic conditions. The

GLACIERS

FIGURE 13.18 Map showing glacier distribution in South America. (After Global Land Ice Measurements from Space (GLIMS); *http://www.flag.wr.usgs.gov/GLIMS/glimshome.html.*)

sudden advance of southern glaciers at about 14 ka highlights the importance of moisture on glacier behavior, when there was no major cooling but rather an increase in wet conditions. Glaciers terminating into the sea, instead, behaved in closer unison with the northern ice sheets, as indicated by the steady recession of southern tidewater glaciers from 18 to 15 ka when sea level rose and more intense iceberg calving occurred.

14,000–10,000 Years BP An overall climatic amelioration led to the recession of most glaciers, punctuated by short-lived, local advances. Significant changes in flora occurred, with tree lines rising on mountain slopes, and peat bogs developing in southern islands before 11 ka. Although not as well defined and as yet not as well understood, a slight glacial readvance occurred between 11 and 10 ka comparable to that of the Younger Dryas of Europe.

10,000 Years BP–Present   During the first 5000 years of the Holocene, most of the lowlands and plateaus were freed of glacial ice, and full interglacial conditions were established, with the hypsithermal period being achieved about 8–6 ka. This high-temperature period had some significant effects on the drier parts of the Southern Hemisphere, such as a water level drop of about 55 m in Lake Titicaca in Peru and Bolivia, and perhaps the drying out of some Patagonian lakes. Sea level also rose a few meters, as indicated by shoreline features. At about 5 ka, the climate deteriorated again and some glaciers readvanced throughout the Southern Hemisphere, lake levels rose, peatlands expanded, and variations in ecozones occurred. The later part of the Holocene has been marked by variable climate.

ENDNOTE

1. During these times the Mediterranean Sea was cut off from the Atlantic Ocean. It dried out and a considerable amount of salt and other evaporites was deposited throughout the basin.

CHAPTER 14

Pre-Quaternary Glaciations

INTRODUCTION

For the last billion years or so, Earth's climate has fluctuated between warmer and colder phases (Fig. 14.1A). The warmer phases, called greenhouse periods, had equable conditions from equator to pole, and were sometimes almost completely ice free (Fig. 14.1B). For example, in the mid-Mesozoic greenhouse, temperate conditions extended to high latitudes toward both poles. The cold phases, called icehouse periods, had more extreme conditions. There were large differences in temperature, and a greater contrast in climate, between the poles and the tropics. Ice covered much of the polar to mid-latitude regions, and large continental ice sheets and glaciers periodically advanced and retreated from them. Reliable records of major glaciations start in the Proterozoic (about 2.6 Ga; Fig. 14.1A,B). During the last billion years, major icehouse periods have recurred approximately every 300 Ma. These were in the late Proterozoic (Neoproterozoic) (which lasted from about 800 to 550 Ma), late Paleozoic (340 to 200 Ma), and the Tertiary-Quaternary (40 Ma to Present). During the Phanerozoic (last 600 Ma), other relatively smaller glaciations covered parts of continents as well during the Late Ordovician (~450 Ma) in Africa, and Late Devonian (~380 Ma) in parts of South America.

BASIS OF INTERPRETATION

The Quaternary glaciation, being the most recent, is used to interpret ancient glaciations. Older glaciations, though, have their own peculiarities due to tectonic settings, evolutionary stages of flora and fauna at the time, and so on. The older the events, the less complete and less reliable the preserved evidence is.

Ancient glacial or otherwise cold conditions are inferred from physical, biological, and chemical criteria preserved in sediments. Landforms are normally not preserved and cannot provide information about old glaciers. Exceptions are some exhumed Carboniferous eskers and drumlins in Brazil and Ordovician roches moutonnées and similar features in Saharan Africa. Striated pavements, if accompanied by other evidence of cold and glacial settings, may indicate glaciation. Diamictites are the best sedimentary indicators of past glacial conditions, particularly when they contain flat-iron-shaped, faceted, striated clasts and are relatively thick and widespread; and are even better if they rest on striated surfaces or boulder pavements. Other indicators of frigid (though not necessarily glacial) conditions are (1) fine,

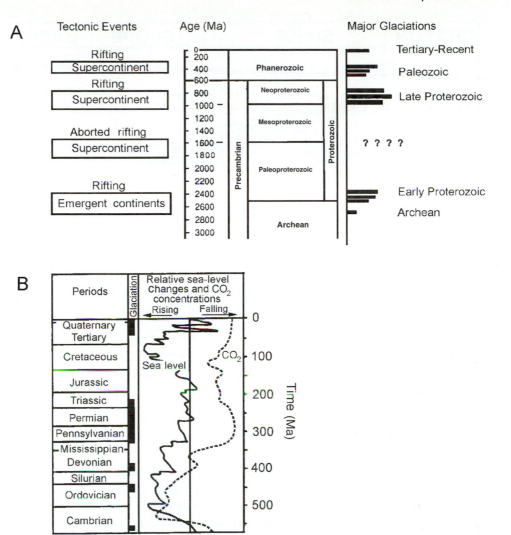

FIGURE 14.1   Diagrams showing occurrence of major glaciations on Earth and associated changes in sea level and other parameters. *(A)* Overall occurrence in relation to periods when continents were joined (supercontinentality). *(B)* Glaciations of the last 600 Ma (Phanerozoic) and their relation to sea level and carbon dioxide. (After Plumb 1991; Eyles and Young 1994.)

massive to laminated deposits with isolated pebbles (lonestones), or pebble clusters, possibly dropped by floating ice; and (2) multiple-composition (polymictic) conglomerates or sandstone with unweathered fresh mineral grains such as feldspar. Cold climates, possibly related to glacial conditions, can also be inferred from some fossil assemblages such as the Carboniferous marine *Euridesma* (clam) assemblage, and from oxygen isotope analyses of fossil shells.

Most of the features mentioned, however, can be formed in cold, nonglacial settings as well. Furthermore, slumping in subaerial or subaqueous environments, particularly in tectonically active regions, may form local diamictite. Indeed, isolated features can lead to erroneous conclusions. It is their assemblage that gives confidence to the glacial interpretation.

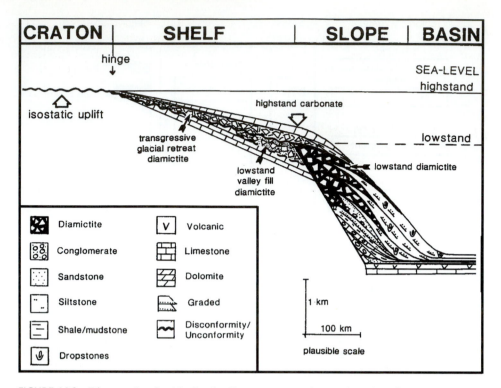

FIGURE 14.2 Diagram showing idealized sedimentary successions and erosions (unconformities) on a shelf and continental slope during the advance and retreat of a grounded glacier. (After Brookfield 1994.)

Preservation of the sedimentary record poses a problem in the interpretation of ancient glaciations, particularly in establishing their areal extent. Most of what was formed on raised uplifting lands has been eroded. The best glacigenic and associated cold-climate successions are preserved in subsiding basins, such as continental rifts, shelves, and oceanic settings, where ongoing tectonic subsidence provided accommodation space for thick sediment accumulations. As in any type of sedimentary accumulation, these glacial successions are incomplete. They may contain unconformities, and may be interstratified with materials of various and warmer settings. Where the successions are best preserved, it should be possible, albeit seldom tried, to establish the behavior of the feeding glaciers, and/or the effects of glacially induced eustasy and isostasy on sedimentation. For example, during lowstand, detrital sediments (including reworked carbonates) derived from the glacier itself or by erosion of exposed shelves are deposited on the basin slope and floor (Fig. 14.2). During interglacial or postglacial highstand, sediments include more slowly deposited carbonates and mature (better sorted, rounded) clastics on shallower, ice-free shelves and shores (Fig. 14.2). As with any other rapidly deposited sediments, glacigenic successions may undergo postdepositional deformation, which complicates their interpretation. For example, basin slopes with unstable deposits may undergo soft sediment deformation and slumps, often on a large scale, which are difficult to recognize in normal outcrops. The Storegga slide, off western Norway, has a headwall 290 km long, extends 800 km into the basin, and contains individual slabs up to 200 m thick and 30 km long.

Notwithstanding these difficulties, the major features of ancient glaciations can usually be reconstructed. Understanding them is important, not only for reconstructing Earth's history, but also for predicting what may occur in the future, as Quaternary glaciation will eventually end.

ARCHEAN GLACIATIONS

Archean glaciations are almost impossible to decipher in the limited, deformed, and metamorphosed remnants of these ancient deposits now existing on Earth. A possible exception is the late Archean (2 to 3 Ga) polymictic conglomerates of the gold-bearing Witwatersrand Supergroup in South Africa (Fig. 14.3A), which may be proglacial in origin.

EARLY PROTEROZOIC GLACIATIONS

Glacial deposits are recorded from the early Proterozoic (Paleoproterozoic) but not from the middle Proterozoic (Mesoproterozoic) (Fig. 14.1A). Earliest Paleoproterozoic glacial deposits are common and widely distributed in Canada (Ontario, Quebec, Northwest Territories), the United States (primarily in Michigan and Wyoming), and possibly Finland, India, South Africa, and western Australia (Fig. 14.3A).

Some of the best early Proterozoic glacial records occur in northern Ontario. They consist of glaciomarine diamictites and rhythmites with locally numerous lonestones, of about 2.3 Ga (Fig. 14.4). These cold-climate deposits occur at different stratigraphic levels separated by sediments locally highly weathered. This indicates multiple glaciations alternating with glacier-free periods.

LATE PROTEROZOIC GLACIATIONS

Extensive glaciations occurred during the late Proterozoic (Neoproterozoic) over all continents (Figs. 14.1A, 14.3A). Glacial periods alternated with nonglacial ones that were marked by intense weathering. Late Proterozoic glacial deposits are peculiar in that they occur in rifts and are associated with carbonates on continents that, as determined from paleomagnetic measurements, were located at low latitudes (0–20° N) during the glaciation.

Various hypotheses have been advanced to explain this apparently anomalous low latitude-equatorial glaciation, but, first, the data need critical scrutiny. The information is mixed. (1) Most diamictites are now confirmed as tillites. (2) The associated carbonates are often detrital, reworked from exposed interglacial carbonate platforms by glaciers during low glacial sea levels. In any case, carbonates do not necessarily indicate warm conditions. Recent carbonate sediments are now accumulating in sediment-starved areas on top of submerged tills in the cold Norwegian, Barents, and Arctic Seas. (3) Paleomagnetic data taken after 1950 are reliable and confirm the low-latitude position of the glacigenic deposits. However, the paleomagnetism itself may have been reset during rapid drift of the continents from high-

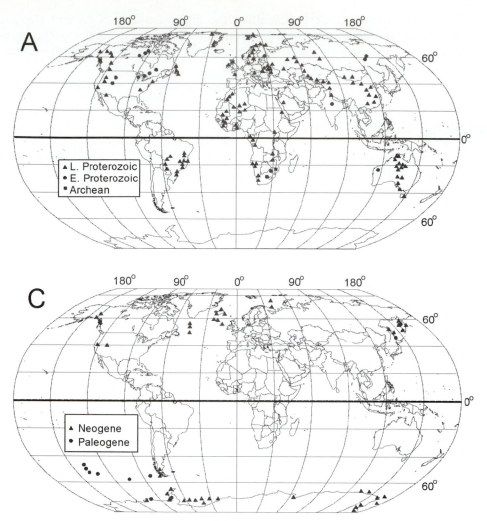

FIGURE 14.3 Maps showing locations where the record of pre-Pleistocene glaciations is found: *(A)* Archean, Proterozoic; *(B)* lower Paleozoic; *(C)* Carboniferous, Permian; *(D)* Paleogene, Neogene. (After Hambrey and Harland 1981.)

er to lower latitudes. That is, the low-latitude paleomagnetic indications may have nothing to do with the position of the continents during the glaciations. Assuming that the low-latitude location of the glacigenic sediment is correct, then those glaciations may have occurred either because of a global refrigeration or because of the high obliquity of the angle of Earth's rotation. At the time, the rotational axis was on the order of 50° (at the moment the obliquity of the axis of rotation is about 23°). More detailed work on the various continents is required to establish which hypothesis is most likely. For example, the first (global refrigeration) would imply that the glaciations were contemporaneous on all continents; the second (high obliquity) would imply diachronous glaciations as different continents drifted across the cold, glacial-forming regions. Evidence from Australia seems to favor high obliquity for the late Proterozoic. Isotopic $\delta^{13}C$ drops to a very negative value during glaciation, suggesting a complete cessation of biological production in the oceans. This, and

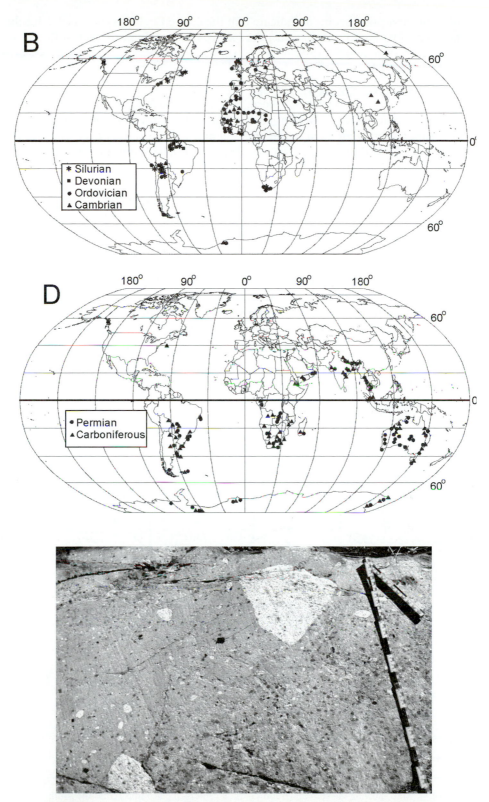

FIGURE 14.4 Photograph of an early Proterozoic diamictite at Elliot Lake, in Northern Ontario.

other observations, seems to indicate that the entire Earth was covered in ice during these late Proterozoic glaciations: the *Snowball Earth* hypothesis (Hoffman, Kaufman, Halversan, and Schrag 1998).

LATE PALEOZOIC GLACIATIONS

Several glaciations have been recorded throughout the Phanerozoic, some localized, others involving most continents. Significant glaciations occurred during the Cambrian, Late Devonian, Late Ordovician; major, widespread, long-lasting glaciations occurred during the upper Paleozoic (Carboniferous–early Permian) and the Cenozoic icehouse periods (Figs. 14.1B, 14.3B–D). The Mesozoic (Triassic, Jurassic, and Cretaceous) is usually considered to be a long-lasting, warm, greenhouse age, though even for these times some cold-climate deposits have been found: for example, in the Aptian stage of the Cretaceous of Eurasia. This would imply that relatively small glaciers might have existed in high-latitude areas.

During the Paleozoic, the centers of glaciation changed position through time. Glaciation occurred during the Cambrian in China, during the Ordovician in northern Africa and in parts of South America, during the Devonian in southwestern and northern South America, and during the Carboniferous in most continents of the Southern Hemisphere plus the Arabian Peninsula, India, and adjacent areas (Fig. 14.5). This reflects, for the most part, polar wandering and drifting of the con-

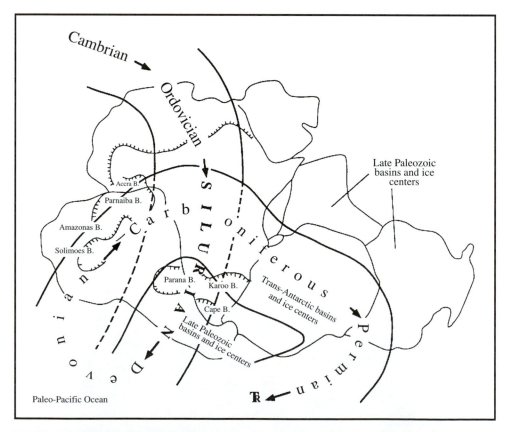

FIGURE 14.5 Map showing the spiral migration path of centers of glaciation across the southern continents, here represented in the joined position they had about 270 Ma, that is, *Gondwanaland.* (After Caputo and Crowell 1985; Eyles and Young 1994.)

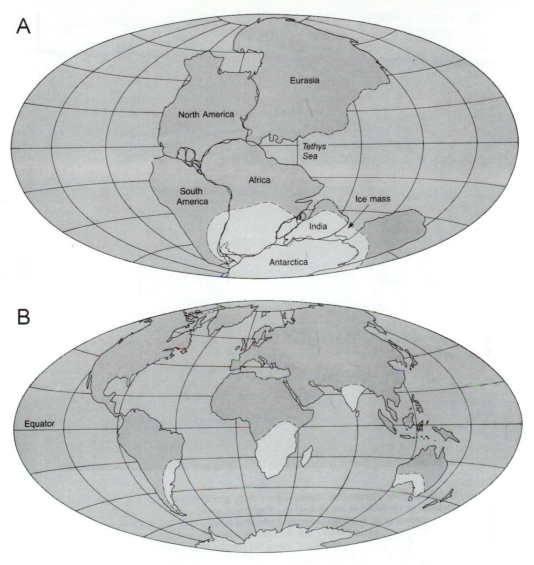

FIGURE 14.6   Maps showing *(A)* maximum extent of the Carboniferous-Permian glaciation in Gondwanaland; *(B)* present-day distribution of Carboniferous-Permian glacial deposits. (From Tarbuck and Lutgens 1999.)

tinents into and away from mid-high latitudes where glaciers could develop. The Carboniferous–lower Permian glaciation occurred when the continents were close together (Fig. 14.6A). Since then the continents have spread apart, carrying piggyback the geological record of the ancient glaciation.

The Carboniferous–lower Permian glaciation is as important as the Quaternary one and is analyzed here in some detail. Like the Quaternary ones, the Carboniferous–lower Permian glaciations had several centers. The older, Lower Carboniferous (~350 Ma) ones were located in southwestern South America (Bolivia, Argentina, Paraguay), and the younger ones were recorded in latest lower Permian mountain glaciers of South Africa, Australia, and the ice sheets of Antarctica. For 100 Ma the glaciers advanced and retreated many times. The climax of the

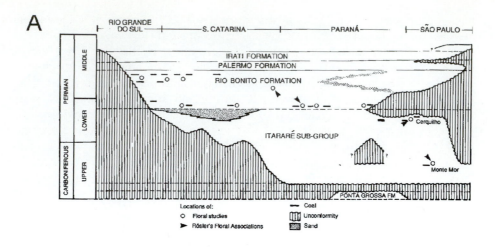

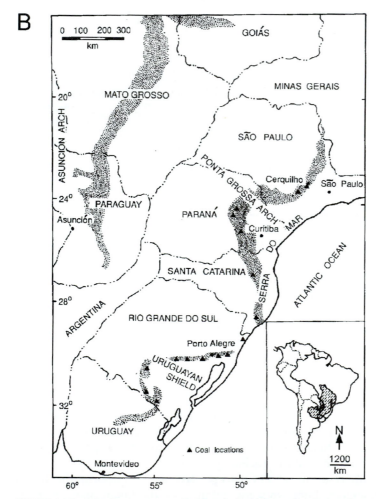

FIGURE 14.7 Distribution of coals in the Paraná Basin of Brazil. *(A)* Stratigraphic cross section across the Paraná Basin. *(B)* Map of the Paraná Basin with location of major coal occurrences (triangles). At the margins of the basin, the coals lie directly on glaciated surfaces, whereas at the center they lie on later-emersed, marine deposits. (From Martini and Rocha-Campos 1991.)

glaciation occurred in Late Carboniferous (~300 Ma) and covered great parts of the high-latitude and polar regions of Gondwanaland (Fig. 14.6A).

Although they developed on a supercontinent in the Southern Hemisphere, the Carboniferous to lower Permian glaciations are the most similar to those of the Quaternary because they occurred after major land plants had evolved. Thick organic deposits (peat, coal) developed during and after glaciation. Vast peatlands formed in the southern glaciated lands either directly on glacigenic deposits or, toward the centers of the basin, on marine or lacustrine materials after they emerged. Cold-climate coal-forming peat formed if the deglaciated land was immediately available for vegetation growth, as may have been the case around the continental margins of basins. Temperate to warm-climate peat developed on land that emerged later due to basin fill and tectonics (Fig. 14.7). In cases where vegetation growth followed soon after deglaciation, there is a good pollen and macrofauna record of the changing climate, from cold to warm conditions, during the growth of the peats (now coals) (Fig. 14.8). Although the Carboniferous plant associations differed from modern ones (flowering plants had not evolved, and floras were dominated by ferns and seed ferns during the Carboniferous), there are good indications that similar Carboniferous and Quaternary ecological niches were occupied by plants with similar adaptations. Thus, the Carboniferous *Botrychiopsis* assemblage is probably equivalent to recent tundra assemblages, the *Glossopteris* assemblage is equivalent to taiga assemblages, and the *Gangamopteris-Glossopteris* flora is equivalent to boreal to temperate flora of today (Figs. 14.9, 14.10). During the end of the

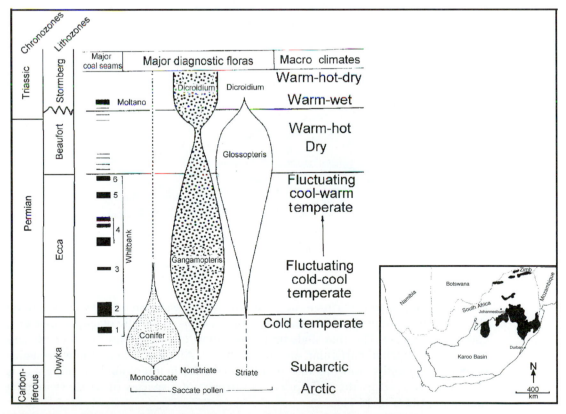

FIGURE 14.8   Diagram showing vertical changes in pollen in South Africa coals, indicating that only the lowermost ones are true cold-climate deposits. (After Falcon 1989.)

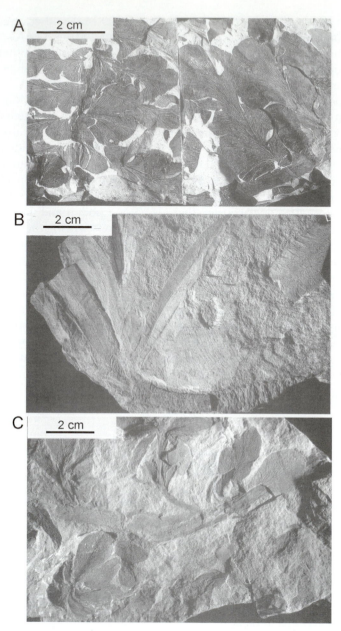

FIGURE 14.9 Carboniferous plants: *(A) Botrychiopsis; (B) Glossopteris; (C) Gang-amopteris.* (From White 1988.)

Carboniferous–lower Permian deglaciation, rapid widespread transgression on shelves fostered deposition of relatively thin, widespread, locally black, organic-rich clays (now shales). This thin shale unit covers most of eastern South America, central-eastern Brazil in particular, and South Africa. Although the unit may be slightly time transgressive over this vast area, its formation in intercommunicating seas is also demonstrated by the widespread occurrence of a species of small, swimming marine reptile (Fig. 14.11).

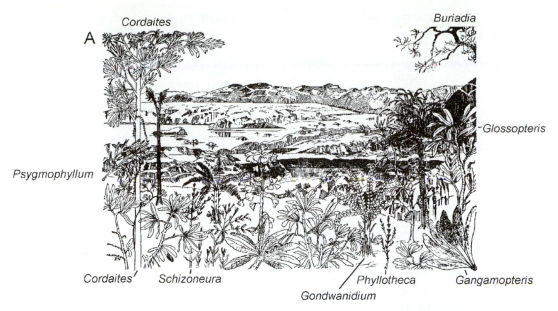

FIGURE 14.10   Diagram showing a reconstructed Carboniferous cold-climate landscape. (After Seward 1959.)

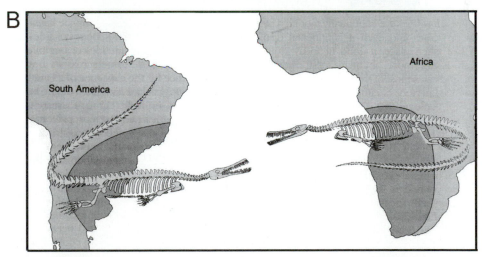

FIGURE 14.11   *(A)* Late Paleozoic postglacial marine reptile *(Mesosaurus)*. *(B)* Distribution of *Mesosaurus* in South America and South Africa. (From Tarbuck and Lutgens 1999.)

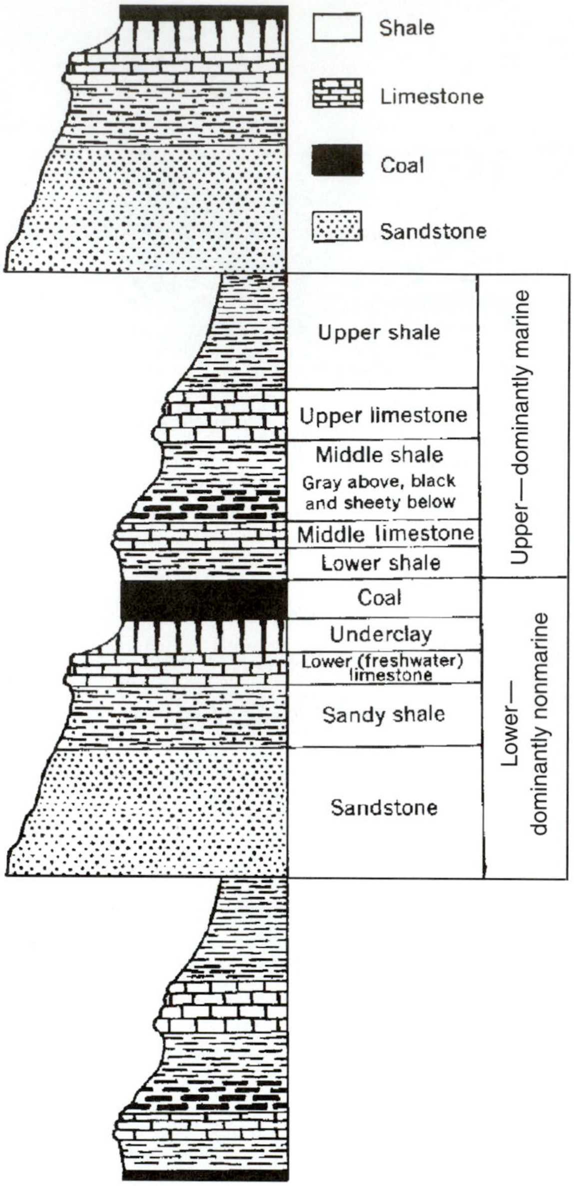

Shale

Limestone

Coal

Sandstone

Upper shale

Upper limestone

Middle shale
Gray above, black
and sheety below

Middle limestone

Lower shale

Coal

Underclay

Lower (freshwater)
limestone

Sandy shale

Sandstone

Upper—dominantly marine

Lower—
dominantly nonmarine

FIGURE 14.12 Idealized, vertical, lithological succession of a Carboniferous cyclothem of central North America, characterized by basal fluvial to deltaic clastic sediments capped by peat/coal, in turn capped by dominantly marine deposits, all recording an overall transgression. These successions are stacked in a stratigraphic column, sharply bounded by rapid regressions marked by the abrupt transition from open marine shale to coastal or fluvial sandstone. (After Weller 1960; Kay and Colbert 1965.)

A further global effect of the large Carboniferous-Permian glaciations is recorded in the rhythmic nearshore sedimentation in the Northern Hemisphere. Changing sea level and climatic fluctuations controlled thick, coal-bearing, cyclic, sedimentary deposits, called *cyclothems,* in various parts of the globe (Fig. 14.12). They record recurring rapid transgressions and regressions.

CHAPTER 15

Causes of Glaciation

Earth's climate changes over time due to variations in total heat budget (which causes changes over the whole Earth) and to variations in heat budget distribution (which redistributes heat over Earth's surface), either separately or in combination. Long-term changes produce gradual cooling toward prolonged icehouse conditions lasting many million years every 300 million years or so, if the late Proterozoic (750–550 Ma, midpoint 650 Ma), Carboniferous-Permian (320–270 Ma, midpoint 300 Ma), and Quaternary (1.5–2 Ma) are representative. Shorter glacial/interglacial and stadial/interstadial fluctuations occurred within each icehouse period. For example, there were at least 17 glaciations in the late Proterozoic, in the Carboniferous-Permian, and in the Quaternary.

So, theories of glaciation must explain both the long-term changes between icehouse and greenhouse conditions, and the shorter fluctuations within icehouse periods of glacial/interglacial and stadial/interstadial phases.

There are three fundamental problems in developing hypotheses for both long-term changes and shorter fluctuations. First is the problem of distinguishing cause and effect: for example, is the rise in carbon dioxide a cause of deglaciation or an effect of it? Second is the problem of separating dependent and independent variables: for example, the interactions of changes in atmospheric gases, oceanic circulation, continental ice-sheet areas, planetary and local reflectivity (albedo), and so on. The various positive and negative feedbacks of these systems give many "chicken-egg" dilemmas: which components are responsible for, and which are the result of, particular changes. Third is the problem of working out the order of events in one area and determining whether they correspond (or not) to similar events elsewhere: for example, are the major Proterozoic glacial/interglacial phases synchronous worldwide?

Here, we discuss possible causes of both long- and short-term fluctuations separately, and then note the possible effects and feedbacks of combinations. In both cases, various factors may reinforce or annul each other (Table 15.1). Causes are both external, operating from outside the Earth system, and internal, operating within the Earth system (Fig. 15.1).

LONG-TERM CHANGES

Each icehouse starts with prolonged gradual cooling, culminates in a long period of fluctuating polar ice sheets, and ends with relatively rapid global warming. Four main factors may affect Earth's heat budget and cause climate to vary:

TABLE 15.1   Hypotheses for Changing Climatic Conditions on Earth

Macro-factors	Consequences
Extraterrestrial	
Change in solar luminosity	Long-term reduction (30% more in Precambrian than now)
	Short-term reduction of energy related to sunspot activities
Planet crosses dusty interstellar areas	Reduction of solar energy arriving on Earth
Changes in galactic year	Reduction of solar energy arriving on Earth
Development of icy rings around equator (Saturn-like)	Reduction of solar energy arriving on Earth
Inclination of rotation axis (obliquity)	Differential insolation at mid-latitudes
Wobbling of the rotation axis (precession)	Differential insolation at mid-latitudes
Terrestrial	
Plate tectonics	Agglomeration of continents and supercontinent formation
	Rising average land altitude
	Lowering average sea level
	Increase in weathering and potential decrease in pCO_2 ($\sim CO_2$ concentration)
Plate tectonics	Dispersal of continents
	Rising of average sea level
	Decrease in weathering, potential increase in volcanism, and rise in pCO_2
Plate tectonics	Positioning of continents in mid-high latitudes
Tectonics	Uplift of continental areas
Volcanism	Increased dust concentration in atmosphere
Volcanism	Increased atmospheric concentration of CO_2 (pCO_2) and methane
Sedimentation (carbonates, peat, coal)	Decrease in pCO_2
Weathering of silici-clastic rocks	Decrease in pCO_2
Weathering of carbonates and organic deposits	Increase in pCO_2
Organism activities: Photosynthetic microbiota	Increase in pCO_2
Burning of fossil fuels: humans	
Change in atmospheric and marine currents	Jet streams
	Gulf Stream
	N. Atlantic Deep Water current
	El Niño and La Niña

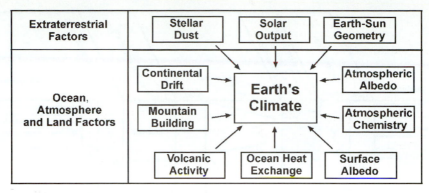

FIGURE 15.1   Diagram of major factors that influence Earth's climate. (After *http://www.geog.ouc.bc.ca/physygeog/contents/7y.html.*)

(1) changes in radiation received from the Sun; (2) changes in heat from Earth's interior; (3) changes in the proportion of greenhouse gases in the atmosphere; and (4) changes in the distribution of land and sea and their effects on oceanic and atmospheric circulation. Changes due to the Milankovitch effect occur in both icehouse and greenhouse periods and cannot explain how major cooling starts (Box 15.1).

1. Long-term change in radiation received from the Sun may be due to either changes in the Sun's output, or changes in radiation received on Earth.

Long-term fluctuations in the Sun's radiation output are possible, though current hypotheses provide no mechanism for this, and also indicate that the Sun has heated up through time. During the Archean (before 2500 Ma), the Sun's luminosity was about 30% less than today. But, paradoxically, the Archean has no good evidence of glaciation, probably because the atmosphere was richer in greenhouse gases such as CO_2, and perhaps because continents were too small to support large ice sheets.

Reduction in radiation received on Earth could also be due to variations in interstellar dust or ice. The solar system rotates around the center of the Milky Way galaxy about every 300 million years (the galactic year), passing through stationary nebulae of hydrogen-rich particles as it does so. These could reduce the amount of radiation received on Earth, and cause gradual cooling. This hypothesis requires more testing by further observations on nebulae density, and by calculations on whether they can affect the energy received by Earth enough to drastically influence its climate. In any case, this periodicity can be matched with major icehouse conditions only back as far as 900 Ma. Even for this period it cannot account for the smaller glaciations such as those of the Ordovician, unless the cyclicity is 150 Ma (half a galactic year) (Fig. 15.2). Also, older Precambrian glaciations show no comparable periodicity.

2. Changes in heat flow from Earth's interior are primarily dependent on variations in the activity and extent of volcanoes. These are directly related to the lengths of volcanic ridges and subduction zones and the rate of plate movements. Times of reduced plate movements should precede icehouse periods. There is some support for this. Both the late Proterozoic and Carboniferous-Permian ice ages followed the assembly of large supercontinents, and reduction in plate boundary activity and extent. But the Quaternary glaciation followed a time of rapid plate movements and multiple plate boundaries in the Tertiary. On the other hand, volca-

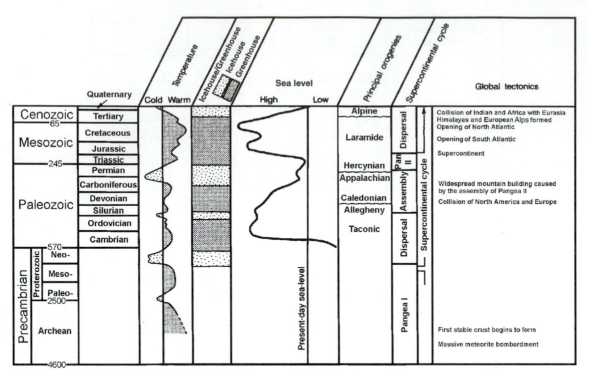

FIGURE 15.2   Diagram showing cold and warm ages in relation to major events during Earth's history. (After Doyle et al. 1994.)

noes also increase atmospheric dust concentrations, which reflects radiation and reduces solar input.

3. Changes in the proportion of greenhouse gases are caused by variations in volcanic activity, by photosynthesis, by deposition of limestones and gas hydrates, and by weathering of siliceous rocks (Fig. 15.3). The most important atmospheric gases that trap long-wave energy, and thus heat up Earth's surface, are water vapor, carbon dioxide, and methane. Much attention has been placed lately on the effect of carbon dioxide (CO_2) on global warming. Carbon dioxide concentration in the atmosphere is affected by many factors, such as the following.

a. *Volcanoes* pump out carbon dioxide; so icehouse conditions should correspond with reduced volcanism.

b. *Photosynthesis* concentrates the carbon from carbon dioxide into organic matter, which can be buried (as in peat and coal). The evolution of abundant multicellular algae precedes the late Proterozoic ice age, and the evolution of complex land plant communities precedes the Carboniferous-Permian ice age. However, there is no corresponding change during the Tertiary cooling, and furthermore no major organic deposits in the late Proterozoic.

c. *Limestones* stored CO_2 in geologic deposits, removing it from the hydrosphere and atmosphere. Whether this is an amount significant enough to affect the overall concentration of carbon dioxide in the atmosphere and thus the climate is debatable. There are indications that extensive shelf seas with thick limestone deposits preceded the late Proterozoic, the Ordovician, the Carboniferous-Permian, and the Tertiary cooling (the latter in the Creta-

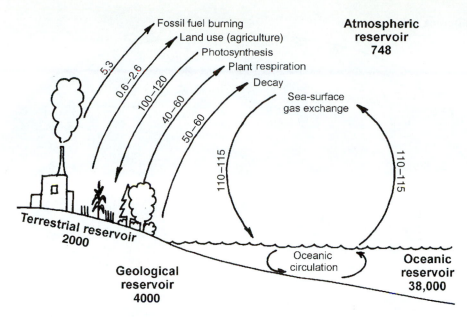

FIGURE 15.3   Major carbon reservoirs in gigatons (1980 estimates), and exchange between reservoirs usually as CO_2. Biological pumping is the mechanism that carries carbon into deep water from the surficial zones. Gas hydrates in deep and/or cold oceanic sediments are not included here but may be as much as 15,000 gigatons. (After Post et al. 1990.)

ceous). However, cooling in both the Tertiary and Carboniferous seem to have lasted long after limestone deposition had decreased. Furthermore, other times of extensive limestone deposition — for example, in the Silurian and mid-Jurassic — occurred during hothouse periods.

d. *Weathering of rocks* also changes the amount of CO_2 in the atmosphere. Weathering of siliciclastic rocks removes carbon dioxide, while weathering of carbonates and organic materials like peat and coal adds carbon dioxide to the atmosphere. Rapid plate movements and continental collisions raise mountains and promote increased weathering of siliciclastic rocks. Indeed, in the last 800 Ma or so there seems to be a correlation between cooling of the atmosphere and rapid drift and collision of continents.

4. Changes in the distribution of land and sea affect not only oceanic and atmospheric circulations but also the possibility of developing polar ice sheets. There is a good correlation between glacial periods and times when polar regions are occupied by continents (such as the present Antarctica) or by enclosed seas (present Arctic Ocean). High-latitude continents allow snow to accumulate at low altitudes and build up to form ice sheets. The snow and ice increase albedo and reflect a greater proportion of incoming solar radiation back into space, thus lowering Earth's heat budget. The present north-south alignment of major seas and continents also limits free oceanic circulation in the Northern Hemisphere and isolates the present Arctic Ocean. Conversely, free latitudinal circulation around Antarctica isolates this continent from any southerly warm water flow. In contrast, most greenhouse periods have highly interconnected seas, as in the Cretaceous (Fig. 15.4).

Redistribution of landmasses may be the fundamental control on the main factors that affect climate and determine icehouse and greenhouse conditions. As-

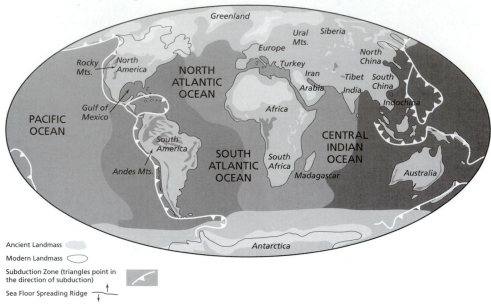

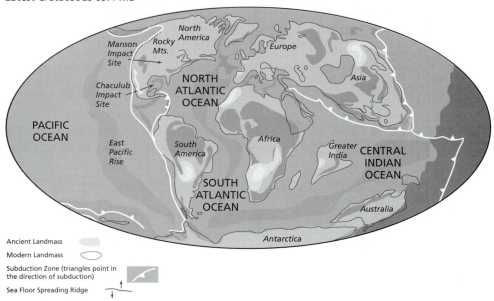

FIGURE 15.4   Paleogeographic maps of Earth for the late Cretaceous and Pleistocene. (From Scotese 1997.)

sembly of supercontinents occurs approximately every 400 Ma, when continental collisions generate high and extensive mountain ranges (Fig. 15.5). In these ranges, faster weathering of siliciclastic igneous and metamorphic rocks reduces atmospheric CO_2 concentrations, which causes climatic cooling. Breakup and dispersal of continents forms new oceanic basins, changes ocean and atmospheric circulations, and increases volcanism and mantle degassing (increasing CO_2 and warming Earth). All things being equal, icehouse conditions should develop during assembly

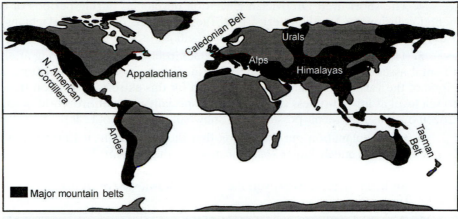

Mountains	Orogeny	Age
Alps and Himalayas	Alpine	Tertiary
North American Cordillera/Andes	Laramide	Mesozoic
Caledonia Belt/Appalachians	Hercynian Appalachian Caledonian Alleghenian Taconic	Paleozoic
Urals	Uralian	Early Paleozoic
Tasman Belt	Tasmanian	Late Paleozoic

FIGURE 15.5 Major mountain chains of the world developed at different times. (After unknown; see website for update.)

of continents and greenhouse conditions during breakup and dispersal of continents. However, explaining the major glaciations of Earth simply with geographic landmass redistribution and its associated effects is difficult, as can be appreciated by analyzing some of the major climatic events, such as the following, backward through the ages.

1. The Tertiary cooling and Quaternary icehouse occurred with a large continent (Antarctica) over the South Pole, but during continental dispersal. Indeed, full glacial conditions existed in Antarctica when that continent became separated from South America and Australia and the circum-Antarctic marine current could develop. This current isolated the Antarctic continent from any influence of tropical marine currents from the north, and rapid cooling occurred.

2. Comparably, the mid-Mesozoic hothouse developed and forests grew in Antarctica when a large landmass (Gondwanaland) was over the South Pole and the Gondwanaland and Laurasia supercontinents were in the early stages of breakup and dispersal.

3. The very long Carboniferous-Permian icehouse fits best with the model of major glaciation taking place after major continental collision. It developed as the Gondwanaland supercontinent was formed and lay astride the South Pole. As in the Quaternary case, the Carboniferous-Permian glaciers were accompanied by large forests and peatlands in the tropical zones, and, upon deglaciation, large mid- to high-latitude areas were covered by cold-climate peatlands and forests (taiga). These plants and the related peat have affected the concentration of carbon dioxide in the atmosphere, but this most likely was not a determinant factor in establishing or ending the icehouse age.

4. The short-lived Ordovician glaciation developed during periods of continental collisions in North America (the Taconics, parts of the Appalachians) and Europe (Caledonian/Hercynian orogenetic belts).

5. The late Proterozoic (Neoproterozoic) icehouse developed long after continental assembly on a low-lying supercontinent, which was actively rifting apart. None of the various explanations put forward for this extreme glaciation (reaching to sea level at low latitudes) is satisfactory, especially if the entire world ocean was frozen over, as now seems possible (the Snowball Earth hypothesis).

    a. One explanation put forward is that the inclination of Earth's rotational axis was much larger (~54° instead of the present 23°), so that tropical areas underwent the same seasonal changes as polar regions, and received a lot less radiation. However, there is no known way of tilting Earth's axis to this extent.

    b. A second explanation uses rapid drifting of continents into and out of polar regions, so that the tropical latitudes obtained from paleomagnetism of beds associated with tills are due to overprinting in tropical regions, before or after glaciation. If this were true, glaciation would not have occurred in tropical regions but in mid- to high-latitude zones as usual.

    c. A third explanation is that enhanced weathering on an equatorial continent, and therefore removal of atmospheric CO_2, would cause cooling and the formation of a sea ice cover at the poles. Further cooling would continue until glaciers began developing on the equatorial continents.

6. The Mesoproterozoic supercontinent shows no evidence of glaciation, although much weathering of siliciclastic material probably occurred and solar radiation was lower than at present. To explain this, an exceptionally high rate of CO_2 outgassing from Earth's interior has to be invoked to compensate and maintain apparent hothouse conditions.

There is no really satisfactory single hypothesis to explain long-term changes in Earth's heat budget and major glaciations. Different combinations of conditions may lead to different results, or, vice versa, apparently similar combinations have drastically different results. Perhaps future work will find out that fluctuations in output from the Sun are primary causes of major climatic changes.

SHORT-TERM FLUCTUATIONS

Once an icehouse age begins, shorter and smaller fluctuations cause ice sheets and glaciers to repetitively advance and retreat, with regular alternations between glacial and interglacial periods. Within glacial periods, glaciers expand (stadial stages) and contract (interstadial stages), and within these, colder and warmer times also occur. There are a number of possible causes, both external and internal to Earth, for these cycles.

Fluctuations in the Sun's Output

There is a decade-long, recurring reduction in solar energy during increased sunspot activities. This affects Earth's climate, but observations have been made only within the last few centuries, and the effects of sunspot activity cannot be determined from the geological record.

Fluctuations in Insolation in Different Parts of Earth

In the Quaternary, the oscillation between warm and cold periods initially had a periodicity of 41,000 years; after about 900,000 years the periodicity changed to 100,000 years. These regular fluctuations are attributed to regular changes in the amount of solar radiation Earth receives, due to regular changes in the geometry of Earth's orbit and its attitude in this orbit. Originally considered by Croll in the nineteenth century, this astronomical explanation was best expressed mathematically by Milankovitch in 1941. The Milankovitch hypothesis explains the various climatic alternations by astronomical variations (Box 15.1).

The Milankovitch effect is not strong enough to cause an icehouse, but it can modulate glaciations by forcing changes in atmospheric and marine currents, particularly when continental masses occur at mid-latitudes where the effect of changing insolation is at its maximum. The Milankovitch effect is ubiquitous throughout Earth's history and has been recorded in sediments of various environments, from ocean basin settings to inland loess and other settings. The various periodicities combine to give an insolation variation graph, which fits actual changes during the Quaternary reasonably well (Box 15.1).

Fluctuations in Greenhouse Gases

Variations in greenhouse gases, like CH_4 (methane) and CO_2, can be caused by various feedback (or autocyclic) mechanisms, some of which have already been considered for long-term fluctuations of climate. On a short time scale, removal of atmospheric CO_2 due to continental weathering and/or photosynthesis by plants cools the earth, allowing continental glaciers to grow and cover the land. This protects the land from weathering and reduces the amount and rate of plant photosynthesis; and so the rate of CO_2 removal may drop below the rate of addition from volcanoes and other sources. The resulting increase in CO_2 in the atmosphere may cause warming and ultimately glacier collapse. The land exposed by glacier retreat (and not inundated by rising sea level) starts weathering again and is colonized by plants. And the whole cycle repeats. However, there are several difficulties with this hypothesis. One is related to the fact that CO_2 is not the most important gas in the atmosphere that affects retention of the long-wave energy causing heating of Earth's surface (greenhouse effect). Its influence may be dwarfed by that of water vapor, which is much more abundant and has similar effects on the energy waves. The other difficulty relates to the fact that even if the concentration of CO_2 in the atmosphere is a determinant factor in changing the climate, the oceans are the largest CO_2 reservoir at the surface of Earth, and gas hydrates are the largest reservoir of carbon. Both these reservoirs regulate the atmospheric CO_2 concentration and may relegate the effects of weathering and photosynthesis to a secondary role.

Changes in Atmospheric Circulation Patterns

Air and marine currents may be affected by many factors, such as sea level change, tectonic uplift or subsidence of parts of the continents and sea floor, and changes in latitudinal temperature gradients. Reorganization of jet streams in the atmosphere or of the large marine currents may lead to drastic variations in climate in certain regions, with the possible local nucleation of glaciers. A case in point is Iceland, which has experienced numerous glacial and interglacial periods, even during historical times, due to shifting of the Gulf Stream.

BOX 15.1   **Milankovitch Effect**

Planet Earth travels through space displaying several movements.

1. Its speed of rotation around its axis has been slowing through geological time. For example, in Devonian times (about 380 Ma), a day was shorter, making a year about 32 days longer. This is indicated by the growth rings of solitary corals in which each large annual growth ring contains 397 smaller daily growth rings. During the Cenozoic, the speed of rotation has not changed, but the axis has, both in inclination (tilt) about the plane of the orbit and in orientation in respect to a fixed star.

a. The tilt of the axis of rotation changes about $1\frac{1}{2}°$ on either side of the present angle of $23\frac{1}{2}°$ over a period of about 41 ka. This is called obliquity (Fig. Box 15.1.1A).

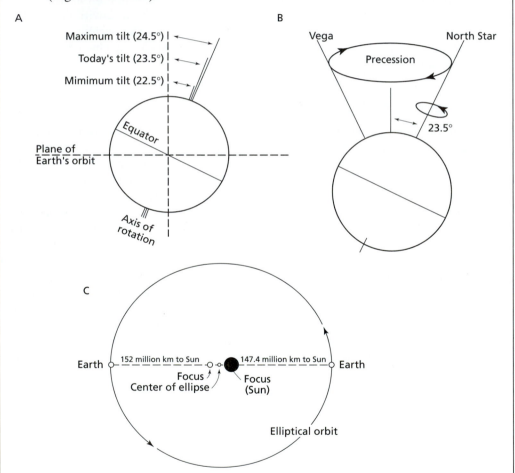

FIGURE BOX 15.1.1   Diagrams showing Earth's orbital movements. *(A)* Tilt of rotation axis. *(B)* Wobbling of the rotation axis. *(C)* Variation in eccentricity of the orbit around the Sun. (After Tarbuck and Lutgens 1999.)

b. The axis of rotation does not always point at the same distant point in space; rather, it wobbles like a spinning top, completing a circle over a period of about 24 ka. This phenomenon is called axial precession. It is caused by the differential attraction of the Sun and the Moon on the terrestrial equatorial bulge (Fig. Box 15.1.1B).

BOX 15.1 *(continued)*

2. Earth also rotates around the Sun on an orbit that varies from nearly circular to markedly elliptical, over a period of about 96 ka. This is called the eccentricity of the orbit. The Sun is located at one focus of the ellipsoidal orbit; the other focus is empty (Fig. Box 15.1.1C).

The various motions of Earth along the orbit lead different parts of the planet to receive different amounts of solar energy (insolation) through the year. In the Northern Hemisphere, maximum insolation occurs when Earth is tilted toward the Sun (summer solstice, nowadays on June 21) and lower insolation during the winter solstice (now December 21) (Fig. Box 15.1.2). At these times, the daylight time is, respectively, the longest and the shortest of the year. An equinox occurs when the number of day and night hours is the same, now on May 20 and September 22. Note that at present the summer solstice occurs when Earth is farthest from the Sun (aphelion) and the winter solstice when Earth is the closest to the Sun (perihelion). The dates of the solstices and equinoxes, though, change in time due to the wobbling of Earth's rotational axis, such that about 11.5 ka the summer solstice occurred at perihelion and the winter solstice at aphelion.

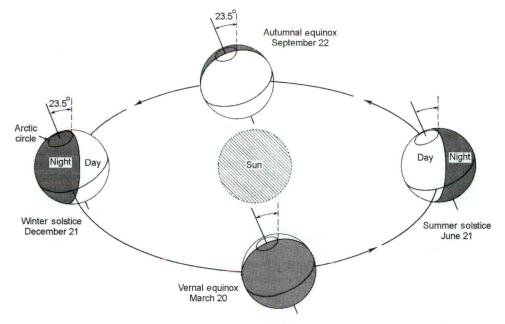

FIGURE BOX 15.1.2 Diagram showing approximate position of Earth in its orbit around the Sun, and its solstices and equinoxes for the Northern Hemisphere. (After Strahler 1972.)

All this led to a small variation in insolation, with some past northern winters colder and summers warmer than the present. The opposite of all this holds true for the Southern Hemisphere. So on the whole, Earth receives an equal average annual amount of heat, which does not vary much from year to year, only over the long term (several hundreds of million years).

Everything else being equal, it is not so much the total amount of solar energy received on Earth that determines changes in overall climate, but rather the distribution of this energy on the planet. The variable distribution in energy caused by

(continues)

BOX 15.1 (*continued*)

the obliquity, precession, and eccentricity has been considered a possible contributor to initiation of glaciation. Milankovitch was the first to calculate in detail the contribution of the various factors for the Late Pleistocene. The sum contribution for the last 130 ka is most important, and it has a cyclic pattern at mid-latitudes (60° N; Fig. Box 15.1.3A,B). It is at those latitudes in the Northern Hemisphere where glaciers, such as those of North America, recurringly nucleated during the Late Pleistocene. There is, therefore, a good correlation between cold climatic conditions of the isotope stages 2, possibly 3 in northern Europe, 4, 5b, and 5c and the recurring low insolation at mid-latitudes. However, it is still debated whether this statistical correlation indicates a cause-effect relation or whether other factors enter into play. The contention by some is that the variation in insolation is small and possibly insufficient in itself to start a glaciation, but that it may trigger variation in air and marine currents that would change the path and intensity of heat exchange between the poles and the equator. These effects would ultimately lead to a glaciation. But factors other than astronomical may influence these fluid currents as well, and may either reinforce or negate the astronomical cycles.

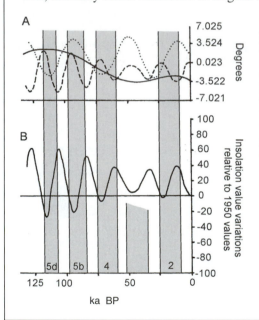

FIGURE BOX 15.1.3 Diagram showing variations in Earth's orbital movements and insolation in the Northern Hemisphere related to isotope stages. *(A)* Variation in eccentricity (continuous line), obliquity (stippled line), and precession (dashed line) for the last 130 ka. *(B)* Calculated cyclical summer insolation at about latitude 65° N (variations from AD 1950). Note that the lower insolation conditions correlate reasonably well with the precession (dashed line) and cold stages (dark columns). (After Dawson 1992; data from Berger 1978a, 1978b; Bradley 1985; Briggs 1970; Larsen and Sejrup 1990.)

Changes in Oceanic Circulation Patterns

Major changes in oceanic circulation patterns have a profound climatic effect. For example, a change or shutoff of the warm, northward-flowing, surficial Gulf Stream would drastically reduce the salinity of the northern seas. This would lead to the shutoff of the cold, more saline, deep, southward-flowing North Atlantic Deep Water current (Fig. 8.5). This would in turn trigger a lower rate of exchange of CO_2 between the oceans and the atmosphere, with a reduction in atmospheric CO_2, a drop in temperature, and an increased rate of sea and glacier ice formation. Such changes in marine currents may occur very rapidly, and they are reflected in rapid climatic changes. An example of this is the cold spell (Younger Dryas) experienced globally, but particularly in Europe and North America, from about 13 to 10 ka, which may be in part associated with changes in the Gulf Stream and the North Atlantic Deep Water flows.

Tectonic Uplift

Tectonics may raise or lower large parts of the continents, and ice caps and glaciers may form in suitable high-altitude areas supplied with enough precipitation. Tectonic uplift in mountain belts may raise large parts of continents, like plateaus, above the snowline, even in tropical areas. Large plateaus usually have glaciers and ice caps only at their uplifted mountainous margins, which intercept precipitation, whereas their internal regions, albeit cold, do not receive sufficient precipitation for glaciers to develop. For example, during the Quaternary, the Tibetan Plateau had glaciers only around its mountainous edges, and added very little to the total glacier volume of Earth.

Another situation develops in rifted areas: for example, around the margins of some continents. Whereas the central parts of the rift form depressions, the flanks are usually uplifted to high elevations. One hypothesis, based on the preservation of many ancient glacial deposits in rifts, suggests that glaciation can develop on the uplifted rift shoulders, and the glaciers flowed and deposited sediments into the rift basin. Dating of the various glacigenic sediments and rifting events is not accurate enough to properly evaluate this hypothesis. Good correlation has not always been found between glaciation and rifting events. For example, major Proterozoic glaciations occurred during pre-rift stages, others during syn-rift stages, and others still during post-rift stages (Fig. 15.6).

Local Autocyclic Changes

As cold climate and particularly cold summers set in, and glaciers start to form, the albedo of Earth's surface increases, and much radiant energy is reflected away. In this case the system feeds on itself: the colder and more snowy it becomes, the more glaciers grow.

Surging of ice streams over seas or large lakes may affect the stability of ice sheets as well. Such surges form ice tongues subjected to rapid calving. This may

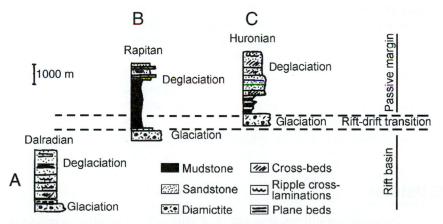

FIGURE 15.6   Diagram showing setting of three major Neoproterozoic successions containing glacial beds in relation to rift and post-rift stages. *(A)* Dalradian unit in Scotia and Ireland. *(B)* Raptian unit in the American northern Canadian Cordillera. *(C)* Huronian unit in central eastern Canada. (After Young 1997.)

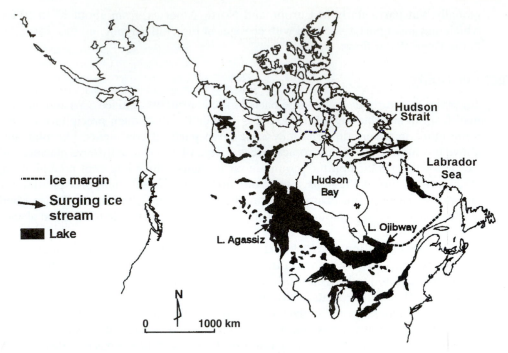

FIGURE 15.7   Map showing location of possible surges of the Laurentide Ice Sheet through the Hudson Strait into the Labrador Sea, leading to a rapid drawdown and collapse of the glacier. (After Andrews 1987; Andrews 1997.)

lead to drawdown of ice from the main body of the glacier, which thins and can eventually collapse. This may have occurred for the Laurentide Ice Sheet when rapid flow of the ice stream exiting the Hudson Strait led to opening of the strait and the drawdown of the glacier toward Hudson Bay and the Labrador Sea (Fig. 15.7).

Land-locked polar ice sheets such as those of Greenland and eastern Antarctica are not directly affected by small (order of a few to 10°C) variations in global temperature, because their average temperature is well below the freezing point, reaching average lows of −52°C. However, global climatic changes have an indirect effect on polar glaciers that ground in deep (up to 500 m) seas, like the present West Antarctic Ice Sheet. This is because as lower-latitude glaciers melt and sea level rises, the polar glaciers may float and icebergs of enormous extent may develop, removing great quantities of ice at once. This occurred in the West Antarctic Ice Sheet in 1998 when a large (approximately 150 km long and 50 km wide) slab of ice (flat-topped iceberg) detached from the ice shelf glacier (Ronne Ice Shelf). A similar sized iceberg detached in 1987 from the Ross Ice Shelf. This process may lead to drastic reduction and eventual collapse of a glacier with sea termination.

CONCLUSIONS

1. Icehouse conditions develop gradually, after overall cooling over about 50 Ma, and end relatively rapidly. We think we understand the cooling, but not the warming. For example, drifting of the continents away from high-latitude areas is one of the best explanations we have so far for warming. But this is far too slow a

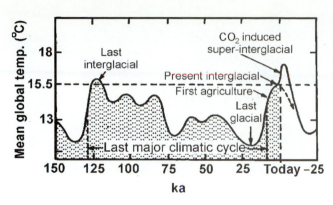

FIGURE 15.8   Diagram showing major climatic cycles for the last 150 ka and possible changes in the next 25 ka. A future glaciation is inevitable, although it may be temporarily delayed by a human-produced high concentration of greenhouse gases in the atmosphere. (From Imbrie and Imbrie 1979; after Mitchell 1977.)

process to explain the relatively rapid end of the Carboniferous-Permian and late Proterozoic ice ages.

2. No two ice ages are the same, and so no one overall hypothesis can explain them all. Though the roughly 300 Ma periodicity may be due to astronomical variations, the development of each ice age is probably due to a concatenation of circumstances.

3. The so-called cyclicity of glaciations is really based on only three examples: the late Proterozoic, Carboniferous-Permian, and Quaternary. Most features identified as cyclic are not so internally, nor do they recur in pre-established time intervals. Exceptions may be astronomical events, which transcend our time-space dimension, but these cannot be properly tested.

4. The current icehouse condition is unlikely to disappear as long as overall extraterrestrial and continental characteristics continue to be favorable for it. We are living in an icehouse stage during an interglacial period. The anthropogenic increase in greenhouse gases, such as CO_2 (Fig. Box 13.1.3), in the atmosphere may temporarily raise the global temperature and lead to a warm super-interglacial, but ultimately renewed glaciation should start in the not-so-distant future (Fig. 15.8).

Useful North American Addresses and Websites

United States

United States Geological Survey
U.S. Department of the Interior
U.S. Geological Survey
Reston, VA 20192
http://www.usgs.gov/

NASA
http://www.nasa.gov/
A good source for all kinds of Earth images.

Links to glacial sites
http://skua.gps.caltech.edu/hermann/
glaciolinks.html
This site may not be around forever but it has a
lot of useful links.

State Surveys

Geological Survey of Alabama
P.O. Box O
Tuscaloosa, AL 35486
205-349-2852
http://www.archives.state.al.us/agencies/
geologic.html

Alaska Geological and Geophysical Survey
794 University Ave., Suite 200
Fairbanks, AK 99709
907-451-5000
Fax: 907-451-5050
Milton A. Wiltse, Director and State Geologist
http://www.dggs.dnr.state.ak.us/

Arizona Geological Survey
416 W. Congress St., Suite 100
Tucson, AZ 85701
520-770-3500
Fax: 520-770-3505
http://www.azgs.state.az.us/

Arkansas Geological Commission
Vardelle Parham Geology Center
3815 W. Roosevelt Rd.
Little Rock, AR 72204
501-296-1877
Fax: 501-663-7360
http://www.state.ar.us/agc/agc.htm

California Division of Mines and Geology
655 S. Hope St., #700
Los Angeles, CA 90017
213-239-0878
Fax: 213-239-0894
http://www.consrv.ca.gov/dmg/

Colorado Geological Survey
1313 Sherman St., Room 715
Denver, CO 80203
Fax: 303-866-2461
http://www.dnr.state.co.us/edo/survey.html

Connecticut Department
of Environmental Protection
79 Elm St.
Hartford, CT 06106-5127
860-424-3555
http://dep.state.ct.us/cgnhs/index.htm

Delaware Geological Survey
University of Delaware
Delaware Geological Survey Building
Newark, DE 19716-7501
302-831-8262
http://www.udel.edu/dgs/dgs.html

Florida Geological Survey
903 W. Tennessee St.
Tallahassee, FL 32304-7700
850-488-9380
Fax: 850-488-8086
http://www.dep.state.fl.us/geo/

Georgia Geologic Survey Branch
Dr. William H. McLemore, State Geologist
19 Martin Luther King Jr. Dr., Room 400
Atlanta, GA 30334
404-656-3214
http://ganet.org/dnr/environ/branches/
geosurv/gsnar.htm

U.S. Geological Survey's
Hawaiian Volcano Observatory
808-967-7328
http://www.hvo.wr.usgs.gov/

Idaho Geological Survey
Morrill Hall, Room 332
University of Idaho
Moscow, ID 83844-3014
208-885-7991
http://www.uidaho.edu/igs/igs.html

Illinois State Geological Survey
615 E. Peabody
Champaign, IL 61820
217-333-4747
http://www.isgs.uiuc.edu/

Indiana Geological Survey
611 N. Walnut Grove
Bloomington, IN 47405-2208
812-855-7636
Fax: 812-855-2862
http://adamite.igs.indiana.edu/index.htm

Iowa Geological Survey Bureau
109 Trowbridge Hall
Iowa City, IA 52242-1319
319-335-1575
http://www.igsb.uiowa.edu/

Kansas Geological Survey
University of Kansas
1930 Constant Ave.
Lawrence, KS 66047-3726
785-864-3965
http://www.kgs.ukans.edu/kgs.html

Kentucky Geological Survey
228 Mining and Mineral Resources Bldg.
University of Kentucky
Lexington, KY 40506-0107
606-257-3896
http://www.uky.edu/KGS/home.htm

Louisiana Geological Survey
P.O. Box G, University Station
Baton Rouge, LA 70893
225-388-5320
Fax: 225-388-3662
http://www.leeric.lsu.edu/lgs/

Maine Geological Survey
22 State House Station
Augusta, ME 04333
207-287-2801
http://www.state.me.us/doc/nrimc/mgs/mgs.htm

Maryland Geological Survey
2300 St. Paul St.
Baltimore, MD 21218
410-554-5500
http://mgs.dnr.md.gov/

Massachusetts Office of Environmental Affairs
100 Cambridge St., 20th Fl.
Boston, MA 02202
617-727-9800, ext. 218
http://www.magnet.state.ma.us/envir/eoea.htm

Michigan Geological Survey Division
P.O. Box 30256
735 E. Hazel St.
Lansing, MI 48909-7756
517-334-6943
http://www.deq.state.mi.us/gsd/

Minnesota Geological Survey
2642 University Ave. West
St. Paul, MN 55114-1057
612-627-4782
Fax: 612-627-4778
http://www.geo.umn.edu/mgs/

Mississippi Office of Geology
P.O. Box 20307
Jackson, MS 39289
601-961-5523

Missouri Department of Natural Resources
Division of Geology and Land Survey
P.O. Box 250
Rolla, MO 65402
573-368-2125
http://www.dnr.state.mo.us/dgls/homedgls.htm

The Montana Bureau of Mines and Geology
1300 W. Park St.
Butte, MT 59701-8997
406-496-4167
Fax: 406-496-4451
http://mbmgsun.mtech.edu/

Nebraska Conservation and Survey Division
Nebraska Hall
University of Nebraska–Lincoln
Lincoln, NE 68588-0517
402-472-3471
Fax: 402-472-4608
http://csd.unl.edu/csd.html

Nevada Bureau of Mines and Geology
Mail Stop 178
University of Nevada
Reno, NV 89557-0088
775-784-6691
http://www.nbmg.unr.edu/

New Hampshire Department
of Environmental Services — PIP Unit
6 Hazen Dr.
Concord, NH 03301
603-271-3503
http://www.des.state.nh.us/geo1link.htm

New Jersey Geological Survey
P.O. Box 427
29 Arctic Pkwy.
Trenton, NJ 08625-0427
609-777-1038
http://www.state.nj.us/dep/njgs/

New Mexico Bureau of Mines
and Mineral Resources
801 Leroy Pl.
Socorro, NM 87801-4796
505-835-5490
http://geoinfo.nmt.edu/

New York State Geological Survey
New York State Museum Office
of Communications
3040, Cultural Education Center
Albany, NY 12230
http://www.nysm.nysed.gov/geology.html

North Carolina Geological Survey
Division of Land Resources
1612 Mail Service Center
512 N. Salisbury St.
Raleigh, NC 27611-7687
919-715-9718
http://www.geology.enr.state.nc.us/

North Dakota Geological Survey
600 E. Boulevard Ave.
Bismarck, ND 58505-0840
614-265-6988
http://www.state.nd.us/ndgs/

Ohio Geological Survey
4383 Fountain Square Dr.
Columbus, OH 43224-1362
614-265-6576
Fax: 614-447-1918
http://www.dnr.state.oh.us/odnr/geo_survey/

Oklahoma Geological Survey
100 E. Boyd, Room N-131
Norman, OK 73019-0628
405-360-2886
http://www.ou.edu/special/ogs-pttc/

Oregon Department of Geology
and Mineral Industries
800 NE Oregon St., Suite 965
Portland, OR 97232
503-731-4444
http://sarvis.dogami.state.or.us/homepage/

Pennsylvania Geological Survey
Harrisburg Office:
P.O. Box 8453
1500 N. 3rd St., 2nd Fl.
Harrisburg, PA 17105-8453
717-787-2169
Fax: 717-783-7267
http://www.dcnr.state.pa.us/topogeo/
indexbig.htm

Rhode Island Geological Survey
University of Rhode Island
Department of Geosciences
Kingston, RI 02881
401-874-2265
Fax: 401-874-2190
http://www.uri.edu/cels/gel/rigs.html

South Carolina Geological Survey
C. W. Clendenin Jr., State Geologist
5 Geology Rd.
Columbia, SC 29212
803-896-7708
Fax: 803-896-7695
http://www.dnr.state.sc.us/geology/
geohome.htm

South Dakota Geological Survey
Akeley Science Center
University of South Dakota
Vermillion, SD 57069
605-677-5227
http://www.sdgs.usd.edu/

Tennessee Department of Environment and
Conservation
Division of Geology
401 Church St.
Nashville, TN 37243-0445
615-532-1516
http://www.state.tn.us/environment/tdg/
index.html

Texas Bureau of Economic Geology
University Station, Box X
Austin, TX 78713-8924
512-471-1534
Fax: 512-471-0140
http://www.utexas.edu/research/beg/

Utah Geological Survey
1594 W. North Temple
P.O. Box 146100
Salt Lake City, UT 84114-6100
888-882-4627
http://www.ugs.state.ut.us/

Vermont Geological Survey
103 S. Main St., Laundry Bldg.
Waterbury, VT 05671-0301
802-241-3608
http://www.anr.state.vt.us/geology/
vgshmpg.htm

Virginia Department of Mines, Minerals and
Energy
Office of Information Services
P.O. Drawer 900
Big Stone Gap, VA 24219
540-523-8146
http://www.mme.state.va.us/

Washington Department of Natural Resources
Geology and Earth Sciences Division
1111 Washington St. SE, Room 148
P.O. Box 47007
Olympia, WA 98504-7007
360-902-1450
Fax: 360-902-1785
http://www.wa.gov/dnr/htdocs/ger/ger.html

West Virginia Geological and Economic Survey
Mont Chateau Research Center
P.O. Box 879
Morgantown, WV 26507-0879
304-594-2331
http://www.wvgs.wvnet.edu/

Wisconsin Geological
and Natural History Survey
3817 Mineral Point Rd.
Madison, WI 53705-5100
608-263-7389
Fax: 608-262-8086
http://www.uwex.edu/wgnhs/

Wyoming State Geological Survey
P.O. Box 3008
Laramie, WY 82071
307-766-2286
http://www.wsgsweb.uwyo.edu/

Canada

Geological Survey of Canada
601 Booth St.
Ottawa, ON K1A 0E8
613-996-3919
Fax: 613-943-8742
http://www.nrcan.gc.ca/gsc/index_e.html

Provincial Surveys

Alberta Geological Survey
4th Fl., Twin Atria
4999-98 Ave.
Edmonton, AB T6B 2X3
780-422-1927
Fax: 780-422-1459
http://www.ags.gov.ab.ca/AGS_GENERAL/
AGSLOCATION.HTP

British Columbia and Yukon Chamber
of Mines
840 W. Hastings St.
Vancouver, BC V6C 1C8
604-681-5328
Fax: 604-681-2363
http://www.bc-mining-house.com/

Manitoba Industry Trade and Mines
http://www.gov.mb.ca/em/index.html

New Brunswick Department
of Natural Resources and Energy
P.O. Box 6000 (Hugh John Flemming
Forestry Complex)
Fredericton, NB E3B 5H1
506-453-2614
Fax: 506-457-4881
http://www.gov.nb.ca/dnre/minerals/index.htm

Geological Survey of Newfoundland
and Labrador
Department of Mines and Energy
P.O. Box 8700
St. John's, NF A1B 4J6
709-729-6193
Fax: 709-729-4491
http://zeppo.geosurv.gov.nf.ca/

Nova Scotia Department of Natural Resources
Minerals and Energy Branch
P.O. Box 698
Halifax, NS B3J 2T9
902-424-5935
Fax: 902-424-7735
http://www.gov.ns.ca/natr/meb/

Ontario Ministry of Northern Development
and Mines
Mines and Minerals Division
933 Ramsey Lake Rd.
Willet Green Miller Centre, Level B-6
Sudbury, ON P3E 6B5
705-670-5877
Fax: 705-670-5818
http://www.gov.on.ca/MNDM/MINES/
mmdhpge.htm

Géologie Québec
5700 4ème Ave. Ouest, local A-208
Charlesbourg, QUE
418-627-6274
Fax: 418-643-2816
http://www.geologie-quebec.gouv.qc.ca/english/

Saskatchewan Energy and Mines
2101 Scarth St.
Regina, SK S4P 3V7
306-787-2526
Fax: 306-787-7338
http://www.gov.sk.ca/enermine/

Glossary of Selected Terms

A-horizon — The uppermost layer of a soil, containing organic material and leached minerals. This is the layer predominantly used by plants and animals, including humans, to sustain life.

ablation — All processes by which snow and ice are lost from a glacier. These processes include melting, evaporation (sublimation), wind erosion, and calving.

ablation till — A general term for loose material deposited during the downwasting of nearly static glacial ice, either contained within or accumulated on the surface of the glacier.

ablation zone — The zone on the surface of a glacier with net ablation.

active layer — The top layer of ground subject to annual freezing-and-thawing in areas underlain by permafrost.

alas — A depression with steep sides and a flat, grass-covered floor, found in thermokarst terrain and produced by thawing of extensive areas of very thick ice-rich permafrost.

albedo — The percent of incoming radiation that is reflected by a natural surface such as the ground, ice, snow, or water. An extremely reflective surface such as ice or snow has a high albedo.

alpine glacier — Any glacier in a mountain range that is largely confined by the surrounding topography. An alpine glacier usually originates in a cirque and may flow down into a valley previously carved by a stream.

angstrom — a unit of length equal to 10^{-10} meter, or one hundred-millionth of a centimeter.

angular unconformity — An unconformity in which the bedding planes of the rocks above and below are not parallel.

aquiclude — A geologic formation that acts as a barrier to groundwater flow due to its low hydraulic conductivity.

aquifer — A rock or sediment that transmits groundwater easily.

aquifuge — A material that contains no interconnected pores and therefore neither absorbs nor transmits water.

arete — A narrow, jagged mountain crest sculpted by alpine glaciers and formed by backward erosion of adjoining cirque walls.

artesian well — A well that penetrates an aquifer containing water under pressure. Water in an artesian well thus rises above the surrounding water table.

aspect — The direction a slope faces with respect to the compass or to the rays of the Sun.

asthenosphere — The plastic zone of the mantle. This is the zone that deforms and can adjust to changes in load, causing the phenomenon of isostasy.

B-horizon — The intermediate layer in a soil, situated below the A-horizon and containing significant amounts of clays and oxides. This is commonly the layer where material carried downward from the A-horizon accumulates.

barchan dune — A crescent-shaped dune with tips extending downwind, making this side concave and the upwind side convex. The tips travel faster than the main body of the dune because they are not anchored by vegetation. This type of dune forms primarily in desert conditions.

barrier island — A long, narrow, sandy island that is parallel to the shore and that commonly has a sequence of beaches, dunes, vegetated zones, and swampy terrains in a transect from the open ocean to a lagoon.

basal till — Unconsolidated material of mixed composition deposited at the base (bottom) of a glacier.

beach ridge — A low ridge of beach sediments parallel to the shore, formed by the action of waves and currents on the beach and located beyond the present limit of storm waves or tides. Beach ridges may occur alone or in a series of approximately parallel deposits and may represent successive positions of an advancing or rebounding shoreline.

bedload — The sediment that a stream moves along the bottom of its channel by rolling and bouncing.

bergschrund — A wedge-shaped ice crevasse near, but not at the head of, a valley glacier.

biochemical — Refers to a process, sediment, or rock having a biological origin. A common example is limestone, formed from elements extracted from seawater by living organisms.

biostratigraphy — A stratigraphic correlation scheme based on fossil fauna or flora.

bioturbation — Partial or total mixing of unconsolidated sediment by the actions of burrowing or rooting organisms.

blowout — A general term for a depression formed by wind erosion on a sandy deposit, especially where protective vegetation is disturbed or destroyed.

bog — A peatland with primarily sphagnum mosses that are capable of maintaining a perched water table isolated from the mineral substratum.

boulder clay — A clay-rich, stony till.

boulder train — A linear zone containing glacially transported boulders (erratics) on the lee side of a well-defined outcrop from which the erratics are derived.

braided stream — A stream with multiple channels that interweave as a result of repeated bifurcation and convergence of flow around interchannel bars, resembling (in plan view) the strands of a complex braid. Braiding is generally confined to broad, shallow streams of highly variable discharge, high bedload, low sinuosity, and noncohesive bank material.

C-horizon — The lowest layer of a soil, consisting of parent materials and their chemically weathered products.

calving — The breaking off and floating away of icebergs from a glacier with a marine or lacustrine terminus. Calving is a very efficient form of ablation, helping to stabilize the extent of ice sheets (like Antarctica) that might otherwise expand continuously due to a positive mass budget.

capacity — The maximum amount of sediment a stream can carry.

carbon 14 — Naturally occurring radioactive isotope of carbon with a half-life of 5730 years. Carbon 14 is particularly useful in dating carbon-rich material such as wood or shells of less than about 40 ka.

cavitation — The erosive effect of collapsing high-pressure bubbles in water.

chattermark — A small, curved scar on bedrock made by rock fragments carried at the base of a glacier. Each mark is roughly transverse to the direction of flow, and either convex (lunate) or concave (crescentic) in the upflow direction.

chert — A hard sedimentary rock, consisting primarily of interlocking crystals of quartz less than about 30 μm in diameter; it may contain amorphous silica (opal).

chronostratigraphic — Refers to a stratigraphic correlation of sediments that were deposited during a specific time interval.

cirque — A semicircular, concave, bowl-like area (natural amphitheater) with a steep face primarily resulting from the erosive activity of a mountain glacier.

clast — An individual constituent, grain, or fragment of sediment or rock, produced by the mechanical weathering (disintegration) of a larger rock mass.

clay — A large number of hydrous aluminosilicate minerals formed by weathering and hydration of other silicates; also, any mineral fragment smaller than 1/512 mm.

cobble — A clast between 64 and 256 mm in diameter.

competence — A measurement of the largest clast that a flowing fluid (water or wind) can carry. It depends on velocity — for instance, a small but swift river may have greater competence than a larger but slower-moving one.

conduction — The transfer of heat energy, normally through a solid, via the mechanism of atomic or molecular impact without the movement of mass.

confluence — The place where two glaciers (or rivers) join.

conglomerate — A coarse-grained, clastic sedimentary rock composed predominantly of rounded to subangular rock fragments larger than 2 mm, commonly with a matrix of sand and finer material. Cements include silica, calcium carbonate, and iron oxides. If the rock fragments are angular, the rock is called *breccia*.

continuous permafrost — A geographic zone where permafrost occurs everywhere beneath the land surface, but not necessarily beneath bodies of water.

Coriolis force — A force resulting from Earth's rotation. This force diverts ocean and atmospheric currents to the right in the northern hemisphere and to the left in the southern hemisphere.

crag and tail — A streamlined hill formed by a glacier, consisting of a bedrock knob (crag) with a tail of till on the lee side.

crevasse filling — A short, straight ridge of stratified sand and gravel believed to have been deposited in a crevasse of a wasting glacier and left standing after the ice melted.

cross-bedding — Cross-stratification in which the cross-beds are more than 1 cm thick. This is referred to as cross-lamination if the cross-beds are equal to or less than 1 cm thick.

cryoplanation — The smoothing and lowering of a land surface by intensive frost action supplemented by the erosive and transportive actions of running water, moving ice, solifluction, and other agents.

cryoturbation — A collective term used to describe all soil movements due to frost action.

cyclothem — An informal, lithostratigraphic unit typically associated with unstable shelf or interior basin conditions in which alternate marine transgressions and regressions occur.

daughter element — An element or isotope created as the end or transitional product of the radioactive decay of another isotope.

deflation — The sorting out, lifting, and removal of loose, dry, fine-grained particles (clays, silts, and fine sands) by wind.

delta plain — The level or nearly level surface composing the exposed part of a delta. A delta plain is normally a floodplain characterized by repeated channel bifurcation and divergence and by multiple distributary channels.

density current — A subaqueous current that flows on the bottom of a sea or lake because of incoming water that is more dense due to temperature or suspended sediments. See also *turbidity current* and *hyperpycnal flow*.

desert pavement — A natural, residual concentration of wind-polished, closely packed pebbles, boulders, and other rock fragments that mantle a desert surface where wind action and sheetwash have removed all smaller particles. It usually protects the underlying, finer-grained material from further deflation.

detrital sediment — A sediment deposited by a physical process.

diamict — A general term that includes both diamictite (rock) and diamicton (sediments). Diamicton is a generic term for any nonlithified, nonsorted, or poorly sorted sediment containing a wide range of particle sizes. Although similar to till, *diamicton* is used when the genesis of the sediment is uncertain.

diatom — A one-celled aquatic plant that has a siliceous framework. Diatomaceous earth is a geologic deposit composed chiefly or wholly of the remains of diatoms. A diatomite is a chertlike rock formed by the lithification of diatomaceous earth.

discontinuous permafrost — An area where permafrost occurs in some places but not others.

distal — Describes a sediment deposited farthest from its source area.

drift — A general term applied to all material (clay, silt, sand, gravel, and boulders) transported by a glacier and deposited directly by or from the ice, or by glacial meltwater.

dropstone — An oversized stone (compared to the host sediments) in laminated sediment. It may depress the underlying laminae and can be covered by "draped laminae." Most dropstones originate through ice-rafting; also called a "lonestone."

eccentricity — The degree to which Earth's orbit around the Sun varies from a perfect circle. It ranges from about 1% to 5% over a 100,000-year cycle.

elastic limit — The maximum stress that can be applied to a solid substance without resulting in permanent strain (deformation).

end moraine — A ridgelike accumulation of poorly or moderately sorted sediment that is produced at the outer margin of an actively flowing glacier at any given time.

englacial — Describes sediment that is carried in the inner parts of a glacier; also refers to the sediments once they are deposited — for instance, englacial till.

eolian — Pertaining to wind, as in material transported and deposited (eolian deposit) by the wind. These materials include clastic substances such as dune sands, sand sheets, loess deposits, and clay.

eon — The largest division of geologic time, embracing several eras. For example, we are now in the Phanerozoic Eon (600 Ma to present).

epoch — One subdivision of a geologic period, often chosen to correspond to a stratigraphic series.

era — A division of geologic time including several periods, but smaller than an eon. Commonly recognized eras are Precambrian, Paleozoic, Mesozoic, and Cenozoic.

erratic — A rock fragment carried by glacial ice, or by floating ice (ice-rafting), and subsequently deposited at some distance from the outcrop from which it was derived. Erratics generally, though not necessarily, rest on bedrock or sediments of different lithology. They range in size from pebbles to house-sized blocks.

escarpment — A relatively continuous cliff or steep slope produced by erosion or faulting. This term is most commonly applied to cliffs produced by differential erosion.

esker — A long, narrow, sinuous ridge composed of stratified sand and gravel deposited by a subglacial stream flowing in an ice tunnel of a retreating glacier. Eskers range in length from less than a kilometer to hundreds of kilometers, and in height from 3 to 30 meters.

estuary — The seaward end of a river valley where freshwater comes into contact with seawater and where tidal effects are evident.

eustatic change — A sea level change that affects the whole Earth.

exhumed — Describes formerly buried landforms, geomorphic surfaces, or paleosols that have been re-exposed by erosion of the covering material.

facies — A group of characteristics that distinguish one group of beds from another within a stratigraphic unit; also, the sum of all primary lithologic and paleontologic characteristics of a sediment or sedimentary rock that are used to infer its origin and environment.

fen — A peatland with mainly grassy vegetation; it may contain stunted trees.

firn — A transition form between snow and glacial ice resulting from the consolidation, metamorphosis, and melting-refreezing of snow.

fjord — The seaward end of a deep, glacially excavated valley that becomes submerged after the ice melts. Typically it has a shallow sill, or threshold of solid rock or glacial sediment submerged near its mouth.

floodplain — The nearly level plain that borders a stream and is subject to inundation under flood-stage conditions unless protected artificially. It is usually a constructional landform built of sediment deposited during overflow and lateral migration of stream-meander bends.

flow till — A supraglacial till that is modified and transported by mass flow at the terminus of a glacier.

flute — A streamlined groove or ridge parallel to the direction of ice movement, formed in newly deposited till or older drift. Flutes range in height from a few centimeters to 25 meters, and in length from a few meters to 20 kilometers.

foraminifera — Oceanic protozoa, most of which have shells composed of calcite. Foraminiferal ooze is a calcareous sediment composed of the shells of dead foraminifera.

forebulge — A small upwarping of Earth's crust just outside the zone of crustal loading and subsidence, caused by an ice sheet or thrust sheets of mountain chains.

foredune — A coastal dune or dune ridge oriented parallel to the shoreline, occurring at the landward margin of the beach, along the shoreward face of a beach ridge, or at the landward limit of the highest tide, and more or less stabilized by vegetation.

foreset bed — One of the inclined beds found in cross-bedding; also, an inclined bed deposited on the outer front of a delta.

formation — The basic lithostratigraphic unit in the local classification of rocks; also, a body of rock generally characterized by some degree of internal lithologic homogeneity or distinctive lithology that is mappable at Earth's surface (at scales on the order of 1:25,000) or traceable in the subsurface. Formations may be combined into *groups* or subdivided into *members.*

frost boil — A small mound of fresh soil material formed by frost action; also, a type of unsorted circle commonly found in fine-grained sediment underlain by permafrost. Frost boils may also develop in areas affected by seasonal frost.

frost shattering — The mechanical disintegration, splitting, or breakup of a rock or soil caused by the pressure exerted by freezing water in cracks or pores, or along bedding planes.

glacial lake — A lake formed in contact with a glacier.

glacial rebound — The uplift of the crust that takes place after the retreat of a continental glacier as the crust and mantle readjust to the lessened load. See also *isostasy.*

glacial striations (striae) — Scratches left on bedrock and boulders by the abrasive action of clasts carried at the base of the moving glacier. These striations show the direction of movement.

glacier surge — A period of unusually rapid movement of a glacier, sometimes lasting more than a year. This is often attributed to slippage of the glacier at its base.

glaciofluvial deposit — Glacially derived sediment that is sorted and deposited by streams flowing from the melting ice. These deposits are stratified and may occur in the form of outwash plains, valley trains, deltas, kames, and eskers.

glaciolacustrine deposit — Material ranging from fine clay to gravel that is derived from glaciers and deposited in glacial lakes by water or floating ice. Many of these materials are bedded or laminated with varves or rhythmites and may contain dropstones.

glaciomarine deposit — Glacially eroded, terrestrially derived sediments (clay, silt, sand, and gravel) that accumulate on the ocean floor.

graded bedding — A bed in which the coarsest particles are concentrated at the bottom and grade gradually upward into finer materials. The whole graded bed is deposited by a waning current.

gravel — Coarse alluvial sediments, containing mostly particles larger than 2 mm in size.

ground moraine — An extensive, low relief area of till having an uneven or undulating surface and commonly bounded on the distal end by a recessional or end moraine.

grounding line — The point at which a tidewater glacier floats free of its bed. From this point inland it acts on its bed; from this point seaward it floats, and thus it may accelerate, thin, and calve.

half-life — The time period in which half of the atoms of a radioactive element in a sample decay into their direct products.

hanging valley — A tributary valley in which the floor at the lower end is notably higher than the floor of the main valley in the area of junction. It may be occupied by a hanging glacier.

high-center polygon — A patterned ground feature; also, a polygon in which the center is raised relative to its boundary.

Holocene — The later epoch of the Quaternary Period of geologic time, extending from the end of the Pleistocene Epoch (arbitrarily taken to be 10,000 years ago) to the present.

horn — A high, rocky, sharp-pointed, steep-sided mountain peak with prominent faces and ridges, bounded by the intersecting walls of three or more cirques that have been cut back into the mountain by headward erosion of glaciers. The ridge between adjoining cirques is called an *arete.*

hydration — A chemical reaction, usually in weathering, that adds water to a mineral structure.

hyperpycnal flow — A current generated by density differences. It occurs when water flows into a basin with water of marked density difference.

hypsithermal period — A postglacial climatic optimum (the warmest time).

ice cap — A dome-shaped cover of perennial ice and snow, covering the summit area of a mountain mass but allowing nunataks to emerge through it, or covering a flat landmass such as an arctic island. An ice cap spreads outwards in all directions due to its own weight and has an area of less than 25,000 square kilometers.

ice-rafting — The transportation of rock fragments of all sizes on or within icebergs, ice floes, or other forms of floating ice.

ice shelf — A continuous plate of floating ice that often extends seaward from a glacier or ice sheet on the shore.

ice stream — A zone of high velocity within an ice cap or ice sheet.

ice wedge — A generally wedge-shaped body with an apex pointing downward. Ice wedges are composed of foliated or vertically banded ice.

ice-wedge polygon — Patterned ground in areas of ice wedges. These polygons are common in poorly drained areas and may be high-centered or low-centered.

insolation — Solar energy received at Earth's surface.

interstadial — A halt, with minor fluctuations in the advance or retreat of an ice sheet.

isostasy — The mechanism by which areas of the crust are uplifted or subside until their mass is buoyantly supported, or "floats," on the dense mantle beneath.

isotope — One of several forms of an element, all having the same number of protons in the nucleus but differing in their number of neutrons and thus in their atomic weight.

jökulhlaup — An Icelandic term for a glacial outburst flood, especially when an ice dam impounding a glacial lake breaks. Similar breaks drained glacial Lake Missoula in the Late Pleistocene and created the channeled scablands in the Pacific Northwest.

ka — Abbreviation for "thousand years."

kame — A mound, knob, or short irregular ridge, composed of stratified sand and gravel deposited by a subglacial stream as a fan or delta at the margin of a melting glacier, or by a supraglacial stream entering a low place or hole on the surface of the glacier.

kame moraine — (a) An end moraine that contains numerous kames; (b) a group of kames along the front of a stagnant glacier, commonly composed of the slumped remnants of a formerly continuous outwash plain built up over the foot of rapidly wasting or stagnant ice.

katabatic wind — Cold, dry, heavy air flowing downslope at the snout of a glacier, sometimes attaining high speeds.

kettle — A steep-sided, bowl-shaped depression in drift deposits, often containing a lake or swamp. A kettle is formed by the melting of a large, detached block of stagnant ice that had been wholly or partly buried in the drift.

Lake Missoula — A glacial lake in northwest Montana during Pleistocene times that was formed by an ice dam of the Cordilleran Ice Sheet. This dam broke periodically, flooding a portion of current-day Washington State (see *jökulhlaup*).

laminar flow — A fluid flow in which the paths of water molecules (and sediment particles) are straight or gently curved, and parallel.

lateral moraine — A ridgelike moraine carried on and deposited at the side margin of a valley glacier. It is composed chiefly of rock fragments derived from valley walls by glacial abrasion and plucking.

levee — An artificial or natural embankment built along the margin of a watercourse or an arm of the sea.

lodgment till — A basal till often containing fragments oriented with their long axes generally parallel to the direction of ice movement.

loess — Silt-sized particles transported and deposited by wind. Commonly a loess deposit thins and the mean-particle size decreases as distance from the source area increases. Loess sources are primarily either from glacial meltwaters (cold loess) or from nonglacial, arid environments, such as deserts (hot loess).

lonestone — See *dropstone*.

longitudinal dune — A dune with its long axis parallel to the direction of the prevailing wind.

longshore current — A current that moves parallel to a shore, caused by breaking waves that approach the shore obliquely. The current can effectively move sediment along a beach. This movement of sediment is known as *longshore drift*.

low-center polygon — A patterned ground feature; also, a polygon in which the center is depressed relative to its boundary.

Ma — Abbreviation for "million years."

marine limit — The highest postglacial elevation of sea level in a glaciated area.

marsh — Periodically wet or continually flooded areas with the surface not deeply submerged. Covered primarily with sedges, cattails, rushes, or other gramineous plants.

medial moraine — An elongated moraine carried in or upon the middle of a glacier and parallel to its sides, usually formed by the merging of adjacent and inner lateral moraines below the junction of two coalescing valley glaciers. An irregular ridge in the middle of the glacial valley remains when the glacier has disappeared.

Miocene — The epoch of the Tertiary Period of geologic time immediately preceding the Pliocene Epoch (from approximately 23 to 5.2 million years ago).

moulin — A shaft by which supraglacial meltwater enters a glacier to become englacial or subglacial meltwater.

nanometer — 10^{-9} meters.

nivation — The excavation of a shallow depression or hollow on a mountainside through frost action, mass wasting, and water erosion associated with snowbanks or small glaciers.

nunatak — An isolated peak of bedrock that projects prominently above the surface of a glacier and is completely surrounded by glacier ice.

outwash — Stratified sand and gravel "washed out" from a glacier by meltwater streams and deposited in front of the end moraine or the margin of an active glacier to form a low-relief outwash plain. The coarser material is deposited nearer to the ice.

oxidation — A chemical weathering reaction in which electrons are lost from an atom and its charge becomes more positive.

paleosol — An ancient, usually buried soil.

palsa — An elliptical dome-like permafrost mound containing alternating layers of ice lenses and peat or mineral soil, commonly 3–10 m high and 2–25 m long. Palsas occur in subarctic settings.

Pangaea — A large proto-continent from which present continents have been broken off by the mechanism of sea-floor spreading and continental drift.

parabolic dune — A sand dune with a long, scoop-shaped form, convex in the downwind direction so that its horns point upwind. Its ground plan, when perfectly developed, approximates the form of a parabola.

parent isotope — An element that is transformed by radioactive decay into a different (daughter) element.

parent material — The partially weathered mineral or organic matter from which a soil is developed by pedogenic processes.

paternoster lakes — A chain of lakes in a glacial valley.

peat — Variably decomposed, unconsolidated organic matter accumulated in wetlands, frequently of high acidity.

peat plateau — A generally flat-topped expanse of peat, elevated above the general surface of a peatland, and containing segregated ice that may or may not extend downward into the underlying mineral soil.

peatland — A wetland that contains at least 40 cm of peat. All peatlands are wetlands but the inverse is not necessarily true. Types of peatlands are bog, fen, swamp, and marsh.

perched water table — An isolated body of groundwater that is perched above and separated from the main water table by an aquiclude.

periglacial — Pertaining to processes, conditions, areas, climates, and topographic features occurring near glaciers and ice sheets, and influenced by the cold temperature of the ice. The term was originally introduced to designate the climate and related geologic features peripheral to ice sheets of the Pleistocene.

permafrost — Ground, soil, or rock that remains at or below 0°C for at least two years. It is defined on the basis of temperature and does not necessarily contain ground ice.

pingo — A relatively large, conical mound of soil-covered ice (commonly 30–50 m high and up to 400 m in diameter) raised in part by hydrostatic pressure within and below the permafrost of arctic regions.

Pleistocene — The earlier epoch of the Quaternary Period of geologic time, following the Pliocene Epoch and preceding the Holocene Epoch (from approximately 1.6 million to 10,000 years ago).

Pliocene — The last epoch of the Tertiary Period of geologic time, following the Miocene Epoch and preceding the Pleistocene Epoch (from approximately 5.2 to 1.6 million years ago).

plucking — A process of glacial erosion by which blocks of rock are pulled away from fractured bedrock.

pluvial lake — (a) A lake formed in a period of exceptionally heavy rainfall; (b) a lake formed in the Pleistocene Epoch during a time of glacial advance, and now either extinct (relict) or existing as a remnant (lake) similar to Lake Bonneville in the western United States.

pothole — A semispherical hole in the bedrock of a streambed, formed by the drilling action of small pebbles and cobbles in a strong current.

ppm — Abbreviation for "parts per million."

pressure melting point — The temperature and pressure at which a solid such as ice changes into a liquid such as water. The melting point of ice decreases about 0.7°C per vertical kilometer of ice.

proglacial lake — A type of glacial lake that forms just beyond the margin of an advancing or retreating glacier.

proximal — Describes a sediment deposited nearest its source area.

Quaternary — The later period of the Cenozoic Era of geologic time, extending from the end of the Tertiary Period (about 1.6 million years ago) to the present and comprising two epochs, the Pleistocene (Ice Age) and Holocene (Recent).

radiolaria — A class of one-celled marine animals with siliceous skeletons that have existed in the ocean throughout the Phanerozoic Eon. A siliceous deep-sea sediment composed largely of the skeletons of radiolaria is called radiolarian ooze. Radiolarite is the sedimentary rock formed from radiolarian ooze.

raised beach — See *beach ridge.*

recessional moraine — (a) An end or lateral moraine, built during a temporary but significant halt in the final retreat of a glacier; (b) a moraine built during a minor readvance of the ice front during a period of overall recession.

recharge — In hydrology, the replenishment of groundwater by infiltration of meteoric water through the soil.

reef — A ridgelike or moundlike structure, layered or massive, built by sedentary calcareous organisms, especially corals, and consisting mostly of their remains; it is wave-resistant and stands above the surrounding contemporaneously deposited sediment.

regelation — The refreezing of meltwater.

regression — A relative drop in sea level.

relict — Pertaining to features such as landforms, geomorphic surfaces, and paleosols that are mostly products of past environments.

rhythmite — An individual unit in a succession of beds developed by rhythmic sedimentation. The term implies no limit as to thickness or complexity of bedding and it carries no time or seasonal connotation.

ripple mark — An undulating surface of alternating, subparallel, small-scale ridges and depressions, commonly composed of sand. It is produced on land by wind and underwater by currents or wave action. The crests of the ridges are generally at right angles or oblique to the direction of flow.

roche moutonnée — An elongated, protruding knob or hillock of bedrock, sculpted by a glacier so that its long axis is oriented in the direction of ice movement; an upstream (stoss or scour) side is gently inclined, smoothly rounded, and striated; and a downstream (lee or pluck) side is steep and rough.

rock glacier — A mass of poorly sorted angular boulders and fine material, containing either interstitial ice a meter or so below the surface (ice-cemented) or a buried glacier (ice-cored). Rock glaciers occur in permafrost areas and range from a few hundred meters to several kilometers in length.

saltation — The movement of sand grains by short jumps above the ground or streambed that is under the influence of a current too weak to keep them permanently suspended.

sandur (plural: sandar) — An Icelandic term that indicates a sandy-gravelly plain where sediment, mostly gravel, is deposited from glacial meltwater.

saprolite — Soft, friable, chemically weathered bedrock that retains the fabric and structure of the parent rock.

scabland — An elevated, flat-lying area with little if any soil cover, sparse vegetation, and usually deep, dry channels scoured into the surface, especially by glacial meltwaters. An often-quoted example is the channeled scabland of Washington State.

snowline — The elevation above which more snow accumulates during the winter than melts during the subsequent summer.

solifluction — Slow downslope movement of water-saturated regolith. Rates of flow vary widely. The presence of frozen substrate or even freezing-and-thawing is not implied in the original definition, though one component of solifluction can be creep of frozen ground. This term is commonly applied to processes operating in both seasonal frost and permafrost areas.

sorted circle — A type of circular patterned ground with a sorted appearance commonly due to a border of coarse fragments surrounding finer material. It may occur singly or in groups. The diameters of sorted circles range from a few centimeters to more than 10 meters, and their coarse fragment borders may be up to 35 cm high and 8 to 12 cm wide.

sphagnum — A major peat-forming moss in cool to cold climates.

sporadic permafrost — The area near the southern boundary of discontinuous permafrost where permafrost occurs in isolated patches.

stadial — Used in connection with Quaternary glaciations to refer to a relatively cold phase (advance) between warmer phases (retreats).

stagnant ice — Glacial ice that has ceased movement.

stratified drift — Sediments deposited by glacial meltwater that are somewhat sorted and layered.

stress — Force per unit area.

stripe — A type of patterned ground that develops on steep slopes.

subaerial — Describes conditions, processes, and features that exist or operate either above water on the land surface or in the open air.

subaqueous — Describes conditions, processes, and features that exist or operate underwater.

subglacial — Describes conditions, processes, and features beneath a glacier.

sublimation — The change of a material from a solid state directly to a gas state without turning to liquid in between.

supraglacial — Describes conditions, processes, and features at the top surface of a glacier.

suspended load — The fine sediment kept suspended in a stream during transport.

swamp — A peatland with at least 25% of its surface area covered by trees.

talik — A Russian term that indicates the layer of unfrozen ground between frozen layers.

talus — A deposit of rock fragments of any size or shape (usually coarse and angular) derived from and lying at the base of a cliff or very steep rock slope.

tarn — A glacial lake produced by scouring. These are often found in cirques.

temperate glacier — A glacier in which the ice is at or near the pressure melting point throughout.

terminal moraine — An end moraine that marks the farthest advance of a glacier and usually has the form of an arcuate or concentric ridge.

Tertiary — The period of the Cenozoic Era of geologic time, extending from approximately 65 to 1.6 million years ago. Its subdivisions comprise, in order of most recent to oldest, the Pliocene, Miocene, Oligocene, Eocene, and Paleocene Epochs.

thermokarst — Karst-like topographic features, such as depressions and lakes, produced in permafrost regions by local melting of ground ice and subsequent settling of the ground.

tidewater glacier — A glacier that terminates in seawater. Flexing by tidal fluctuation may accelerate calving of icebergs, thus stabilizing the glacier terminus at or just beyond the grounding line.

topset bed — A horizontal sedimentary bed formed at the top of a delta and overlying the foreset beds.

transgression — A relative rise in sea level that causes areas of land to become submerged.

transverse dune — An asymmetric sand dune elongated perpendicular to the prevailing wind direction, having a gentle windward slope and a steep leeward slope standing at or near the angle of repose of sand.

tunnel valley — A valley cut into drift and other loose material, or into bedrock, by a subglacial stream.

turbidite — The sedimentary deposit of a turbidity current, showing typical graded bedding and sedimentary structures.

turbidity current — A mass of water and sediment that flows downslope along the bottom of a sea or lake because it is denser than the surrounding water.

turbulent flow — A high-velocity flow in which streamlines are neither parallel nor straight but curled into small, tight eddies (compare *laminar flow*).

unit cell — The smallest bonded group of atoms in a mineral that can be repeated in three directions to form a crystal.

unsorted circle — A type of circular patterned ground with a nonsorted appearance due to the absence of a border of coarse fragments. Vegetation characteristically outlines the pattern by forming a bordering ridge. The diameters of unsorted circles commonly range from 0.5 to 3 m.

valley train — A long, narrow body of outwash confined within a valley beyond a glacier.

varve — A thin pair of graded glaciolacustrine layers seasonally deposited, with a coarser, thicker summer layer and a finer-grained, thinner winter one.

ventifact — A stone or pebble that has been shaped, worn, faceted, cut, or polished by the abrasive action of windblown sand, usually under arid conditions.

X-ray diffraction — In mineralogy, the process of identifying mineral structures by exposing crystals to X-rays and studying the resulting diffraction pattern.

Younger Dryas — A European term for the late glacial time centered about 10,500 years BP.

Note: Many definitions appearing here were adapted from other glossaries. Some of these are:

Jackson, J. A., and Bates, R. L. (eds.). 1997. *Glossary of geology* (4th ed.). Alexandria, Va.: American Geological Institute. 769 pp.

William W. Locke's online glossary of important terms in glacial geology can be found at http://gemini.oscs.montana.edu/~geol445/hyperglac/glossary.htm

Karen Lempke's nicely illustrated online glossary of alpine glacial landforms can be found at

http://www.uwsp.edu/geo/faculty/lemke/vgd_alpine/glossary.html#erosionallandforms

References

REFERENCES FOR FIGURES

Allard, M. 1996. Geomorphological changes and permafrost dynamics: Key factors in changing arctic ecosystems. An example from Bylot Island, Nunavut, Canada. *Geoscience Canada* 23: 205–212.

*Andersen, B. G., and Borns, H. W., Jr. 1994. *The ice age world.* Oslo: Scandinavian University Press. 208 pp. Very good pictures and diagrams. Brief summary of deglaciation of Europe.

Andrews, J. T. 1987. The late Wisconsin glaciation and deglaciation of the Laurentide Ice Sheet. In: W. F. Ruddiman and H. E. Wright Jr. (eds.), *North America and adjacent oceans during the last deglaciation,* pp. 13–37. Vol. K-3 of *The geology of North America.* Boulder, Colo.: Geological Society of America.

————. 1997. Northern Hemisphere (Laurentide) deglaciation: Processes and responses of ice sheet/ocean interaction. In: I. P. Martini (ed.), *Late glacial and postglacial environmental changes,* pp. 9–27. New York: Oxford University Press.

Baker, V. R. 1973. *Paleohydrology and sedimentology of Lake Missoula flooding in eastern Washington.* Geological Society of America Special Paper 144. 79 pp.

————. 1983. Large-scale palaeohydrology. In: K. I. Gregory (ed.), *Background to palaeohydrology,* pp. 455–478. Chichester: Wiley.

Banks, B. 1989. *Satellite images: Photographs of Canada from space.* Energy, Mines and Resources Canada. Ottawa: Canada Centre for Remote Sensing. 120 pp.

Barnett, P. J. 1985. Glacial retreat and lake levels, North Central Lake Erie Basin, Ontario. In: P. F. Karrow and P. E. Calkin (eds.), *Quaternary evolution of the Great Lakes,* Geological Association of Canada Special Paper 30, pp. 185–194.

Barnhardt, W. A., Gehreis, W. R., Belknap, D. F., and Kelley, J. T. 1995. Late Quaternary relative sea-level change in the western Gulf of Maine: Evidence for a migrating forebulge. *Geology* 23: 317–320.

Bell, R. E., and Karl, D. M. 1999. Evolutionary processes: A focus of decade-long ecosystem study of Antarctic's Lake Vostok. *EOS* 80: 573–579.

*Benn, D. I., and Evans, D. J. A. 1998. *Glaciers & glaciation.* London: Edward Arnold. 734 pp. Up-to-date textbook and references on glacial geo-

morphology for senior classes and a good start for any research and project on this topic.

*Bennett, M. R., and Glasser, N. F. 1996. *Glacial geology: Ice sheets and landforms.* Chichester: Wiley. 364 pp.

Benson, L., and Thompson, R. S. 1987. The physical record of lakes in the Great Basin. In: W. F. Ruddiman and H. E. Wright, Jr. (eds.), *North America and adjacent oceans during the last deglaciation,* pp. 241–260. Vol. K-3 of *The geology of North America.* Boulder, Colo.: Geological Society of America.

Benson, S. 1959. *Physical investigations on the snow and firn of northwest Greenland 1952, 1953, and 1954.* U.S. Snow, Ice and Permafrost Research Establishment, Research Report 26.

Berger, A. L. 1978a. Long-term variations of caloric insolation resulting from the Earth's orbital elements. *Quaternary Research* 9: 139–167.

————. 1978b. Insolation signatures of Quaternary climatic changes. *Il Nuovo Cimento* 2C, 1: 63–87.

Beskow, G. 1930. Erdfliessen und struktuböden. *Geol. Fören. Stockh., Förh.* 52: 622–638.

Black, R. F. 1974. Ice-wedge polygons of northern Alaska. In: D. R. Coates (ed.), *Glacial geomorphology,* pp. 247–275. Binghamton: State University of New York.

Blatt, H., Middleton, G., and Murray, R. 1972. *Origin of sedimentary rocks.* Englewood Cliffs, N.J.: Prentice-Hall. 634 pp. (2d ed., 1980, 782 pp.)

Bluck, B. J. 1967. Sedimentation of beach gravels: Examples from South Wales. *Journal of Sedimentary Petrology* 37: 128–156.

Boothroyd, J. C., and Ashley, G. M. 1975. Processes, bar morphology, and sedimentary structures on braided outwash fans, northeastern Gulf of Alaska. In: A. V. Jopling and B. C. McDonald (eds.), *Glaciofluvial and glaciolacustrine sedimentation,* pp. 193–222. Society of Economic Paleontologists and Mineralogists, Special Publication No. 23.

Boulton, G. S. 1990. Sedimentary and sea level changes during glacial cycles and their control of glacimarine facies architecture. In: J. A. Dowdeswell and J. D. Scoutese (eds.), *Glacimarine environments: Processes and sediments,* pp. 15–52. Geological Society Special Publication, No. 53.

————. 1996. Theory of glacial erosion, transport and deposition as a consequence of subglacial sediment deformation beneath mid-latitude ice sheets. *Journal of Glaciology* 42: 43–46.

*Indicates glacial geology textbook

Bowen, D. Q., Richmond, G. M., Fullerton, D. S., Sibrava, V., Fulton, R. J., and Velichko, A. A. 1986. Correlation of Quaternary glaciations in the Northern Hemisphere. In: V. Sibrava, D. Q. Bowen, and G. M. Richmond (eds.), *Quaternary glaciations in the Northern Hemisphere.* Special issue of *Quaternary Science Reviews* 5: 509–510 and Chart 1.

Bradley, R. S. 1985. *Quaternary paleoclimatology.* Boston: Allen & Unwin. 472 pp.

Brewer, M. 1958. Some results of geothermal investigations of permafrost in northern Alaska. *American Geophysical Union Transactions* 39: 19–26.

Briggs, J. C. 1970. A faunal history of the North Atlantic Ocean. *Systematic Zoology* 19: 19–34.

Brigham-Grette, J. 1996. Geochronology of glacial deposits. In: J. Menzies (ed.), *Past glacial environments,* pp. 377–410. Oxford: Butterworth-Heinemann.

Broecher, W. S., and Denton, G. H. 1990. Urachen der Vereisungszyklen. *Spektrum der Wiss.* 3: 88–89.

Brookfield, M. B. 1994. Problems in applying preservation, facies and sequence models to Sinian (Neoproterozoic) glacial sequences in Australia and Asia. *Precambrian Research* 70: 113–143.

Brown, R. J. E. 1967. Comparison of permafrost conditions in Canada and the U.S.S.R. *Polar Research* 13: 741–751.

———. 1968. Occurrence of permafrost in Canadian peatlands. Proceedings of 3rd International Peat Congress, Quebec, 1968, pp. 174–181.

———. 1970. *Permafrost in Canada: Its influence on northern development.* Toronto: University of Toronto Press. 234 pp.

Buchard, B. 1978. Oxygen isotope palaeotemperatures from the Tertiary period in the North Sea area. *Nature* 275: 121–123.

Calkin, P. E. 1995. Global glacial chronologies and causes of glaciation. In: J. Menzies (ed.), *Modern glacial environments,* pp. 9–75. Oxford: Butterworth-Heinemann.

Caputo, M. V., and Crowell, J. C. 1985. Migration of glacial centers across Gondwana during the Paleozoic Era. *Geological Society of America Bulletin* 96: 1020–1036.

Carson, C. E., and Hussey, K. M. 1962. The oriented lakes of arctic Alaska. *Journal of Geology* 70: 417–439.

Chapman, L. J., and Putnam, D. F. 1966. *The physiography of southern Ontario.* Toronto: University of Toronto Press. 386 pp. Somewhat dated information but still a good summary of the morphology of this area. Good map of the Pleistocene geology.

Chorley, R. J. 1959. The shape of drumlins. *Journal of Glaciology* 3: 339–344.

Clapperton, C. M. 1997. Termination of the Pleistocene and Holocene changes in South America and other glaciated parts of the Southern Hemisphere. In: I. P. Martini (ed.), *Late glacial and postglacial environmental changes,* pp. 61–78. New York: Oxford University Press.

Clark, J. A., Hendriks, M., Timmermans, T. J., Struck, C., and Hilverda, K. J. 1994. Glacial isostatic deformation of the Great Lakes region. *Geological Society of America Bulletin* 106: 19–31.

Clark, P. U. 1997. Sediment deformation beneath the Laurentide Ice Sheet. In: I. P. Martini (ed.), *Late glacial and postglacial environmental changes,* pp. 81–97. New York: Oxford University Press.

Clarke, T. H., and Stearn, C. W. 1960. *The geological evolution of North America.* New York: Ronald Press. 434 pp.

Clayton, C. M. 1997. Termination of the Pleistocene and Holocene changes in South America and other glaciated parts of the Southern Hemisphere. In: I. P. Martini (ed.), *Late glacial and postglacial environmental changes,* pp. 61–78. New York: Oxford University Press.

Colbeck, S. C. 1992. *A review of the processes that control snow friction.* CRREL (Cold Regions Research and Engineering Laboratory), U.S. Army Corps of Engineers, Hanover, Germany. Monograph 92-2. 40 pp.

Colby, B. R. 1963. Fluvial sediments: A summary of source, transportation, deposition, and measurement of sediment discharge. *Geological Survey Bulletin* (U.S.) 1181-A: 1–47.

Colman, S. M., Pierce, K. L., and Birkeland, P. W. 1987. Suggested terminology for Quaternary dating methods. *Quaternary Research* 28: 314–319.

Costa, J. E. 1987. Floods from dam failure. In: V. R. Baker, R. C. Kochel, and P. C. Patton (eds.), *Flood geomorphology,* pp. 439–463. New York: Wiley.

Cowan, W. R., Sharpe, D. R., Feenstra, B. H., and Gwyn, Q. H. J. 1978. Glacial geology of the Toronto-Owen Sound area. In: *Toronto '78 Field Trip Guidebook of Joint Meeting of the Geological Society of America,* pp. 1–16. Geological Association of Canada and Mineralogical Association of Canada.

Cowell, D. W., Wickware, G. M., Boissoneau, A. W., Jeglum, J. K., and Sims, R. A. 1983. Hudson Bay Lowland peatland inventory. *Proceedings of Peatland Inventory Methodology Workshop,* pp. 88–102. Ottawa: Land Resource Research Institute, Agriculture Canada.

Czeppe, Z. 1960. Thermic differentiation of the active layer and its influence upon the frost heave in periglacial regions (Spitsbergen). *Bull. Acad. Pol. Sci.,* classe III, 8: 149–152.

Davis, W. M. 1906. The sculpture of mountains by glaciers. *Scottish Geographical Magazine* 22.

*Dawson, A. G. 1992. *Ice age Earth.* London: Routledge. 293 pp. Well-written, easy to understand book about the last 18,000 years of Earth.

Delorme, L. D., Thomas, R. L., and Karrow, P. F. 1990. Quaternary geology—Waterloo/Burlington and Lake Ontario. In: D. McKenzie (ed.), *Quaternary environs of Lake Erie and Ontario,* Field trip guide, pp. 142–162. Waterloo, Ontario: Quaternary Institute, University of Waterloo.

Derbyshire, E., and Owen, L. A. 1996. Glacioaeolian processes, sediments and landforms. In: J. Menzies (ed.), *Past glacial environments,* pp. 213–237. Oxford: Butterworth-Heinemann.

Dijkmans, J. W. A. 1990. *Aspects of geomorphology and thermoluminescence dating of cold-climate eolian sands.* Utrecht: Geographisch Instituut Rijksuniversiteit. 250 pp.

Dionne, J. C. 1978. Holocene relative sea-level fluctuations in the St. Lawrence estuary, Quebec, Canada. *Quaternary Research* 29: 233–244.

Doyle, P., Bennett, M., and Baxter, A. 1994. *The key to Earth history: An introduction to stratigraphy.* Chichester: Wiley. 231 pp.

Dreimanis, A. 1969. Selection of genetically significant parameters for investigation of tills. *Geografia* 8: 15–20.

———. 1978. Terminology and genetic classifications of tills and moraines currently used in Europe and North America. In: *Ground moraines of Continental deposits,* pp. 12–27. Proceedings Mat. Int. Symp (Moskwa), Geol. Inst. Acad. Sci.

———. 1988. Tills: Their genetic terminology and classification. In: R. P. Goldthwait and C. L. Matsch (eds.), *Genetic classification of glacigenic deposits,* pp. 17–83. Rotterdam: Balkema.

Dunbar, C. O., and Rodgers, J. 1957. *Historical geology.* New York: Wiley. 356 pp.

*Ehlers, J. 1996. *Quaternary geology and glacial geology.* Chichester: Wiley. 578 pp. Very good book for senior people primarily dealing with stratigraphy. Valuable sections on dating Quaternary deposits, methods of study of Quaternary sediments and Quaternary geology of Europe.

Einarsson, Porleifur. 1994. *Geology of Iceland.* Reykjavik: Mál og menning. 309 pp. Good book about Iceland.

*Embleton, C., and King, C. A. M. 1975. *Periglacial geomorphology.* London: Edward Arnold. 203 pp. Classic text, with some dated information.

Eyles, N. 1993. Earth's glacial record and its tectonic setting. *Earth Science Reviews* 35. 248 pp. Very good summary of glacial materials and events on Earth.

Eyles, N., and Eyles, C. H. 1992. Glacial depositional systems. In: R. G. Walker and N. P. James (eds.), *Facies models,* pp. 73–100. Waterloo: Geological Association of Canada.

Eyles, N., and Westgate, J. A. 1987. Restricted regional extent of the Laurentide ice sheets in the Great Lakes basins during early Wisconsin glaciation. *Geology* 15: 537–540.

Eyles, N., and Young, M. A. 1994. Geodynamic controls on glaciation in Earth history. In: M. Deynoux, J. M. G. Miller, E. W. Domack, N. Eyles, I. J. Fairchild, and G. M. Young (eds.), *Earth's glacial record,* pp. 1–28. Cambridge: Cambridge University Press.

Fairbanks, R. G. 1989. A 17,000-year glacio-eustatic sea level record: Influence of glacial melting rates on the Younger Dryas event and deep-oceanic circulation. *Nature* 342: 737–642.

Falcon, R. M. S. 1989. Macro- and micro-factors affecting coal-seam quality and distribution in southern Africa with particular reference to the No. 2 seam, Witbank coalfield, South Africa. *International Journal of Coal Geology* 12: 681–731.

*Flint, R. F. 1957. *Glacial and Pleistocene geology.* New York: Wiley. 553 pp. The best textbook on the subject ever written. It contains dated informa-

tion, but some of its maps and concepts are still used today.

Forester, R. M. 1987. Late Quaternary paleoclimate records from lacustrine ostracods. In: W. F. Ruddiman and H. E. Wright Jr. (eds.), *North America and adjacent oceans during the last deglaciation,* pp. 261–276. In vol. K-3 of *The geology of North America.* Boulder, Colo.: Geological Society of America.

*French, H. M. 1976. *The periglacial environment.* London: Longman. 309 pp.

*———. 1996. *The periglacial environment.* 2d. Ed. Harlow: Addison Wesley Longman. 341 pp. Up-to-date, expanded second edition of the book.

Friedman, G. M., Sanders, J. E., and Kopaska-Merkel, D. C. 1992. *Principles of sedimentary deposits.* New York: Macmillan. 717 pp.

Fuller, M. L. 1914. *The geology of Long Island, New York.* U.S. Geological Survey Professional Paper 82. 231 pp.

Goldwait, R. P. 1951. Development of end moraines in east-central Baffin Island. *Journal of Geology* 59: 567–577.

Goudie, A. 1992. *Environmental change.* Oxford: Clarendon Press. 329 pp.

Gray, D. M., and Male, D. H. 1981. *Handbook of snow: Principles, processes, management, and use.* Toronto: Pergamon Press. 776 pp. Good manual for study of snow as a resource and hazard.

Gustavson, T. C. 1975. Sedimentation and physical limnology in proglacial Malaspina Lake, southeastern Alaska. In: A. V. Jopling and B. C. McDonald (eds.), *Glaciofluvial and glaciolacustrine sedimentation,* pp. 249–263. Society of Economic Paleontologists and Mineralogists, Special Publication No. 23.

Hambrey, M. J., and Harland, W. B. 1981. *Earth's pre-Pleistocene glacial record.* Cambridge: Cambridge University Press. 1004 pp. The most complete reference on occurrence of ancient glaciations; however, some diamictites are now not considered to be glacial in origin.

Han, T. 1997. *The great Quinghai-Xizang Ice Sheet.* Beijing: Geological Publishing House. 96 pp.

Hare, F. K. 1976. Late Pleistocene and Holocene climates: Some persistent problems. *Quaternary Research* 6: 507–517.

Harris, S. E. 1943. Friction cracks and the direction of glacial movement. *Journal of Geology* 51: 244–258.

Hunt, C. B. 1972. *Geology of soils: Their evolution, classification, and uses.* San Francisco: W. H. Freeman. 344 pp.

———. 1974. *Natural regions of the United States and Canada.* San Francisco: W. H. Freeman. 725 pp.

Imbrie, J., and Imbrie, K. P. 1979. *Ice ages: Solving the mystery.* Hillside, N.J.: Enslow. 224 pp. Very interesting historical perspectives.

Jopling, A. V., Irving, W. N., and Beebe, B. F. 1981. Stratigraphic, sedimentologic and faunal evidence for the occurrence of pre-Sangamon artifacts in northern Yukon. *Arctic* 34: 3–33.

Karrow, P. F., and Morgan, A. V. 1975. Quaternary stratigraphy of the Toronto area, pp. 161–179. Geological Association of Canada, Field trips guide-book. Waterloo, Ontario: University of Waterloo.

Karte, J. 1987. Pleistocene periglacial conditions and geomorphology in north central Europe. In: J. Boardman (ed.), *Periglacial processes and landforms in Britain and Ireland,* pp. 67–75. Cambridge: Cambridge University Press.

Kay, M., and Colbert, E. H. 1965. *Stratigraphy and life history.* New York: Wiley. 736 pp.

Kelly, J. T., Dickson, S. M., Belknap, D. F., and Stuckenrath, R., Jr. 1992. Sea-level change and late Quaternary sediment accumulation on the southern Maine inner continental shelf. In: C. Fletcher and J. Wehmiller (eds.), *Quaternary coasts of the United States: Marine and lacustrine systems,* pp. 23–34. Society of Economic Paleontologists and Mineralogists, Special Publication 48.

Kennett, J. P., and Shackleton, N. J. 1975. Laurentide ice sheet meltwater recorded in Gulf of Mexico deep-sea cores. *Science* 188: 147–150.

Komar, P. D. 1988. Sediment transport by floods. In: V. R. Baker, R. C. Kochel, and P. C. Patton (eds.), *Flood geomorphology,* pp. 97–111. New York: Wiley.

Koster, E. A. 1988. Ancient and modern cold-climate aeolian sand deposition: A review. *Journal of Quaternary Science* 3: 69–83.

Kuhle, M. 1997. New findings concerning the Ice Age (Last Glacial Maximum) glacier cover of the East-Pamir, of the Nanga Parbat up to the Central Himalaya and of Tibet, as well as the age of the Tibetan Inland Ice. *GeoJournal* 42: 87–257.

Kupsch, W. O. 1962. Ice-thrust ridges in western Canada. *Journal of Geology* 70: 582–594.

Lachenbruch, A. H. 1962. *Mechanics of thermal contraction cracks and ice-wedge polygons in permafrost.* Geological Society of America Special Paper 70. 69 pp.

Larsen, E., and Sejrup, H.-P. 1990. Weichselian land-sea interactions: Western Norway–Norwegian Sea. *Quaternary Science Reviews* 9: 85–98.

Legget, R. F. 1968. *Soils in Canada.* Royal Society of Canada, Special Publication 3. Toronto: University of Toronto Press.

Leopold, L. B., and Maddock, T. 1953. *The hydraulic geometry of stream channels and some physiographic implications.* U.S. Geological Survey Professional Paper 252. 57 pp.

Leverett, F., and Taylor, F. B. 1915. *The Pleistocene of Indiana and Michigan and the history of the Great lakes.* U.S. Geological Survey Monograph 53. 529 pp.

Lewis, C. F. M. 1970. Recent uplift of Manitoulin Island, Ontario. *Canadian Journal of Earth Sciences* 7: 665–675.

Liu, Tungsheng. (ed.). 1985. *Loess and the environment.* Beijing: China Ocean Press. 251 pp.

Lorimer, R. L. C. 1967. *Edinburgh — Scotland's capital.* Edinburgh: Oliver and Boyd. 192 pp.

*Lowe, J. J., and Walker, M. 1984. *Reconstructing Quaternary environments.* Essex: Longman. 446 pp.

MacDonald, G. M. 1990. Palynology. In: B. G. Warner (ed.), *Methods in Quaternary ecology,* pp. 37–63.

St. John's, Newfoundland: Geoscience Canada, Memorial University.

Mackay, J. R. 1970. Disturbances to the tundra and forest tundra environment of western Arctic. *Canadian Geotechnical Journal* 7: 420–432.

———. 1972. Some observations on growth of pingos. In: D. E. Kerfoot (ed.), *Mackenzie Delta area monograph,* pp. 141–147. 22nd International Geographical Congress, Brock University, Montreal.

———. 1973. The growth of pingos, western Arctic coast, Canada. *Canadian Journal of Earth Sciences* 10: 979–1004.

Maizels, J. 1993. Lithofacies variations within sandur deposits: The role of runoff regime, flow dynamics, and sediment supply characteristics. *Sedimentary Geology* 85: 299–325.

Martini, I. P., and Glooschenko, W. A. 1985. Cold climate peat formation in Canada, and its relevance to lower Permian Coal Measures of Australia. *Earth Science Reviews* 22: 107–140.

Martini, I. P., Kwong, J. K., and Sadura, S. 1993. Sediment ice rafting and cold-climate fluvial deposits: Albany River, Ontario, Canada. In: M. Marzo and C. Puidgefabregas (eds.), *Alluvial sedimentation,* pp. 63–76. IAS Special Publication 17.

Martini, I. P., and Rocha-Campos, A. C. 1991. Interglacial and early post-glacial, Lower Gondwana coal sequences in the Paraná Basin. In: H. Ulbrick and A. C. Rocha-Campos (eds.), *Gondwana Seven — Proceedings of the Seventh International Gondwana Symposium,* pp. 317–337. Instituto de Geociências, Universidade de São Paulo, São Paulo, Brazil.

Martinson, D. G., Pisias, N. G., Hays, J. D., Imbrie, J., Moore, T. C., and Shackleton, N. J. 1987. Age, dating and orbital theory of the Ice Ages: Development of a high resolution 0- to 300,000-year chronostratigraphy. *Quaternary Research* 27: 1–29.

Miall, A. D. 1978. Lithofacies types and vertical profile models in braided river deposits: A summary. In: A. D. Miall (ed.), *Fluvial sedimentology,* pp. 597–604. Canadian Society of Petroleum Geologists, Memoir 5.

Middleton, G. V., and Southard, J. B. 1978. *Mechanics of sediment movement.* Society of Economic Paleontologists and Mineralogists, Short Course 3. Tulsa, Okla.: Society of Sedimentary Geology.

Miller, K. G., Fairbank, R. G., and Mountain, G. S. 1987. Tertiary oxygen isotope synthesis, sea level history, and continental margin erosion. *Paleoceanography* 2: 1–19.

Mitchell, J. M., Jr. 1977. The changing climate. In: *Energy and climate,* pp. 51–58. Washington, D.C.: Geophysics Study Committee, National Academy of Sciences.

Mitchum, R. M., Jr., Vail, P. R., and Sangree, J. B. (eds.). 1977. *Stratigraphic interpretation of seismic reflection patterns in depositional sequences,* pp. 135–143. American Association of Petroleum Geologists, Memoir 26.

Müller, F. 1959. Beobachtungen über pingos. Medd. om Grønland, 153, No. 3.

Nye, J. F. 1952. A method of calculating the thickness of ice sheets. *Nature* 169: 529–530.

Oceanus Science Slides. 1998. *http://www.joi-odp.org/ joi/Oceanus/Sediments/Sed1.html.*

Østrem, G., Ziegler, T., and Ekman, S. R. 1970. *Slamtansprotundersökelser i norske bre-elver 1969.* (See Østrem, Ziegler, and Ekman 1973 for translation.)

———. 1973. *A study of sediment transport in Norwegian glacial rivers 1969.* Translated by H. Carstens. Publication No. IWR-35. Fairbanks: Institute of Water Resources, University of Alaska. 59 pp.

Pérez Alberti, A. 1993. Xeomorfoloxía. In: A. Peréz Alberti (dir.), *Xeografía de Galicia.* Tomo 3. Ed. Santiago de Compostela: Gran Enciclopedia Gallega Ediciones.

Pérez Alberti, A., and Covelo Abeleira, P. 1996. Recontrución de la dinamica glaciar del Alto Bibei durante el Pleistoceno reciente a partir del estudio de los sedimentos acumulados en Pias (Noroeste de la Peninsula Ibérica). In: A. Peréz Alberti, P. Martini, W. Chesworth, and A. Martines Cortizas (eds.), *Dinamica y evolución de medio Cuaternarios,* pp. 115–129. Santiago de Compostela: Xunta de Galicia.

Péwé, T. L. 1982. *Geologic hazards of the Fairbanks area, Alaska.* Alaska Geological and Geophysical Surveys, Special Report 15. 109 pp. Very good monograph about permafrost response to human activities.

———. 1991. Permafrost. In: G. A. Kiers (ed.), *The heritage of engineering geology: The first hundred years,* pp. 277–298. Geological Society of America Centennial Special Vol. 3.

Piotrowski, J. A. 1993. Tunnel-valley formation in northwest Germany: Geology, mechanisms of formation, and subglacial bed conditions for the Bornhoved tunnel valley. *Sedimentary Geology* 89: 107–141.

Plumb, K. A. 1991. New Precambrian time scale. *Episodes* 14: 409–428.

Post, A., and LaChapelle, E. R. 1971. *Glacier ice.* Toronto: University of Toronto Press. 110 pp. Very good black-and-white picture book.

Post, W. M., Peng, Tsung-Hung, Emanuel, W. R., King, A. W., Dale, V. H., and De Angelis, D. L. 1990. The global carbon cycle. *American Scientist* 78: 310–326.

Powell, R. D. 1981. A model for sedimentation by tidewater glaciers. *Annales Glaciology* 2: 129–134.

Powell, R. D., and Domack, E. 1995. Modern glaciomarine environments. In: J. Menzies (ed.), *Modern glacial environments,* pp. 445–486. Oxford: Butterworth-Heinemann.

Powell, R. D., and Molnia, B. F. 1989. Glaciomarine sedimentary processes, facies and morphology of the south-southeast Alaska shelf and fjords. *Marine Geology* 85: 359–390.

Prest, K. V. 1970. Quaternary geology of Canada. In: R. J. W. Douglas (ed.), *Geology and economic minerals of Canada,* pp. 677–764. Geological Survey of Canada, Economic Geology Report No. 1.

Prothero, D. R. 1994. *The Eocene-Oligocene transition.* New York: Columbia University Press. 291 pp.

Prothero, D. R., and Schwab, F. 1996. *Sedimentary geology.* New York: W. H. Freeman. 575 pp.

Quigley, R. M. 1980. Geology, mineralogy, and geochemistry of Canadian soft soils: A geotechnical perspective. *Canadian Geotechnical Journal* 17: 261–285.

Reeh, N. 1989. Dynamic and climatic history of the Greenland Ice Sheet. In: R. J. Fulton (ed.), *Quaternary geology of Canada and Greenland,* pp. 795–822. Geological Survey of Canada, Geology of Canada, no. 1. (Also vol. K-1 of *The geology of North America.* Boulder, Colo.: Geological Society of America.)

Richmond, G. M., and Fullerton, D. S. 1986. Introduction to Quaternary glaciations in the United States of America. In: V. Sibrava, D. Q. Bowen, and G. M. Richmond (eds.), *Quaternary glaciations in the Northern Hemisphere.* Special issue of *Quaternary Science Reviews* 5: 3–10.

Riley, J. L. 1982. Hudson Bay Lowland floristic inventory, wetlands catalogue and conservation strategy. In: I. P. Martini (ed.), *Scientific studies on Hudson and James Bays. Naturaliste Canadien* 109: 543–555.

Rösler, O. 1979. Plantas fósseis de São Joâo do Triunfo (PR), Formaçâo Rio Bonito, e suas implicações geológicas. 2o Simpósio Regional de Geologia, vol 1., pp. 181–194. Núcleo de São Paulo: Sociedade Brasileira de Geologia.

Rust, B. R., and Romanelli, R. 1975. Late Quaternary subaqueous outwash deposits near Ottawa, Canada. In: A. V. Jopling and B. C. McDonald (eds.), *Glaciofluvial and glaciolacustrine sedimentation,* pp. 177–192. Society of Economic Paleontologists and Mineralogists, Special Publication No. 23.

Rutter, N. W. 1995. Problematic ice sheets. *Quaternary International* 28: 19–37.

Scholle, P. A. 1978. *A color guide to carbonate rock constituents, texture, cements, and porosities.* American Association of Petroleum Geologists, Memoir 27. 241 pp.

Schweger, C. E. 1989. Paleoecology of the western Canadian ice-free corridor. In: R. J. Fulton (ed.), *Quaternary Geology of Canada and Greenland,* pp. 491–498. Geological Survey of Canada, Geology of Canada, no. 1.

Scotese, C. R. 1997. Paleogeographic atlas, *http://www .scotese.com/earth.htm.*

Seward, A. C. 1959. *Plant life through the ages.* New York: Hafner. 603 pp. Dated but still most informative book.

Shackleton, N. J., Berger, A., and Peltier, W. A. 1990. An alternative astronomical calibration of the lower Pleistocene time scale based on ODP Site 677. *Transactions of the Royal Society of Edinburgh: Earth Sciences* 81: 251–261.

Sharp, R. P. 1960. *Glaciers.* Condon lectures, Oregon State System of Higher Education. 78 pp.

Shenk, H. G., and Muller, S. W. 1941. Stratigraphic terminology. *Geological Society America Bulletin* 52: 1419–1426.

Shoezov, P. E. 1959. *Osnovy geokriologij. Part I.* Akademia Nauk. S.S.S.R.

Shumskii, P. A. 1964. *Principles of structural glaciology.* New York: Dover. 497 pp. Dated book, but with some extremely valuable sections on petrography and other properties of ice.

Singvhi, A. K. 1992. *Thar Desert in Rajasthan.* Geological Society of India, Bangalore. 191 pp.

Sissons, J. B. 1971. The geomorphology of central Edinburgh. *Scottish Geographical Magazine* 87: 185–196.

Smalley, I. J. 1966. The properties of glacial loess and the formation of loess deposits. *Journal Sedimentary Petrology* 36: 669–676.

Smalley, I. J., and Unwin, D. J. 1968. The formation and shape of drumlins and their distribution and orientation in drumlin fields. *Journal of Glaciology* 7: 377–390.

St. Onge, D. A. 1987. The Sangamonian stage and the Laurentide Ice Sheet. *Geographie Physique et Quaternaire* 41: 189–198.

Stearn, C. W., Caroll, R. L., and Clark, T. H. 1979. *Geological evolution of North America.* New York: Wiley. 566 pp.

Strahler, A. N. 1972. *Planet earth.* New York: Harper & Row. 438 pp.

*Sudgen, D. E., and John, B. S. 1976. *Glacier and landscape: A geomorphological approach.* London: Edward Arnold. 376 pp. One of the best textbooks on glacial geomorphology, now with some dated information.

Sundborg, A. 1956. The River Klarälven, a study of fluvial processes. *Geograf. Annaler* 38: 125–316.

Tarbuck, E. J., and Lutgens, F. K. 1999. *Earth: An introduction to physical geology.* Upper Saddle River, N.J.: Prentice-Hall. 634 pp.

Teller, J. T. 1995. History and drainage of large ice dammed lakes along the Laurentide Ice Sheet. *Quaternary International* 28: 83–92.

Tufnell, L. 1984. *Glacier hazards.* London: Longman. 97 pp.

Velichko, A. A., Isayeva, L. L., Makeyev, V. M., Matishov, G. G., and Faustova, M. A. 1984. Late Pleistocene glaciation of the Arctic Shelf, and the reconstruction of Eurasian ice sheets. In: A. A. Velichko (ed.), *Late Quaternary environments of the Soviet Union,* pp. 25–44. London: Longman.

Viles, H., and Spencer, T. *Coastal problems: Geomorphology, ecology and society at the coast.* London: Edward Arnold. 350 pp.

Walcott, R. I. 1970. Isostatic response to loading of the crust in Canada. *Canadian Journal of Earth Sciences* 7: 716–726.

Wang, Y., and Bhan, D. 1985. *Model atlas of surface textures of quartz sand.* Beijing: Science Press. 63 pp.

Washburn, A. L. 1967. Instrumental observations of mass wasting in the Mester Vig district, Northeastern Greenland. *Medd. om Grønland* 166. 328 pp.

———. 1973. *Periglacial processes and environments.* London: Edward Arnold. 157 pp. A must-read book on this subject.

Weertman, J. 1961. Mechanism for the formation of inner moraines found near the edge of cold ice caps and ice sheets. *Journal of Glaciology* 3: 965–978.

Weller, J. M. 1960. *Stratigraphic principle and practice.* New York: Harper & Row. 725 pp.

*West, R. G. 1968. *Pleistocene geology and biology.* London: Longman. 377 pp. A now dated book, but with a still valuable analysis of biology and glacial settings.

White, M. E. 1988. *Greening of Gondwana.* French Forest, Australia: Reed Books. 256 pp. Very good picture book of paleoflora.

Whiteman, C. A. 1995. Process of terrestrial deposition. In: J. Menzies (ed.), *Modern glacial environments,* pp. 293–308. Oxford: Butterworth-Heinemann.

Wilson, R. 1976. *The land that never melts, Auyuittaq National Park.* Ottawa, Ontario: Parks Canada, Department of Indian and Northern Affairs. 212 pp.

Woldstedt, P. 1954. Die Klimakurve des Tertiärs und Quartärs in Mitteleuropa. *Eiszeitalter und Gegenvart* Bd. 4/5: 5–9.

Young, G. M. 1997. Tectonic and glacioeustatic controls on postglacial stratigraphy: Proterozoic examples. In: I. P. Martini (ed.), *Late glacial and postglacial environmental changes,* pp. 249–267. New York: Oxford University Press.

Young, G. M., and Nesbitt, H. W. 1985. The Gowganda Formation in the southern part of the Huronian outcrop belt, Ontario, Canada: Stratigraphy, depositional environments, and regional tectonic significance. *Precambrian Research* 29: 265–301.

Zoltai, S. C., and Pollett, F. C. 1983. Wetlands in Canada: Their classification, distribution, and use. In: A. J. P. Gore (ed.), *Mires: A. Swamp, bog, fen and moor; B. Regional studies,* pp. 245–268. Amsterdam: Elsevier.

ADDITIONAL GENERAL REFERENCES

Ahlmann, H. W. 1948. *Glaciological research on the North Atlantic coasts.* Royal Geographical Society Research, Series 1.

Ashley, G. M., Shaw, J., and Smith, N. D. 1985. *Glacial sedimentary environments.* Society of Economic Paleontologists and Mineralogists, Short Course 16. Tulsa, Okla.: Society of Sedimentary Geology.

Bentley, W. A., and Humphrey, W. J. 1962. *Snow crystals.* New York: Dover. 226 pp. (Reprint of the 1931 edition published by McGraw and Hill.)

Boulton, G. S. 1987. A theory of drumlin formation by subglacial deformation. In: J. Menzies and J. Rose (eds.), *Drumlin symposium,* pp. 25–80. Rotterdam: Balkema.

Brodzikowski, K., and van Loon, A. J. 1991. *Glacigenic sediments.* Amsterdam: Elsevier. 674 pp.

Canada National Hydrological Research Institute. 1978. *Snow crystals.* Canada National Hydrological Research Institute, Paper 1.

*Croot, D. G. 1988. *Glaciotectonics: Forms and processes.* Rotterdam: Balkema. 212 pp.

Denton, G. H., and Hughes, T. J. (eds.). 1981. *The last great ice sheets.* New York: Wiley. 484 pp. Much information and innovative ideas; for senior people, a bit difficult to read.

Dionne, J.-C. 1976. Le Glaciel. *La Revue de Géographie de Montréal* 30: 1–236.

*Drewry, D. 1986. *Glacial geologic processes.* London: Edward Arnold. 276 pp. Well-written book about processes.

*Embleton, C., and King, C. A. M. 1968. *Glacial and periglacial geomorphology.* London: Edward Arnold. 608 pp. Classic text, with some dated information.

*———. 1975. *Glacial geomorphology.* London: Edward Arnold. 573 pp. Classic text, with some dated information.

*Erickson, J. 1996. *Glacial geology.* New York: Facts on File. 248 pp.

*Eyles, N. 1983. *Glacial geology: An introduction for engineers and earth scientists.* Oxford: Pergamon Press. 409 pp.

Frakes, L. A., Francis, J. E., and Syktus, J. L. 1992. *Climate modes of the Phanerozoic.* Cambridge: Cambridge University Press. 274 pp.

Giardino, J. R., Shroder J. F., Jr., and Vitek J. D. 1987. *Rock glaciers.* Boston: Allen & Unwin. 355 pp.

*Hambrey M. 1994. *Glacial environments.* Vancouver: UBC Press. 296 pp. Good summary of glaciomarine studies.

Hoffman, P. F., Kaufman, A. J., Halversan, R. P., and Schrag, D. P. 1998. A Neoproterozoic snowball Earth. *Science* 281: 1324–1346.

*Hooke, R. LeB. 1998. *Principles of glacier mechanics.* Upper Saddle River, N.J.: Prentice-Hall. 284 pp. Senior classes textbook on glaciology.

John, B. 1979. *The world of ice: The natural history of the frozen regions.* London: Orbis. 120 pp. Very good photographs and ideas.

LaChapelle, E. R. 1970. *Field guide to snow crystals.* Seattle: University of Washington Press. 101 pp.

*Menzies, J. (ed.). 1995. *Modern glacial environments.* Oxford: Butterworth-Heinemann. 619 pp. Collections of very good, up-to-date papers: a must-read for senior people and a starting point for project search.

*———. 1996. *Past glacial environments.* Oxford: Butterworth-Heinemann. 598 pp. Collections of very good, up-to-date papers: a must-read for senior people and a starting point for project search.

*Paterson, W. S. B. 1994. *The physics of glaciers.* Oxford: Pergamon-Elsevier Science. 480 pp. A classic, readable text on glaciology.

*Pouder, E. R. 1965. *The physics of ice.* Oxford: Pergamon Press. 151 pp. Good, readable book.

Reading, H. G. (ed.). 1996. *Sedimentary environments.* Oxford: Blackwell Science. 688 pp.

Ritchie, J. C. 1987. *Postglacial vegetation of Canada.* Cambridge: Cambridge University Press. 178 pp.

Ritter, D. F. 1986. *Process geomorphology.* Dubuque, Iowa: Wm. C. Brown. 579 pp.

Shaw, J., Kvill, D. M., and Rains, B. 1989. Drumlins and catastrophic subglacial floods. *Sedimentary Geology* 62: 177–202.

Walker, R. G., and James, N. P. (eds.). 1992. *Facies models.* Waterloo: Geological Association of Canada. 409 pp.

*Williams, P. J., and Smith, M. W. 1989. *The frozen Earth: Fundamentals of geocryology.* Cambridge: Cambridge University Press. 306 pp.

Index